P9-DUH-706

ORGANIC SPECTROSCOPY

Organic Spectroscopy

L.D.S. Yadav
Professor
Department of Chemistry
University of Allahabad
Allahabad-211 002, India

Kluwer Academic Publishers
BOSTON DORDRECHT LONDON

Anamaya Publishers
New Delhi

A C.I.P. catalogue record for the book is available from the Library of Congress

ISBN 1-4020-2574-2

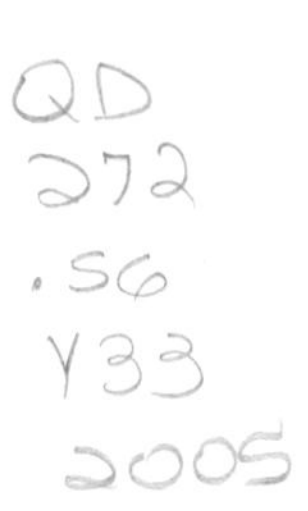

Copublished by Kluwer Academic Publishers,
P.O. Box 17, 3300 AA Dordrecht, The Netherlands
with Anamaya Publishers, New Delhi, India

Sold and distributed in North, Central and South America by
Kluwer Academic Publishers,
101 Philip Drive, Norwell, MA 02061, U.S.A.

In all other countries, except India, sold and distributed by
Kluwer Academic Publishers,
P.O. Box 322, 3300 AH Dordrecht, The Netherlands

In India, sold and distributed by
Anamaya Publishers, F-230, Lado Sarai, New Delhi-110 030, India

Printed in India.

Preface

Nowadays spectoscopy is being used as the most popular technique for structure determination and analysis. Thus, the knowledge of spectroscopy has become necessary for all the students of organic chemistry. The present book is an attempt to give the students the benefit of my over three decades experience of teaching and research. The book deals with UV, visible, IR, Raman, ^{1}H NMR, ^{13}C NMR, ESR and mass spectroscopy along with an introduction to the subject, and spectroscopic solution of structural problems.

The subject matter has been presented in a comprehensive, lucid and systematic manner which is easy to understand even by self-study. I believe that learning by solving problems gives more competence and confidence in the subject. Keeping this in view, sufficient number of solved and unsolved problems are given in each chapter. The answers to the unsolved problems at the end of the book are to check the solution worked out by the students.

The book contains sufficient spectral data in the text, tables and figures. In addition, a large number of spectra of various compounds have been incorporated for the students to see how the spectra actually appears. The spectral data and spectra together would help the reader familiarize with the interpretation.

In compiling this book I have drawn information from various sources available, e.g. review articles, reference work on spectroscopy, spectral catalogues and numerous books. Although individual acknowledgment cannot be made, I feel great pleasure in recording my indebtedness to all the contributors to the above sources.

I express my heartfelt gratitude to Prof. H.P. Tiwari, former Head, Department of Chemistry, University of Allahabad, who was kind enough to spare time from his busy schedule to go through the book and grateful to Prof. J.D. Pandey, former Head, Department of Chemistry, University of Allahabad, for his valuable discussion, especially on Raman Spectroscopy. I express my deep sense of gratitude to Prof. J.P. Sharma for his discussion and suggestions on various aspects of the subject and sincerely thank Prof. J.S. Chauhan and Prof. K.P. Tiwari for their gracious help and encouragement. I am also thankful to all my dear colleagues, particularly Prof. J. Singh, Dr. A.K. Jain, Dr. R.K.P. Singh and Dr. I.R. Siddiqui for their readily available help in many ways, and to the research scholars Mr. B.S. Yadav and Mr. V.K. Rai who proof read the entire manuscript.

I highly appreciate the work of publishing staff of M/s Anamaya Publishers, especially that of Mr. Manish Sejwal, who handled the project promptly and intelligently.

I hope that the book will be useful and successful in its objectives and will gratefully acknowledge any suggestions and comments from the readers for further improvements.

L.D.S. Yadav

Contents

Organic Spectroscopy

1

Introduction to Spectroscopy (Spectrometry)

1.1 Spectroscopy and Electromagnetic Radiations

Organic chemists use spectroscopy as a necessary tool for structure determination. Spectroscopy may be defined as the *study of the quantized interaction of electromagnetic radiations with matter*. Electromagnetic radiations are produced by the oscillation of electric charge and magnetic field residing on the atom. There are various forms of electromagnetic radiation, e.g. light (visible), ultraviolet, infrared, X-rays, microwaves, radio waves, cosmic rays etc.

1.2 Characteristics of Electromagnetic Radiations

All types of radiations have the same velocity (2.998×10^{10} cm/s in vacuum) and require no medium for their propagation, i.e. they can travel even through vacuum. Electromagnetic radiations are characterized by frequencies, wavelengths or wavenumbers.

Frequency ν is defined as the *number of waves which can pass through a point in one second*, measured in cycles per second (cps) or hertz (Hz) (1 Hz = 1 cps).

Wavelength λ is defined as the *distance between two consecutive crests C* or *troughs T* (Fig. 1.1) measured in micrometer (μm) or micron (μ) (1 μm = 1 μ = 10^{-6} m), nanometer (nm) or millimicron (mμ) (1 nm = 1 mμ = 10^{-9} m) and angstrom (Å) (1 Å = 10^{-10} m).

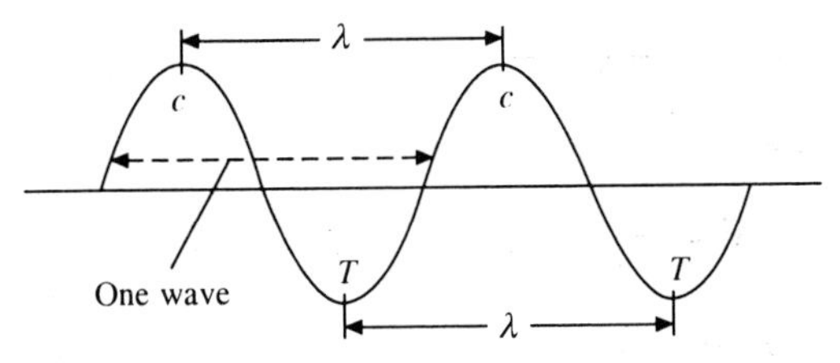

Fig. 1.1 Wavelength λ of an electromagnetic radiation

Wavenumber $\bar{\nu}$ is defined as the *number of waves which can pass through per unit length* usually 1 cm. It is the reciprocal of wavelength expressed in centimeter (cm^{-1}), i.e.

$$\bar{\nu} = \frac{1}{\lambda \text{ (in cm)}}$$

By their definitions, frequency and wavelength are inversely proportional, i.e.

$$\nu \propto \frac{1}{\lambda} \quad \text{or} \quad \nu = \frac{c}{\lambda}$$

where c is velocity of light (2.998×10^{10} cm/s).

Electromagnetic radiation is energy. When a molecule absorbs radiation, it gains energy, and on emitting radiation, it losses energy. The emission or absorption of electromagnetic radiations is quantized and each quantum of radiation is called a quantum or photon. Energy E for a single photon

$$E = h\nu = \frac{hc}{\lambda}$$

where h is Planck's constant (6.626×10^{-27} erg s).

The higher the frequency (or the shorter the wavelength) of the radiation, the greater is its energy.

Energy for a mole of photons. One mole of photons is one Einstein.

$$E = Nh\nu = \frac{Nhc.}{\lambda \text{ (in cm)}} = Nh\bar{\nu}\, c \text{ erg}$$

where N is Avogadro's number (6.023×10^{23} molecules/mole).

Or $$E = \frac{Nhc}{\lambda \text{ (in cm)}} = \frac{2.86 \times 10^{-3}}{\lambda \text{ (in cm)}} = \bar{\nu} \times 2.86 \times 10^{-3} \text{ kcal/mole}$$

(4.184×10^{7} erg = 4.184 J = 1 cal; 1 eV = 23.06 kcal/mole).

1.3 Solved Problems

1. Convert the following wavelengths into the corresponding wavenumbers in cm^{-1}:

(i) 5 μ and (ii) 10^4 nm

Solution

(i) Wavenumber $\bar{\nu} = \frac{1}{\lambda \text{ (in cm)}}$

The given λ = 5 and $\mu = 5 \times 10^{-4}$ cm

Hence $$\bar{\nu} = \frac{1}{5 \times 10^{-4}} = \frac{1 \times 10^4}{5} = 2000 \text{ cm}^{-1}$$

(ii) Wavenumber $\bar{\nu} = \frac{1}{\lambda \text{ (in cm)}}$

The given $\lambda = 10^4$ nm $= 10^4 \times 10^{-7} = 10^{-3}$ cm (1 nm $= 10^{-7}$ cm)

Therefore $$\bar{\nu} = \frac{1}{10^{-3}} = 1 \times 10^3 = 1000 \text{ cm}^{-1}$$

2. Convert wavenumber 1755 cm^{-1} into the corresponding wavelength in μm.

Solution

Wavelength λ (in cm) $= \frac{1}{\bar{v}}$

Hence $$\lambda = \frac{1}{1755} = 0.00057 \text{ cm}$$

$$= 0.00057 \times 10^4\ \mu m\ (1\text{ cm} = 10^4\ \mu m) = 5.7\ \mu m$$

3. Calculate the frequency of the electromagnetic radiations corresponding to the wavelengths (i) 2000 Å and (ii) 4 μm.

Solution

(i) Frequency $v = \frac{c}{\lambda}$

The given wavelength 2000 Å = 2000×10^{-8} cm (1 Å = 10^{-8} cm)

Hence $$v = \frac{2.998 \times 10^{10}\ \text{cm/sec}}{2000 \times 10^{-8}\ \text{cm}} = 1.499 \times 10^{15}\ \text{per sec}$$

$$= 1.499 \times 10^{15}\ \text{Hz (cps) or } 1.499 \times 10^{9}\ \text{MHz}$$

(ii) $v = \frac{c}{\lambda}$

The given wavelength 4 μm = 4×10^{-4} cm (1 μm = 10^{-4} cm)

Therefore $$v = \frac{2.998 \times 10^{10}\ \text{cm/sec}}{4 \times 10^{-4}\ \text{cm}} = 74.95 \times 10^{12}\ \text{per sec}$$

$$= 74.95 \times 10^{12}\ \text{Hz (cps) or } 74.95 \times 10^{6}\ \text{MHz}$$

4. Calculate the wavelength in Å of an electromagnetic radiation having frequency 7×10^{14} Hz.

Solution

$$\lambda = \frac{c}{v} = \frac{2.998 \times 10^{10}\ \text{cm/sec}}{7 \times 10^{14}\ \text{Hz (cps)}} = 0.4283 \times 10^{-4}\ \text{cm or } 4283\ \text{Å}\ (1\text{ cm} = 10^8\ \text{Å})$$

5. Calculate the energy associated with an ultraviolet radiation having wavelength 250 nm. Give the answer in kcal/mole and also in kJ/mole.

Solution

$$E = \frac{Nhc}{\lambda\ \text{(in cm)}} = \frac{2.86 \times 10^{-3}}{\lambda\ \text{(in cm)}}\ \text{kcal/mole}$$

The given λ = 250 nm = 250×10^{-7} cm (1 nm = 10^{-7} cm)

Hence $$E = \frac{2.86 \times 10^{-3}}{250 \times 10^{-7}} = 114.4\ \text{kcal/mole}$$

$$= 114.4 \times 4.184 = 478.65\ \text{kJ/mole}\ (1\text{ kcal} = 4.184\text{ kJ})$$

6. Calculate the energy associated with an ultraviolet radiation having wavelength 286 mm. Give the answer in kcal/mole and also in kJ/mole.

Solution

$$E = \frac{2.86 \times 10^{-3}}{\lambda \text{ (in cm)}} \text{ kcal/mole}$$

The given $\lambda = 286 \text{ m}\mu = 286 \times 10^{-7}$ cm ($1 \text{ m}\mu = 10^{-7}$ cm).

Therefore

$$E = \frac{2.86 \times 10^{-3}}{286 \times 10^{-7}} = 100 \text{ kcal/mole or } 100 \times 4.184 \text{ kJ/mole}$$
$$= 418.4 \text{ kJ/mole}$$

1.4 Electromagnetic Spectrum

Electromagnetic spectrum covers a very wide range of electromagnetic radiations from cosmic rays (having wavelengths in fractions of an angstrom) to radio waves (having wavelengths in meters or even kilometers) at the other end. The arrangement of all types of electromagnetic radiations in order of their wavelengths or frequencies is known as the *complete electromagnetic spectrum* (Fig. 1.2).

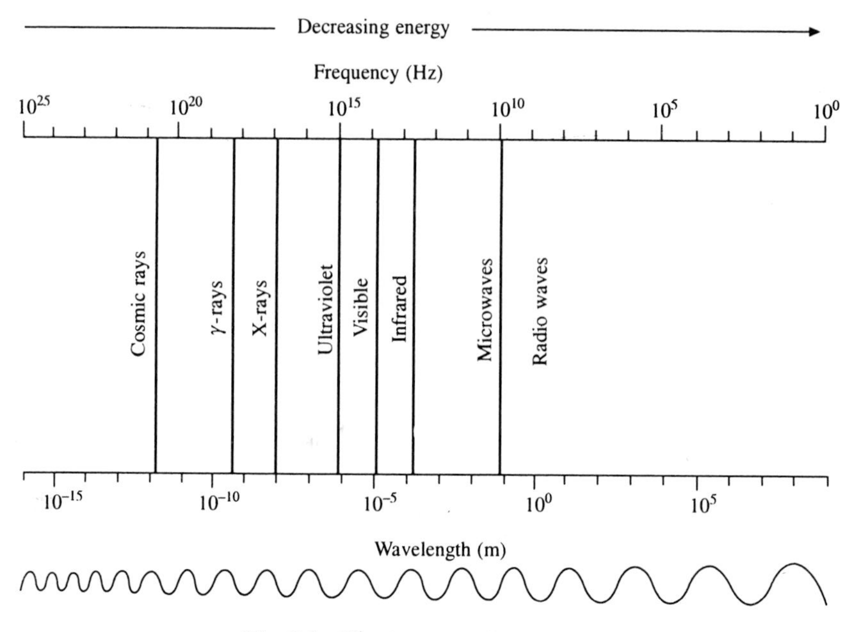

Fig. 1.2 Electromagnetic spectrum

The visible region (400-800 nm) represents only a small portion of the electromagnetic spectrum. Similarly, ulraviolet (UV), X-rays, γ-rays, cosmic rays, infrared (IR), microwaves and radio waves are the other important regions of the electromagnetic spectrum. The apprcximate wavelengths and frequencies of various regions of the electromagnetic spectrum are given in Fig. 1.2. Except the visible region, various regions overlap. The regions of greatest interest to organic chemists are 200-400 nm (ultraviolet), 400-800 nm (visible) and 2.5-15 μ (infrared).

1.5 Absorption and Emission Spectra

When electromagnetic radiations are passed through an organic compound, they may be absorbed to induce electronic, vibrational and rotational transitions in the molecules. The energy required for each of these transitions is quantized. Thus, only the radiation supplying the required quantum (photon) of energy is absorbed and the remaining portion of the incident radiation is transmitted. The wavelengths or frequencies of the absorbed radiations are measured with the help of a spectrometer. Generally, a spectrometer records an absorption spectrum as a plot of the intensity of absorbed or transmitted radiations versus their wavelengths or frequencies. Such spectra which are obtained by absorption of electromagnetic radiations are called *absorption spectra* (Fig. 1.3). UV, visible, IR and NMR spectra are examples of absorption spectra. Absorption band in an absorption spectrum can be characterized by the wavelength at which maximum absorption occurs and the intensity of absorption at this wavelength.

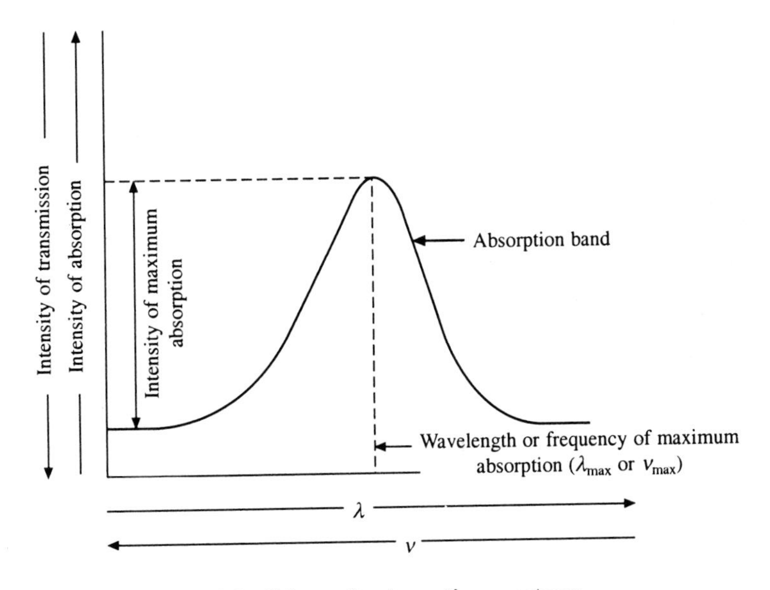

Fig. 1.3 Schematic absorption spectrum

The spectra which are obtained by emission of electromagnetic radiations from the excited substances are known as *emission spectra*, like atomic emission spectra. The excitation is caused by heating the substance to a high temperature either thermally or electrically. The excited substance emits certain radiations when it comes to the ground state and a spectrometer records these radiations as an emission spectrum.

PROBLEMS

1. What is spectroscopy? Explain absorption and emission spectra.

2. Describe important characteristics of electromagnetic radiations. Give expression for calculating the energy for a mole of photons.

3. Arrange the following electromagnetic radiations in order of their decreasing wavenumbers:

Radio waves, UV, visible, IR, X-rays and γ-rays.

4. Write short notes on:

(a) Electromagnetic spectrum, (b) Absorption of radiations and (c) Electromagnetic radiations.

5. Draw a typical absorption spectrum.

6. (a) Convert the following wavelengths in terms of wavenumbers in cm^{-1}:

(i) 2.5 μ (ii) 285 nm (iii) 2.98 μm

(b) The wavelength range of visible radiation is 4000 to 8000 Å. Calculate the corresponding frequency range in MHz.

7. (a) Calculate the energy associated with an infrared radiation having wavelength 4.0 μ. Give your answer in kcal/mole.

(b) Calculate the frequency range in cycles per second of the near-ultraviolet wavelength range 200 to 400 nm.

8. (a) Calculate the energy associated with a radiation having wavelength 6000 Å. Give your answer in kcal $mole^{-1}$ and also in kJ $mole^{-1}$.

(b) The most useful region for infrared spectroscopy is 2.5 to 15 μ. Convert this region into the corresponding wavenumber range in cm^{-1}.

(c) The energy difference between the two electronic states is 23.06 kcal $mole^{-1}$. What will be the frequency of the radiation absorbed when the electronic transition occurs from the lower energy state to the higher energy state?

References

1. A.J. Baker and T. Cairns, Spectroscopic Techniques in Organic Chemistry, Heyden London, 1965.
2. E.F.H. Britlain, W.O. George and C.H.J. Wells, Introduction to Molecular Spectroscopy, Academic Press, London, 1970.
3. J.C.D. Brand and G. Eglinton, Applications of Spectroscopy to Organic Chemistry, Oldbourne Press, London, 1965.
4. J.R. Dyer, Applications of Absorption Spectroscopy of Organic Compounds, Prentice-Hall, Englewood Cliffs, N.J., 1965.
5. W.G. Richards and P.R. Scott, Structure and Spectra of Atoms, Wiley Eastern Ltd., 1978.

2

Ultraviolet (UV) and Visible Spectroscopy

2.1 Introduction

Ultraviolet and visible spectroscopy deals with the recording of the absorption of radiations in the ultraviolet and visible regions of the electromagnetic spectrum. The ultaviolet region extends from 10 to 400 nm. It is subdivided into the near ultraviolet (quartz) region (200–400 nm) and the far or vacuum ultraviolet region (10–200 nm). The visible region extends from 400 to 800 nm.

The absorption of electromagnetic radiations in the UV and visible regions induces the excitation of an electron from a lower to higher molecular orbital (electronic energy level). Since UV and visible spectroscopy involves electronic transitions, it is often called *electronic spectroscopy*. Organic chemists use ultraviolet and visible spectroscopy mainly for detecting the presence and elucidating the nature of the conjugated multiple bonds or aromatic rings.

2.2 Absorption Laws and Molar Absorptivity

A UV-visible spectrophotometer records a UV or visible spectrum (Fig. 2.1) as a plot of wavelengths of absorbed radiations versus the intensity of absorption in terms of absorbance (optical density) A or molar absorptivity (molar extinction coefficient) ε as defined by the Lambert-Beer law. According to Lambert's law, the fraction of incident monochromatic radiation absorbed by a homogeneous medium is independent of the intensity of the incident radiation while Beer's law states that the absorption of a monochromatic radiation by a homogeneous medium is proportional to the number of absorbing molecules. From these laws, the remaining variables give the following equation which expresses the Lambert-Beer law

$$\log_{10} \frac{I_0}{I} = A = \varepsilon c l \qquad (2.1)$$

where I_0 is the intensity of incident radiation, I the intensity of radiation transmitted through the sample solution, A the absorbance or optical density, ε the molar absorptivity or molar extinction coefficient, c the concentration of solute (mole/litre) and l the path length of the sample (cm).

The molar absorptivity of an organic compound is constant at a given wavelength. The intensity of an absorption band in the UV or visible spectrum is usually expressed as the molar absorptivity at maximum absorption, ε_{max} or $\log_{10} \varepsilon_{max}$. The wavelength of the maximum absorption is denoted by λ_{max}.

When the molecular weight of a sample is unknown, or when a mixture is being examined, the intensity of absorption is expressed as $E_{1\,cm}^{1\%}$ (or $A_{1\,cm}^{1\%}$) value, i.e. the absorbance of a 1% solution of the sample in a 1 cm cell

$$E_{1\,cm}^{1\%} = \frac{A}{cl}$$

where c is the concentration in g/100 ml and l the path length of the sample in cm.

This value is easily related to ε by the expression

$$10\varepsilon = E_{1\,cm}^{1\%} \times \text{Molecular weight}$$

2.3 Instrumentation

The desired parameter in spectroscopy is absorbance, but it cannot be directly measured. Thus, a UV-visible spectrophotometer compares the intensity of the transmitted radiation with that of the incident UV-visible radiation. Most UV-visible spectrophotometers are double-beam instruments and consist of a radiation source, monochromator, detectors, amplifier and recording system as shown in Fig. 2.1.

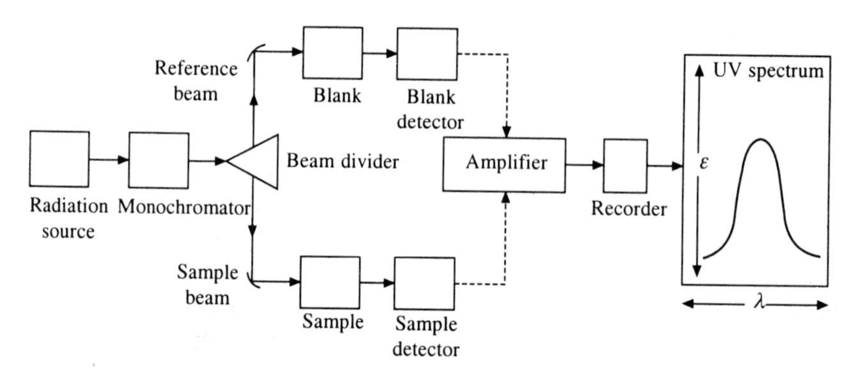

Fig. 2.1 Schematic diagram of a double-beam UV-visible spectrophotometer

(i) Radiation Source

The hydrogen-discharge lamp is the most commonly used source of radiation in the UV region (180-400 nm). A deuterium-discharge lamp is used in its place when more (3-5 times) intensity is desired. A tungsten-filament lamp is used when absorption in the visible region (400–800 nm) is to be determined.

(ii) Monochromator

It disperses the radiations obtained from the source into their separate wavelengths. The most widely used dispersing element is a prism or grating made up of quartz because quartz is transparent throughout the UV range. Glass strongly absorbs ultraviolet radiation, hence it cannot be used in this region. Glass can be satisfactorily used in the visible region. The dispersed radiation is divided by the beam divider into two parallel beams of equal intensity; one of which passes through a transparent cell containing the sample solution and the other through an identical cell containing the solvent. The former is called *sample beam* and the latter *reference beam*.

(iii) Detectors

These have photocells or photomultiplier tubes which generate voltage proportional to the radiation energy that strikes them.

(iv) Amplifier

The spectrophotometer has balancing electronic amplifier which subtracts the absorption of the solvent from that of the solution electronically.

(v) Recorder

It automatically records the spectrum as a plot of wavelengths of absorbed radiations against absorbance A or molar absorptivity ε.

2.4 Sample Handling

UV-visible spectra are usually recorded either in very dilute solutions or in the vapour phase. Quartz cells of 1 cm square are commonly used for recording the spectra. The sample is dissolved in some suitable solvent which does not itself absorb radiation in the region under investigation. Commonly used solvents are cyclohexane, 1,4-dioxane, water, and 95% ethanol. The chosen solvent should be inert to the sample. Generally, 1 mg of the compound with a molecular weight of 100–200 is dissolved in a suitable solvent and made up to, e.g. 100 ml and only a portion of this is used for recording the spectrum.

2.5 Theory (Origin) of UV-Visible Spectroscopy

UV-visible absorption spectra originate from electronic transitions within a molecule. These transitions involving promotion of valence electrons from the ground state to the higher-energy state (excited state) are called *electronic excitations* and are caused by the absorption of radiation energy in the UV-visible regions of the electromagnetic spectrum. Since various energy levels of molecules are quantized, a particular electronic excitation occurs only by the absorption of specific wavelength of radiation corresponding to the required quantum of energy.

2.6 Electronic Transitions

According to molecular orbital theory, the excitation of a molecule by the absorption of radiation in the UV-visible regions involves promotion of its electrons from a bonding, or non-bonding (n) orbital to an antibonding orbital. There are σ and π bonding orbitals associated with σ^* and π^* antibonding orbitals, respectively. Non-bonding (n or p) orbitals are not associated with antibonding orbitals because non-bonding or lone pair of electrons present in them do not form bonds. Following electronic transitions are involved in the UV-visible region (Fig. 2.2):

(i) $\sigma \rightarrow \sigma^*$ (iii) $n \rightarrow \sigma^*$ (iii) $\pi \rightarrow \pi^*$ (iv) $n \rightarrow \pi^*$

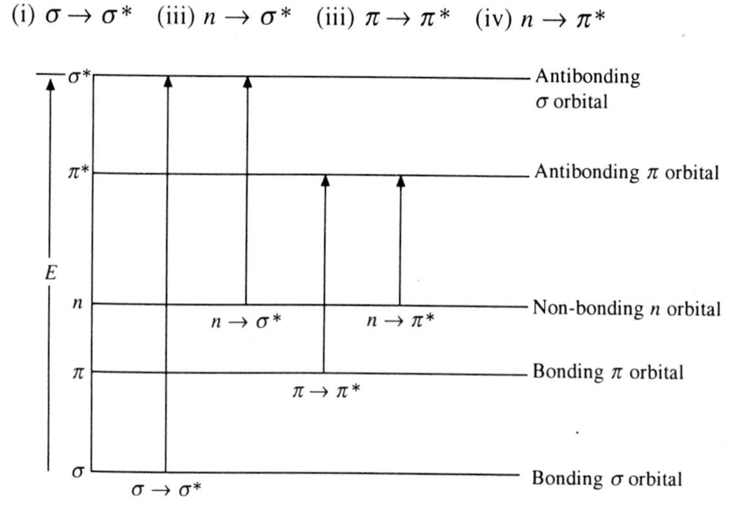

Fig. 2.2 Relative energies of electronic transitions

The usual order of energy required for various electronic transitions is

$$\sigma \rightarrow \sigma^* > n \rightarrow \sigma^* > \pi \rightarrow \pi^* > n \rightarrow \pi^*$$

Now we will discuss the electronic transitions involved in UV-visible spectroscopy.

(i) $\sigma \rightarrow \sigma^*$ Transition

The transition or promotion of an electron from a bonding sigma orbital to the associated antibonding sigma orbital is $\sigma \rightarrow \sigma^*$ transition. It is a high energy process because σ bonds are generally very strong. Thus, these transitions do not occur in normal UV-visible regions (200-800 nm). For example, in alkanes only $\sigma \rightarrow \sigma^*$ transition is available and they absorb high energy UV radiation around 150 nm; ethane shows λ_{max} 135 nm. The region below 200 nm is called *vacuum UV region*, since oxygen present in air absorbs strongly at ~200 nm and below. Similarly, nitrogen absorbs at ~150 nm and below. Thus, an evacuated spectrophotometer is used for studying such high energy transitions (below 200 nm). However, this region is less informative

$$\text{>C—C<} \xrightarrow{\sigma \rightarrow \sigma^*} \text{>C:C<}$$

(ii) $n \rightarrow \sigma^*$ Transition

The transition or promotion of an electron from a non-bonding orbital to an antibonding sigma orbital is designated as $n \rightarrow \sigma^*$ transition. Compounds containing non-bonding electrons on a heteroatom are capable of showing absorption due to $n \rightarrow \sigma^*$ transitions. These transitions require lower energy than $\sigma \rightarrow \sigma^*$ transitions. Some organic compounds undergoing $n \rightarrow \pi^*$ transitions are halides, alcohols, ethers, aldehydes, ketones etc. For example, methyl chloride shows λ_{max} 173 nm, methyl iodide λ_{max} 258 nm, methyl alcohol λ_{max} 183 nm and water λ_{max} 167 nm

$$-\overset{|}{\underset{|}{C}}-C-\ddot{\underset{..}{X}}: \xrightarrow{n \rightarrow \sigma^*} -\overset{|}{\underset{|}{C}}\dot{-}\dot{\underset{..}{X}}:$$

In alkyl halides, the energy required for $n \rightarrow \sigma^*$ transition increases as the electronegativity of the halogen atom increases. This is due to comparatively difficult excitation of non-bonding (n or p) electrons on increase in the electronegativity. The difficult excitation means less probability of transition. The molar extinction coefficient ε increases as the probability of the transition increases. Thus, methyl iodide shows λ_{max} 258 nm, ε_{max} 378 and methyl chloride λ_{max} 173 nm, ε_{max} ~100. Since iodine is less electronegative than chlorine, $n \rightarrow \sigma^*$ transition is more probable at low energy process in methyl iodide than in methyl chloride. Similarly, amines show both higher λ_{max} and ε_{max} than alcohols owing to lesser electronegativity of nitrogen than oxygen, e.g. trimethylamine λ_{max} 227 nm, ε_{max} 900 and methyl alcohol λ_{max} 183 nm, ε_{max} 150. Protonated trimethylamine does not show absorption due to $n \rightarrow \sigma^*$ transition, because it has no non-bonded electrons.

(iii) $\pi \rightarrow \pi^*$ Transition

The transition or promotion of an electron from a π bonding orbital to a π antibonding orbital is designated $\pi \rightarrow \pi^*$ transition. These type of transitions occur in compounds containing one or more covalently unsaturated groups like C=C, C=O, NO_2 etc. $\pi \rightarrow \pi^*$ transitions require lower energy than $n \rightarrow \sigma^*$ transitions. In unconjugated alkenes, this transition occurs in the range 170–190 nm; ethylene shows λ_{max} 171 nm. Similarly, unconjugated carbonyl compounds show $\pi \rightarrow \pi^*$ transition in the range 180–190 nm; acetone shows λ_{max} 188 nm.

$$>C=C< \xrightarrow{\pi \rightarrow \pi^*} -\overset{|}{\underset{|}{C}}\dot{-}\dot{C}<$$

(iv) $n \rightarrow \pi^*$ Transition

The transition or promotion of an electron from a non-bonding orbital to a π antibonding orbital is designated $n \rightarrow \pi^*$ transition. This transition requires lowest energy.

$$>C=\ddot{O}: \xrightarrow{n \rightarrow \pi^*} >C\dot{=}\dot{O}:$$

Saturated aldehydes and ketones show both types of transitions, i.e. low energy $n \rightarrow \pi^*$ and high energy $\pi \rightarrow \pi^*$ occurring in the regions 270–300 nm and 180–190 nm, respectively; acetone shows $n \rightarrow \pi^*$ at λ_{max} 279 nm. The band due to $\pi \rightarrow \pi^*$ transition is more intense, i.e. it has high value of ε than the less intense bands due to $n \rightarrow \pi^*$ transition. In addition, carbonyl compounds also exhibit high energy $n \rightarrow \sigma^*$ transition around 160 nm; acetone shows λ_{max} 166 nm.

Because of different structural environments, identical functional groups in different compounds do not necessarily absorb at exactly the same wavelength.

2.7 Formation of Absorption Bands

Since the energy required for each electronic transition is quantized, the UV-visible spectrum is expected to exhibit a single, discrete line corresponding to each electronic transition. In practice, broad absorption bands are usually observed. In a molecule, each electronic energy level (either in ground state or in excited state) is accompanied by a large number of vibrational (v_0, v_1, v_2 etc.) and rotational (r_0, r_1, r_2 etc.) energy levels which are also quantized (Fig. 2.3). In complex molecules having many atoms, there are still a large number of closer vibrational energy levels.

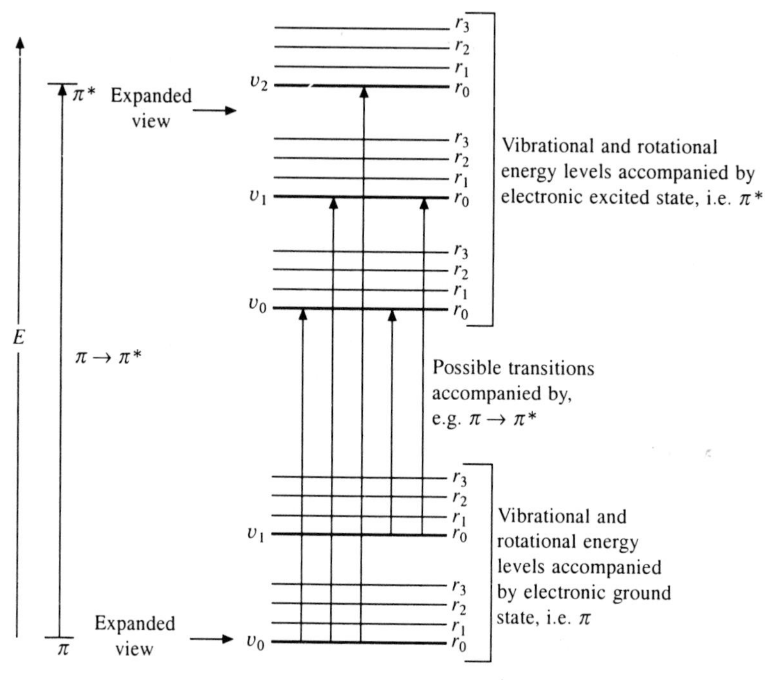

Fig. 2.3 Schematic energy level diagram of a diatomic molecule

The radiation energy passed through a sample is sufficient to induce various electronic transitions as well as transitions in accompanying vibrational and rotational energy levels. However, these transitions have very small energy differences, but the energy required to induce an electronic transition is larger than that required to cause transitions in the accompanying vibrational and rotational energy levels. Thus, the electronic absorption is superimposed upon the accompanying vibrational and rotational absorptions resulting in the formation of broad bands. More clearly, not only a single but a large number of very close wavelengths are absorbed and the closeness of the resulting discrete spectral lines causes them to coalesce to give broad absorption bands in case of complex molecules.

2.8 Designation of Absorption Bands

UV-visible absorption bands may be designated by the type of electronic transition from which they originate, e.g. $\sigma \rightarrow \sigma^*$ band, $\pi \rightarrow \pi^*$ band etc., or by the letter designation. The following letter designation was proposed because more than one band may arise due to the same type of electronic transition.

K-Bands (Conjugated; German: *Konjugierte*). These bands originate from $\pi \rightarrow \pi^*$ transitions in compounds having a $\pi — \pi$ conjugated system (a system having at least two multiple bonds separated by only one single bond), e.g. 1,3-butadiene shows *K*-band at λ_{max} 217 nm, ε_{max} 21,000 and acrolein at λ_{max} 210 nm, ε_{max} 11,500. Aromatic compounds having a chromophoric substituent also exhibit *K*-bands in their UV spectra, e.g. acetophenone shows *K*-band at λ_{max} 240 nm, ε_{max} 13,000 and styrene at λ_{max} 244 nm, ε_{max} 12,000. Usually, *K*-band have high molar absorptivity, $\varepsilon_{max} > 10^4$.

R-Bands (Radical-like; German: *Radikalartig*). These bands originate from $n \rightarrow \pi^*$ transitions of a single chromophoric group, e.g. carbonyl or nitro group. *R*-bands have low molar absorptivity, $\varepsilon_{max} < 100$, and are also called forbidden bands. For example, acetone shows an *R*-band at λ_{max} 279 nm, ε_{max} 15; acrolein at λ_{max} 315 nm, ε_{max} 14 and acetophenone at λ_{max} 319 nm, ε_{max} 50.

B-Bands (Benzenoid bands). These bands originate from $\pi \rightarrow \pi^*$ transitions in aromatic or heteroaromatic compounds. For example, benzene shows a *B*-band at λ_{max} 256, ε_{max} 200 and acetophenone at λ_{max} 278 nm, ε_{max} 1100.

E-Bands (Ethylenic bands). Similar to *B*-bands, these are characteristic of aromatic and heteroaromatic compounds and originate from $\pi \rightarrow \pi^*$ transitions of the ethylenic bonds present in the aromatic ring. *E*-band which appears at a shorter wavelength and is usually more intense is called E_1-band. The low intensity band of the same compound appearing at a longer wavelength is called E_2-band†. For example, benzene exhibits E_1- and E_2-bands near λ_{max} 180 nm, ε_{max} 60,000 and λ_{max} 200 nm, ε_{max} 7900, respectively.

For more examples of *K*-, *R*-, *B*- and *E*-bands, see Tables 2.2, 2.6 and 2.7. A compound may exhibit more than one band in its UV spectrum, either due to the presence of more than one chromophore or due to more than one transition of a single chromophore. A typical UV spectrum is given in Fig. 2.4.

A band may be submerged under some more intense band. Thus, in certain

---

†In another notation it is designated *K*-band but we shall use the designation E_2-band for this.

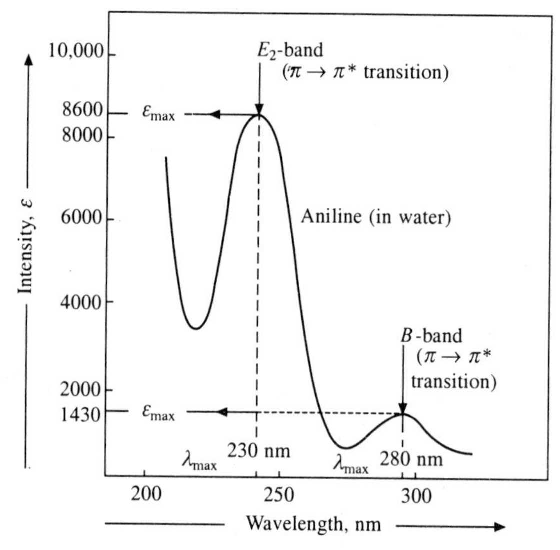

Fig. 2.4 UV spectrum of aniline showing E_2- and B-bands

cases, all the expected bands may not be observed. For example, *B*-bands are sometimes buried under *K*-bands (see Tables 2.6 and 2.7).

2.9 Transition Probability: Allowed and Forbidden Transitions

On exposure to UV or visible radiation, a molecule may or may not absorb the radiation, i.e. it may or may not undergo electronic excitation. The molar absorptivity at maximum absorption

$$\varepsilon_{max} = 0.87 \times 10^{20} \cdot P \cdot a$$

where *P* is the transition probability with values from 0 to 1 and *a* the target area of the absorbing system, i.e. a chromophore.

A chromophore with a length of the order of 10 Å or 10^{-7} cm and with unit probability will have ε_{max} value of ~10^5. Thus, there is a direct relationship between the area of a chromophore and its absorption intensity (ε_{max}). Transitions with ε_{max} values > 10^4 are called *allowed transitions* and are generally caused by $\pi \rightarrow \pi^*$ transitions, e.g. in 1,3-butadiene, the absorption at 217 nm, ε_{max} 21,000 results from the allowed transition. Transitions with ε_{max} values < 10^4 are called *forbidden transitions*. These are generally caused by $n \rightarrow \pi^*$ transitions, e.g. in carbonyl compounds, the absorption near 300 nm with ε_{max} values 10–100 results from the forbidden transition.

In addition to the area of a chromophore, there are also some other factors which govern the transition probability. However, the prediction of their effects on the transition probability is complicated because they involve geometries of the lower and higher energy molecular orbitals as well as the symmetry of the molecule as a whole. Symmetrical molecules have more restrictions on their transitions than comparatively less symmetrical molecules. Consequently, symmetrical molecules like benzene have simple electronic absorption spectra as compared to less symmetrical molecules. There are very less symmetry restrictions for a highly unsymmetrical molecule, thus it will exhibit a complex electronic absorption spectrum.

2.10 Certain Terms Used in Electronic Spectroscopy: Definitions

Chromophore

A covalently unsaturated group responsible for absorption in the UV or visible region is known as a *chromophore*. For example, C=C, C≡C, C=O, C≡N, N=N, NO_2 etc. If a compound absorbs light in the visible region (400–800 nm), only then it appears coloured. Thus, a chromophore may or may not impart colour to a compound depending on whether the chromophore absorbs radiation in the visible or UV region.

Chromophores like C=C or C≡C having π electrons undergo $\pi \to \pi^*$ transitions and those having both π and non-bonding electrons, e.g. C=O, C≡N or N=N, undergo $\pi \to \pi^*$, $n \to \pi^*$ and $n \to \sigma^*$ transitions. Since the wavelength and intensity of absorption depend on a number of factors, there are no set rules for the identification of a chromophore. Characteristics of some common unconjugated chromophores are given in Table 2.1.

Table 2.1 Characteristics of some common unconjugated chromophores

Chromophore	Example	λ_{max} (nm)	ε_{max}	Transition	Solvent
>C=C<	Ethylene	171	15,530	$\pi \to \pi^*$	Vapor
—C≡C—	Acetylene	150	~10,000	$\pi \to \pi^*$	Hexane
		173	6000	$\pi \to \pi^*$	Vapor
>C=O	Acetaldehyde	160	20,000	$n \to \sigma^*$	Vapor
		180	10,000	$\pi \to \pi^*$	Vapor
		290	17	$n \to \pi^*$	Hexane
	Acetone	166	16,000	$n \to \sigma^*$	Vapor
		188	900	$\pi \to \pi^*$	Hexane
		279	15	$n \to \pi^*$	Hexane
—COOH	Acetic acid	204	60	$n \to \pi^*$	Water
—$CONH_2$	Acetamide	178	9500	$\pi \to \pi^*$	Hexane
		220	63	$n \to \pi^*$	Water
—COOR	Ethyl acetate	211	57	$n \to \pi^*$	Ethanol
—NO_2	Nitromethane	201	5000	$\pi \to \pi^*$	Methanol
		274	17	$n \to \pi^*$	Methanol
>C=N—	Acetoxime	190	5000	$n \to \pi^*$	Water
—C≡N	Acetonitrile	167	Weak	$\pi \to \pi^*$	Vapor
—N=N—	Azomethane	338	4	$n \to \pi^*$	Ethanol

Auxochrome

A covalently saturated group which, when attached to a chromophore, changes both the wavelength and the intensity of the absorption maximum is known as *auxochrome*, e.g. NH_2, OH, SH, halogens etc. Auxochromes generally increase the value of λ_{max} as well as ε_{max} by extending the conjugation through resonance. These are also called *colour enhancing groups*. An auxochrome itself does not show absorption above 200 nm. Actually, the combination of chromophore and auxochrome behaves as a new chromophore having different values of λ_{max} and ε_{max}. For example, benzene shows λ_{max} 256 nm, ε_{max} 200, whereas aniline shows λ_{max} 280 nm, ε_{max} 1430 (both increased). Hence, NH_2 group is an auxochrome which extends the conjugation involving the lone pair of electrons on the nitrogen atom resulting in the increased values of λ_{max} and ε_{max}.

Absorption and Intensity Shifts

Bathochromic Shift or Effect. The shift of an absorption maximum to a longer wavelength (Fig. 2.5) due to the presence of an auxochrome, or solvent effect is called a *bathochromic shift* or *red shift*. For example, benzene shows λ_{max} 256 nm and aniline shows λ_{max} 280 nm. Thus, there is a bathochromic shift of 24 nm in the λ_{max} of benzene due to the presence of the auxochrome NH_2. Similarly, a bathochromic shift of $n \rightarrow \pi^*$ band is observed in carbonyl compounds on decreasing solvent polarity, e.g. λ_{max} of acetone is at 264.5 nm in water as compared to 279 nm in hexane.

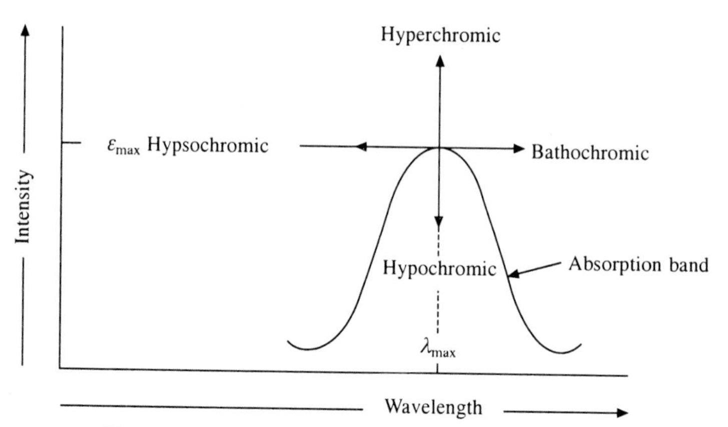

Fig. 2.5 Shifts in absorption position and intensity

Hypsochromic Shift or Effect. The shift of an absorption maximum to a shorter wavelength is called *hypsochromic* or *blue shift* (Fig. 2.5). This is caused by the removal of conjugation or change in the solvent polarity. For example, aniline shows λ_{max} 280 nm, whereas anilinium ion (acidic solution of aniline) shows λ_{max} 254 nm. This hypsochromic shift is due to the removal of $n \rightarrow \pi$ conjugation† of the lone pair of electrons of the nitrogen atom of aniline with the π-bonded

†A system in which an atom having non-bonding electrons is separated from a multiple bond by only one single bond is called *n-p* conjugated system.

system of the benzene ring on protonation because the protonated aniline (anilinium ion) has no lone pair of electrons for conjugation. Similarly, there is a hypsochromic shift of 10–20 nm in the λ_{max} of $\pi \rightarrow \pi^*$ bands of carbonyl compounds on going from ethanol as solvent to hexane, i.e. on decreasing solvent polarity.

Hyperchromic Effect. An effect which leads to an increase in absorption intensity ε_{max} is called *hyperchromic effect* (Fig. 2.5). The introduction of an auxochrome usually causes hyperchromic shift. For example, benzene shows *B*-band at 256 nm, ε_{max} 200, whereas aniline shows *B*-band at 280 nm, ε_{max} 1430. The increase of 1230 in the value ε_{max} of aniline compared to that of benzene is due to the hyperchromic effect of the auxochrome NH_2.

Hypochromic Effect. An effect which leads to a decrease in absorption intensity ε_{max} is called *hypochromic effect* (Fig. 2.5). This is caused by the introduction of a group which distorts the chromophore. For example, biphenyl shows λ_{max} 252 nm, ε_{max} 19,000, whereas 2,2′-dimethylbiphenyl shows λ_{max} 270 nm, ε_{max} 800. The decrease of 18,200 in the value of ε_{max} of 2,2′-dimethylbiphenyl is due to the hypochromic effect of the methyl groups which distort the chromophore by forcing the rings out of coplanarity resulting in the loss of conjugation.

2.11 Conjugated Systems and Transition Energies

When two or more chromophoric groups are conjugated, the absorption maximum is shifted to a longer wavelength (lower energy) and usually to a greater intensity compared to the simple unconjugated chromophore, e.g. ethylene shows λ_{max} 171 nm, ε_{max} 15,530 and 1,3-butadiene λ_{max} 217 nm, ε_{max} 21,000. In conjugated dienes, the π molecular orbitals of the separate alkene groups combine to give two new bonding molecular orbitals designated π_1 and π_2, and two anti-bonding molecular orbitals designated π_3^* and π_4^*. Fig. 2.6 shows the $\pi_2 \rightarrow \pi_3^*$ transition, i.e. the promotion of an electron from a highest occupied molecular orbital (HOMO) to the lowest unoccupied molecular orbital (LUMO) of a conjugated diene, is of very low energy, 1,3-butadiene shows λ_{max} 217 nm

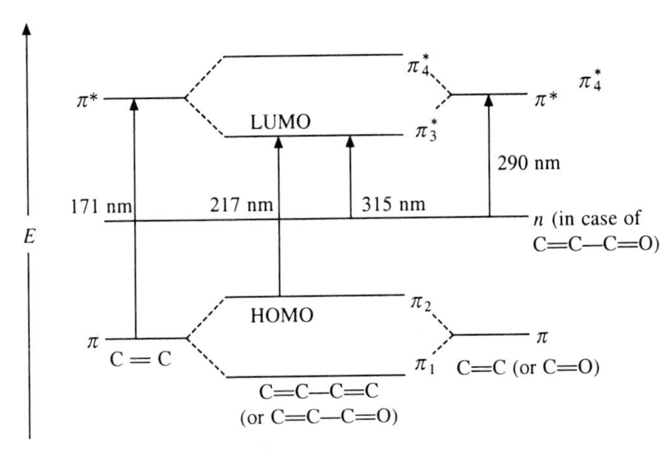

Fig. 2.6 Relative energies of electronic transitions in conjugated systems

which is lower than the $\pi \rightarrow \pi^*$ transition of an unconjugated alkene, ethylene shows λ_{max} 171 nm.

Similarly, in case of other conjugated chromophores, the energy difference between HOMO and LUMO is lowered resulting in the bathochromic shift. Thus, α,β-unsaturated carbonyl compounds show both $\pi \rightarrow \pi^*$ and $n \rightarrow \pi^*$ transitions at longer wavelengths (e.g. acrolein, λ_{max} 210 and 315 nm) compared to unconjugated carbonyl compounds (acetaldehyde, λ_{max} 180 and 290 nm) (Fig. 2.6).

We know that as the energy of electronic transitions decreases, usually their probability increases and so ε_{max} also increases. Since conjugation lowers the energy required for electronic transitions, it increases the value of λ_{max} and usually that of ε_{max} also. Thus, ethylene shows λ_{max} 171 nm, ε_{max} 15,530 and 1,3-butadiene λ_{max} 217 nm, ε_{max} 21,000. For more examples of various conjugated chromophores, see Tables 2.2, 2.6 and 2.7.

Table 2.2 Characteristics of some simple conjugated chromophores

Chromophore	Example	λ_{max} (nm)	ε_{max}	Transition	Solvent	Band
$>C=C-C=C<$	1,3-Butadiene	217	21,000	$\pi \rightarrow \pi^*$	Hexane	*K*
	1,3,5-Hexatriene	258	35,000	$\pi \rightarrow \pi^*$	Hexane	*K*
$>C=C-C\equiv C-$	Vinyl acetylene	219	7600	$\pi \rightarrow \pi^*$	Hexane	*K*
		228	7800	$\pi \rightarrow \pi^*$	Hexane	*K*
$>C=C-C=O$	Acrolein	210	11,500	$\pi \rightarrow \pi^*$	Ethanol	*K*
		315	14	$n \rightarrow \pi^*$	Ethanol	*R*
	Crotonaldehyde	218	18,000	$\pi \rightarrow \pi^*$	Ethanol	*K*
		320	30	$n \rightarrow \pi^*$	Ethanol	*R*
	3-Pentene-2-one	224	9750	$\pi \rightarrow \pi^*$	Ethanol	*K*
		314	38	$n \rightarrow \pi^*$	Ethanol	*R*
$-C\equiv C-C=O$	1-Hexyn-3-one	214	4500	$\pi \rightarrow \pi^*$	Ethanol	*K*
		308	20	$n \rightarrow \pi^*$	Ethanol	*R*
$O=C-C=O$	Glyoxal	195	35	$\pi \rightarrow \pi^*$	Hexane	*K*
		280	3	$n \rightarrow \pi^*$	Hexane	*R*
HOOC—COOH	Oxalic acid	~185	4000	$\pi \rightarrow \pi^*$	Water	*K*
		250	63	$n \rightarrow \pi^*$	Water	*R*

Note: In general, the longer the conjugated system, the higher are the values of λ_{max} and ε_{max}.

Thus, a compound with sufficient conjugation absorbs in the visible region (400–800 nm) and becomes coloured. For example β-carotene, an orange pigment present in carrots, has eleven carbon-carbon double bonds and absorbs in the visible region (λ_{max} 450 nm, ε_{max} 14×10^4) and is coloured.

2.12 Solvent Effects

Since the polarity of a molecule usually changes with electronic transition, the position and the intensity of absorption maxima may be shifted by changing solvent polarity.

(i) $\pi \rightarrow \pi^*$ Transitions (*K*-Bands)

Owing to the non-polar nature of hydrocarbon double bonds, the $\pi \rightarrow \pi^*$ transitions of alkenes, dienes and polyenes are not appreciably affected by changing solvent polarity. The $\pi \rightarrow \pi^*$ transitions of polar compounds, e.g. saturated as well as α,β-unsaturated carbonyl compounds are shifted to longer wavelengths and generally towards higher intensity with increasing solvent polarity. The excited state in this transition is more polar than the ground state, thus, dipole-dipole interaction with a polar solvent lowers the energy of the excited state more than that of the ground state. Thus, there is a bathochromic shift of 10–20 nm in going from hexane as a solvent to ethanol, i.e. on increasing solvent polarity.

(ii) *B*-Bands

These bands also originate from $\pi \rightarrow \pi^*$ transitions, and their position and intensity are not shifted by changing solvent polarity except in case of heteroaromatic compounds which show a marked hyperchromic shift on increasing solvent polarity.

(iii) $\pi \rightarrow \pi^*$ Transitions (*R*-Bands)

It has been found that an increase in solvent polarity usually shifts $n \rightarrow \pi^*$ transitions to shorter wavelengths (higher energy). For example, acetone shows λ_{max} 279 nm in hexane, whereas in water it shows λ_{max} 264.5 nm. This can be explained on the basis that the carbonyl group is more polar in the ground state ($>\overset{\delta+}{C}=\overset{\delta-}{O}$) than in the excited state ($>\overset{\delta-}{C}=\overset{\delta+}{O}$). Thus, dipole-dipole interaction or hydrogen bonding with a polar solvent lowers the energy of the ground state more than that of the excited state resulting in the hypsochromic shift in case of unconjugated as well as conjugated carbonyl compounds with increasing solvent polarity.

(iv) $n \rightarrow \sigma^*$ Transitions

These transitions are affected by solvent polarity, especially by solvents capable of forming hydrogen bond. Alcohols and amines form hydrogen bonds with protic solvents. Such associations involve non-bonding electrons of the heteroatom. The involvement of non-bonding electrons in hydrogen bonding lowers the energy of the *n* orbital, and thus the excitation of these electrons requires greater energy resulting in the hypsochromic shift with increasing polarity.

- Polar solvents stabilize polar groups through dipole-dipole interaction or hydrogen bonding.
- If a chromophore is more polar in the ground state than in the excited state, then polar solvents stabilize the ground state to the greater extent than the excited state. Thus, there is a hypsochromic shift with increasing solvent polarity.
- If a chromophore is more polar in the excited state than in the ground state, then the former is stabilized to a greater extent by polar solvents than the latter. Thus, there is a bathochromic shift with increasing solvent polarity.

It has been found that an increase in solvent polarity usually shifts $n \rightarrow \pi^*$ and $n \rightarrow \sigma^*$ bands to shorter wavelengths, and $\pi \rightarrow \pi^*$ bands of polar compounds to longer wavelengths.

2.13 Woodward-Fieser Rules for Calculating λ_{max} in Conjugated Dienes and Trienes

Woodward (1941) formulated a set of empirical rules for calculating or predicting λ_{max} in conjugated acyclic and six-membered ring dienes. These rules, modified by Fieser and Scott on the basis of wide experience with dienes and trienes, are called Woodward-Fieser rules and are summarized in Table 2.3. First, we discuss the following terms used in Woodward-Fieser rules.

Table 2.3 Woodward-Fieser rules for calculating λ_{max} in conjugated dienes and trienes

Base value for acyclic or heteroannular diene	214 nm
Base value for homoannular diene	253 nm
Increment for each:	
Alkyl substituent or ring residue	5 nm
Exocyclic conjugated double bond	5 nm
Double bond extending conjugation	30 nm
—OR (alkoxy)	6 nm
—Cl, —Br	6 nm
—OCOR (acyloxy)	0 nm
—SR (alkylthio)	30 nm
—NR_2 (dialkylamino)	60 nm
In the same double bond is exocyclic to two rings simultaneously	10 nm
Solvent correction	0 nm
Calculated* λ_{max} of the compound	Total nm

*For $\pi \rightarrow \pi^*$ transition (*K*-band).

(i) Homoannular Dienes

In homoannular dienes, conjugated double bonds are present in the same ring and having *s-cis* (*cisoid*) configuration (*s* = single bond joining the two doubly bonded carbon atoms):

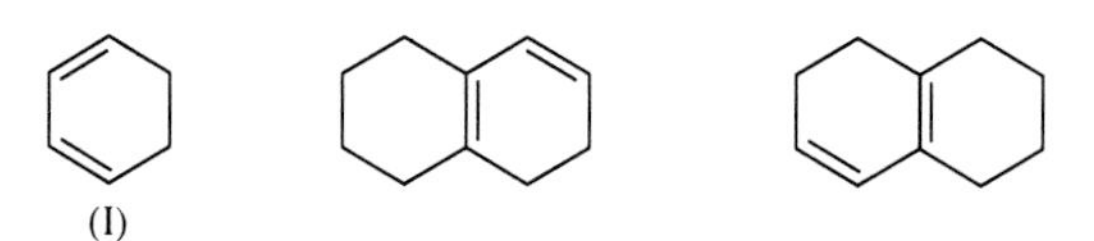

The *s-cis* configuration causes strain which raises the ground state energy level of the molecule leaving the high energy excited state relatively unchanged. Thus, the transition energy is lowered resulting in the shift of absorption position to a longer wavelength. Acyclic dienes exist mostly in the strainless *s-trans* (*transoid*) conformation with relatively lower ground state energy level. Thus, their absorptions appear at shorter wavelengths. For example, 1,3-cyclohexadiene

(I) shows λ_{max} 256 nm, whereas 1,3-butadiene shows λ_{max} 217 nm. Also, due to the shorter distance between the two ends of the chromophore, *s-cis* dienes give lower ε_{max} (~10,000) than that of the *s-trans* dienes (~20,000).

s-trans 1,3-butadiene	⇌	*s-cis* 1,3-butadiene
~ 97.5%		~ 2.5%
λ_{max} 217 nm		

(ii) Heteroannular Dienes

In heteroannular dienes, conjugated double bonds are not present in the same ring and these have *s-trans* (*transoid*) configurations:

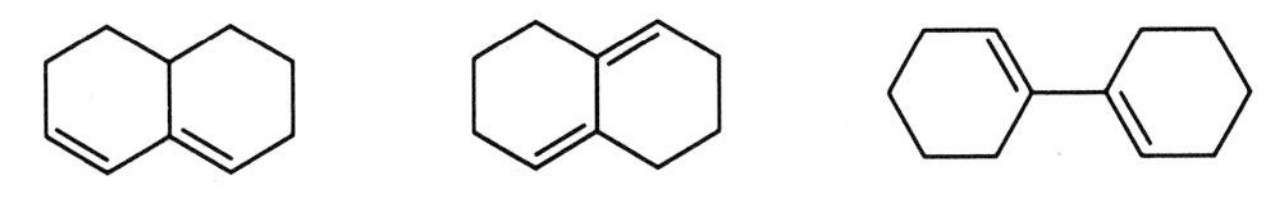

(iii) Exocyclic Conjugated Double Bonds

The carbon-carbon double bonds projecting outside a ring are called *exocyclic double bonds*. For example

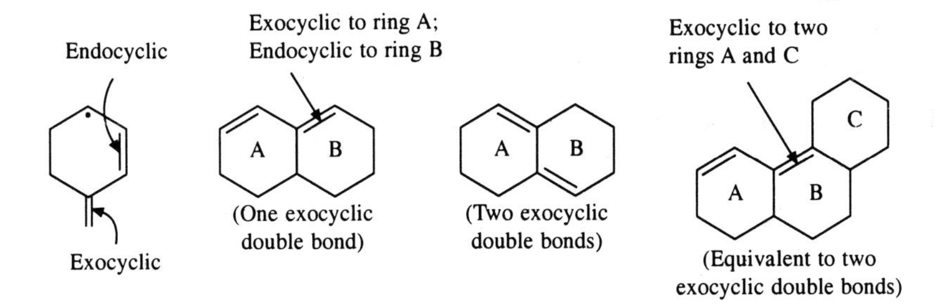

Note that the same double bond may be exocyclic to one ring, while endocyclic to the other and sometimes the same double bond may be exocyclic to two rings simultaneously.

(iv) Alkyl Substituents and Ring Residues

Only the alkyl substituents and ring residues attached to the carbon atoms constituting the conjugated system of the compound are taken into account. Following examples indicate such carbon atoms by numbers and the alkyl substituents and ring residues by dotted lines:

$$CH_3—CH_2 \vdots \overset{4}{C}H=\overset{3}{C}H—\overset{2}{C}H=\overset{1}{C}H \vdots CH_3$$

Two alkyl substituents

$$\text{(cyclohexylidene)} \overset{4}{=}\overset{3}{C}H—\overset{2}{C}H=\overset{1}{C}H_2$$

Two ring residues

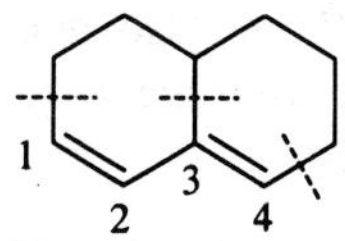

Three ring residues 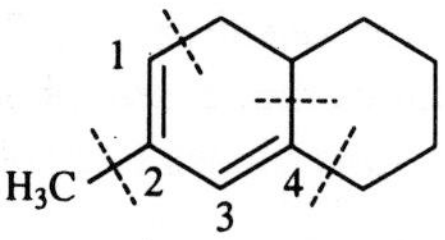 | Three ring residues and one alkyl substituent 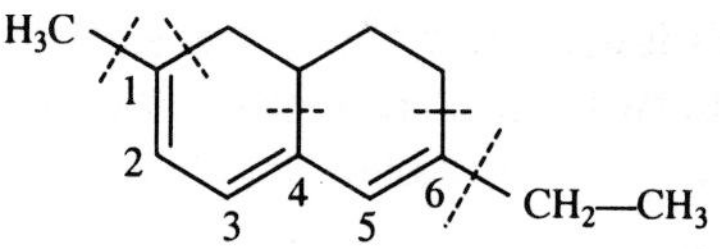 | Three ring residues and two alkyl substituents

In compounds containing both homoanular and heteroannular diene systems, the calculations are based on the longer wavelength (253 nm), i.e. the homoannular diene system.

The calculated and observed values of λ_{max} usually match within ±5 nm as shown in the following examples illustrating the applications of Woodward-Fieser rules (Table 2.3).

Example 1. Using Woodward-Fieser rules, calculate λ_{max} for 2,3-dimethyl-1,3-butadiene.

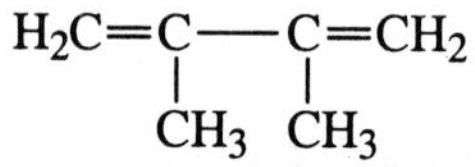

$$H_2C{=}C(CH_3){-}C(CH_3){=}CH_2$$

It is an acyclic diene with two alkyl substituents.
Thus, λ_{max} of this compound is

Base value	214 nm
Two alkyl substituents (2 × 5)	10 nm
Calculated λ_{max}	224 nm
Observed λ_{max}	226 nm

Example 2. Calculate the wavelength of the maximum UV absorption for

Myrcene

Since, it is an acyclic diene with one alkyl substituent, thus

Base value	214 nm
One alkyl substituent	5 nm
Calculated λ_{max}	219 nm
Observed λ_{max}	224 nm

Example 3. Predict the value of λ_{max} for

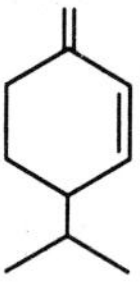

β-phellandrene

This is a heteroannular diene (conjugated double bonds are not in the same ring) with two ring residues and one exocyclic double bond, hence

Base value	214 nm
Two alkyl substituents (2 × 5)	10 nm
One exocyclic double bond	5 nm
Predicted λ_{max}	229 nm
Observed λ_{max}	232 nm

Example 4. Applying Woodward-Fieser rules, calculate the value of absorption maximum for the ethanolic solution of

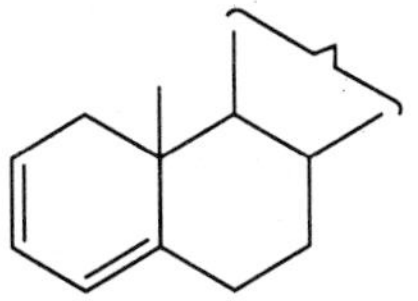

This is a homoannular diene with three ring residues and one exocyclic double bond, thus

Base value	253 nm
Three ring residues (3 × 5)	15 nm
One exocyclic double bond	5 nm
Calculated λ_{max}	273 nm
Observed λ_{max}	275 nm (ε_{max} 10,000)

Example 5. Calculate λ_{max} for the ethanolic solution of

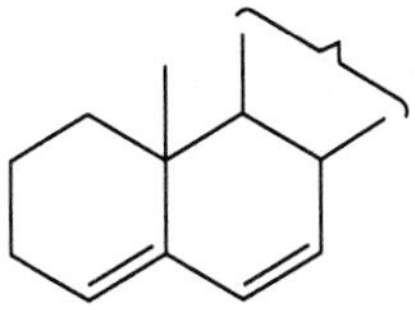

This is a heteroannular diene with three ring residues and one exocyclic double bond, thus

Base value	214 nm
Three ring residues (3 × 5)	15 nm
One exocyclic double bond	5 nm
Calculated λ_{max}	234 nm
Observed λ_{max}	235 nm (ε_{max} 19,000)

Example 6. Hydrogenation of one mole of the triene A with one mole of H_2 gives isomeric dienes having molecular formula $C_{10}H_{14}$. Show how UV spectroscopy and the expected λ_{max} values could distinguish these isomers:

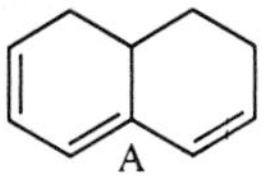

A

The hydrogenation of one mole of the triene A with one mole of H_2 may give three isomeric dienes B, C and D with molecular formula $C_{10}H_{14}$:

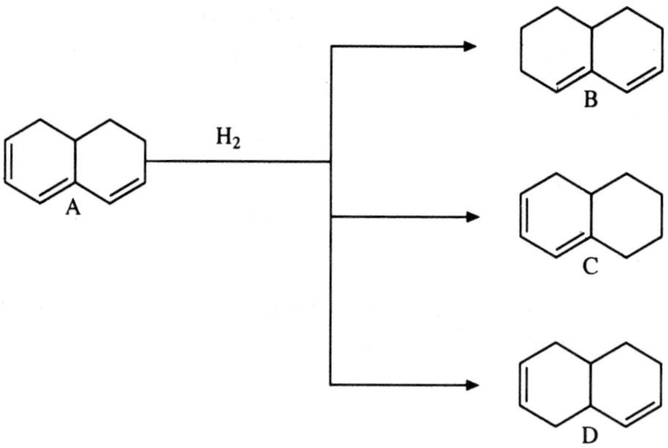

H_2

A B C D

Diene B is a heteroannular conjugated diene with three ring residues and one exocyclic double bond, thus its expected $\lambda_{max} = 214 + 3 \times 5 + 5 = 234$ nm.

Diene C being a homoannular conjugated diene with three ring residues and one exocyclic double bond, thus its expected $\lambda_{max} = 253 + 3 \times 5 + 5 = 273$ nm.

Diene D is an unconjugated diene, hence $\lambda_{max} < 200$ nm.

Thus, by comparing the expected λ_{max} values of the isomeric dienes B, C and D with their observed values of λ_{max}, we can distinguish these isomers.

Example 7. An organic compound can have one of the following structures:

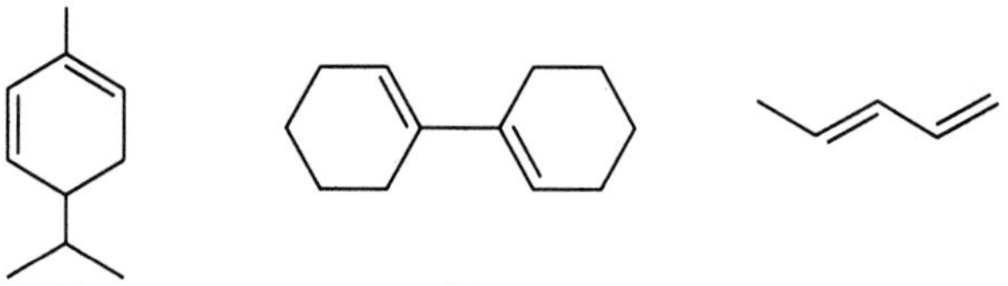

(a) (b) (c)

The λ_{max} of the compound is 236 nm. Which is the most likely structure of the compound? Explain your choice.

Let us calculate the λ_{max} for each of these structures:

Structure (a) is a homoannular diene with two ring residues and one alkyl substituent, hence its calculated $\lambda_{max} = 253 + 3 \times 5 = 268$ nm.

Structure (b) represents a heteroannular diene with four ring residues, hence its calculated $\lambda_{max} = 214 + 4 \times 5 = 234$ nm.

Structure (c) shows an acyclic conjugated diene with one alkyl substituent, hence its calculated $\lambda_{max} = 214 + 5 = 219$ nm.

Since the given λ_{max} of the compound is 236 nm, its most likely structure is (b) because the calculated λ_{max} (234 nm) for this structure is very close to the given value (236 nm).

Example 8. Applying Woodward-Fieser rules, calculate the value of absorption maximum for

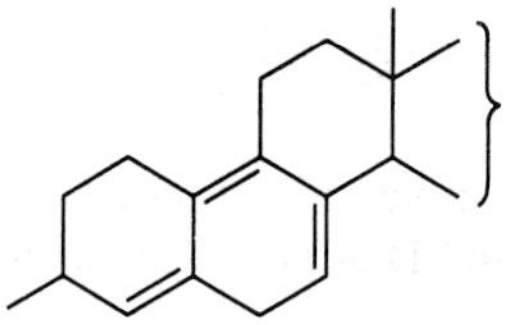

It contains both homoannular and heteroannular diene systems but the calculation of its λ_{max} will be based on the homoannular diene system. There are six ring residues attached to the carbon atoms of the entire conjugated system, one double bond extending conjugation, two exocyclic double bonds and one double bond exocyclic to two rings simultaneously. Thus, λ_{max} of this compound is calculated as:

Base value	253 nm
Six ring residues (6 × 5)	30 nm
One double bond extending conjugation	30 nm
Two exocyclic double bonds (2 × 5)	10 nm
One double bond exocyclic to two rings simultaneously	10 nm
Calculated λ_{max}	333 nm

Example 9. The following dienes have the experimental λ_{max} 243 and 265 nm in ethanol. Giving reasons, correlate the λ_{max} values to the structures (a) and (b):

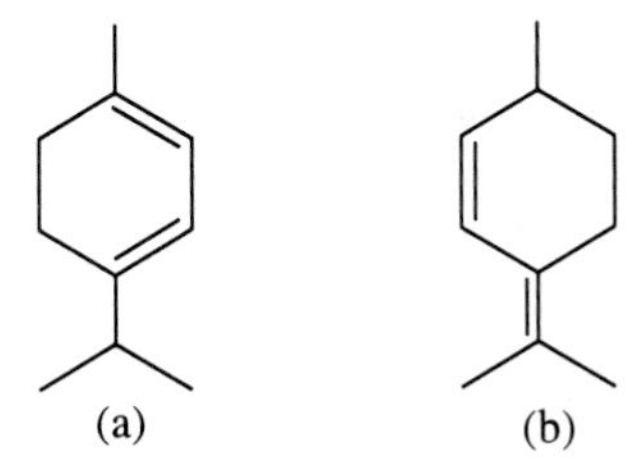

The calculated λ_{max} for (a) (a homoannular diene with two ring residues and two alkyl substituents) is 253 + 2 × 5 + 2 × 5 = 273 nm. Thus, structure (a) is the diene having experimental λ_{max} 265 nm as this value is in fair agreement with the calculated λ_{max} for (a).

The calculated λ_{max} for (b) (a heteroannular diene with two ring residues, two alkyl substituents and one exocyclic double bond) is 214 + 2 × 5 + 2 × 5 + 5 = 239 nm. Thus, the structure (b) is the diene having the experimental λ_{max} 243 nm because this value is in agreement with the calculated λ_{max} for (b).

Example 10. Which of the following compounds is expected to have higher value of λ_{max} and why?

—CH_2—   =CH—CH_3

(a)   (b)

Compound (a) is an unconjugated diene. Hence, it will have λ_{max} < 200 nm. The compound (b) is a heteroannular conjugated diene with two ring residues, one alkyl substituent and one exocyclic double bond. Thus, it is expected to have λ_{max} (214 + 2 × 5 + 5 + 5) = 234 nm, i.e. higher than that of the compound (a).

(v) Exceptions to Woodward-Fieser Rules

Distortion of the chromophore is the most important factor responsible for deviations from the predicted values of λ_{max} for dienes and trienes. Distortion of the chromophore may cause red or blue shifts depending on the nature of the distortion. Thus, the strained diene verbenene (III) has λ_{max} 245.5 nm compared to 232 nm of *β*-phellandrene (II), whereas the calculated value for both is 229 nm. In such bicyclic compounds, 15 nm is added as the ring strain correction to the calculated values of absorption maxima. The diene (IV) which is expected to have λ_{max} 234 nm but actually it is 220 nm (ε_{max} 10,050). This blue shift is due to the distortion of the chromophore resulting in the loss of coplanarity of the double bonds with consequent loss of conjugation. On the other hand, the expected coplanarity of the double bonds in the diene (V) is confirmed by its λ_{max} 243 nm (ε_{max} 15,800) although it still does not agree with the calculated value 234 nm.

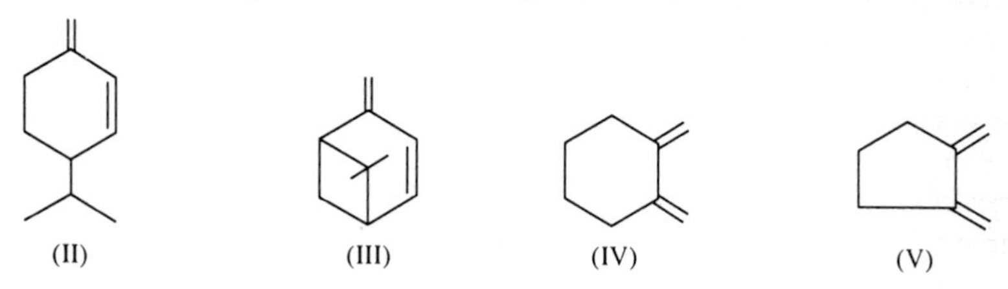

In general, if the strain in the molecule does not considerably affect the coplanarity of the conjugated system, then a bathochromic shift occurs. If the strain forces the chromophore out of coplanarity, then there is loss of conjugation and a hypsochromic shift occurs.

The change of ring size of homoannular dienes from six-membered to any other, say five- or seven-membered, leads to deviations from predicted values due to distortion of the chromophore resulting in the loss of conjugation. Thus, calculations for homoannular dienes are valid only for six-membered rings.

Woodward-Fieser rules work well only for conjugated systems containing up to four double bonds. These rules do not apply satisfactorily to cross (branched)-conjugated svstems like .

2.14 Polyenes and Poly-ynes

As the number of conjugated double bond increases, the values of λ_{max} and ε_{max} also increase and several subsidiary peaks appear. For example, lycopene (the compound responsible for the red colour of tomatoes) has eleven conjugated doublc bonds and absorbs in the visible region (λ_{max} 474 nm, ε_{max} 18.6×10^4). As the chain length in conjugated polyenes increases, the value of λ_{max} for long chains also increases, but not proportionately. This is probably due to the variation in bond lengths between the double and single bonds. In the following cyanine dye analogue, the resonance leads to uniform bond length and bond order along the polyene chain. Since there is no change in bond length in the polyene chain, calculated values of λ_{max} are in agreement with the observed values.

$$Me_2\overset{+}{N}{=}CH{-}(CH{=}CH)_n{-}\ddot{N}Me_2 \longleftrightarrow Me_2\ddot{N}{-}(CH{=}CH)_n{-}CH{=}\overset{+}{N}Me_2$$

In a long chain polyene, change from *trans*- to *cis*-configuration at one or more double bonds lowers both λ_{max} and ε_{max}.

Alkynes, like alkenes, absorb strongly below 200 nm due to $\pi \rightarrow \pi^*$ transitions. When triple bond is conjugated with one or more chromophoric groups as in polyenynes and poly-ynes, radiation of longer wavelength is absorbed. For example, acetylene absorbs at 150 and 173 nm, while vinylacetylene absorbs at 219 and

228 nm. Acetylenic compounds exhibit high intensity bands at shorter wavelength and low intensity bands at longer wavelengths, e.g. —(C≡C)$_3$— shows two bands at λ_{max} 207 nm, ε_{max} 1,35,000 and λ_{max} 306 nm, ε_{max} 120.

2.15 Woodward-Fieser Rules for Calculating λ_{max} in α,β-Unsaturated Carbonyl Compounds

Compounds containing a carbonyl group (C=O) in conjugation with an ethylenic groups (C=C) are called enones. UV spectra of enones are characterized by an intense absorption band (*K*-band) due to $\pi \rightarrow \pi^*$ transition in the range 215-250 nm (ε_{max} usually 10,000–20,000) and a weak *R*-band due to $n \rightarrow \pi^*$ transition in 310-330 nm region (ε_{max} usually 10-100).

Similar to dienes and trienes, there are set rules called Woodward-Fieser rules for calculating or predicting λ_{max} in α,β-unsaturated carbonyl compounds (enones). These rules first framed by Woodward and modified by Fieser and by Scott are given in Table 2.4.

Table 2.4 Woodward-Fieser rules for calculating λ_{max} in α,β-unsaturated carbonyl compounds

$$\delta{-}\overset{\delta}{C}{=}\overset{\gamma}{C}{-}\overset{\beta}{C}{=}\overset{\alpha}{C}{-}C{=}O$$

Base values for		
Acyclic α,β-unsaturated ketones		215 nm
Six-membered cyclic α,β-unsaturated ketones		215 nm
Five-membered cyclic α,β-unsaturated ketones		202 nm
α,β-unsaturated aldehydes		207 nm
Increment for each		
Double bond extending conjugation		30 nm
Alkyl group or ring residue	α	10 nm
	β	12 nm
	γ and higher	18 nm
—OH (hydroxy)	α	35 nm
	β	30 nm
	δ	50 nm
—OAc (acyloxy)	α, β, δ	6 nm
—OMe (methoxy)	α	35 nm
	β	30 nm
	γ	17 nm
	δ	31 nm
—SR (alkylthio)	β	85 nm
—Cl	α	15 nm
	β	12 nm
—Br	α	25 nm
	β	30 nm
—NR_2	β	95 nm
Exocyclic carbon-carbon double bond		5 nm
Homoannular diene component		39 nm
Calculated[†] λ_{max} (in EtOH)		Total nm

[†]For $\pi \rightarrow \pi^*$ transition (*K*-band).

For calculated λ_{max} in other solvents, a solvent correction given in Table 2.5 must be carried out.

Table 2.5 Solvent corrections

Solvent	Correction (nm)
Ethanol	0
Methanol	0
Dioxane	+5
Chloroform	+1
Ether	+7
Hexane	+11
Cyclohexane	+11
Water	–8

Since carbonyl compounds are polar, the positions of the *K*- and *R*-bands of enones are dependent on the solvent. Hence, solvent corrections are required (Table 2.5) to obtain the calculated values of λ_{max} in a particular solvent. The ε_{max} for *cisoid* enones are usually <10,000, while that of transoid are >10,000. The calculated values of λ_{max} are usually within ±5 nm of the observed values as shown in the following examples illustrating the applications of Woodward-Fieser rules (Table 2.4).

Example 1. Applying Woodward-Fieser rules, calculate the λ_{max} for the ethanolic solution of mesityl oxide

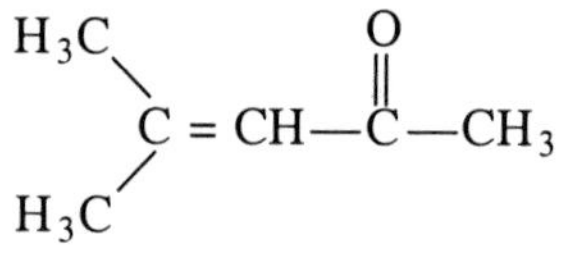

$$(H_3C)_2C{=}CH{-}\overset{O}{\overset{\|}{C}}{-}CH_3$$

This being an acyclic α,β-unsaturated ketone with two β-alkyl substituents, the λ_{max} of this compound is calculated as

Base value	215 nm
Two β-alkyl substituents (2 × 12)	24 nm
Calculated λ_{max}	239 nm
Observed λ_{max}	237 nm

Example 2. Predict the value of λ_{max} (hexane) for

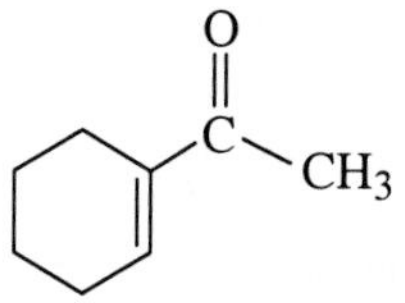

(1-acetylcyclohexene: cyclohexene ring bearing C(=O)CH$_3$)

O, C, CH_3

This is a six-membered cyclic α,β-unsaturated ketone with one α- and one β-alkyl substituents. Hence,

Base value	215 nm
One α-alkyl substituent	10 nm
One β-alkyl substituent	12 nm
Predicted λ_{max} (EtOH)	237 nm
Observed λ_{max} (EtOH)	249 nm
Calculated λ_{max} (hexane)	237 (EtOH)
Solvent correction	+11
	248 nm

Example 3. Calculate λ_{max} (EtOH) for

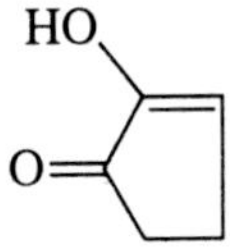

HO

O

It is a five-membered cyclic α,β-unsaturated ketone with one α-hydroxy and one β-ring residue. Thus,

Base value	202 nm
One α-hydroxy groups	35 nm
One β-ring residue	12 nm
Calculated λ_{max} (EtOH)	249 nm
Observed λ_{max} (EtOH)	247 nm

Example 4. What is the expected λ_{max} for

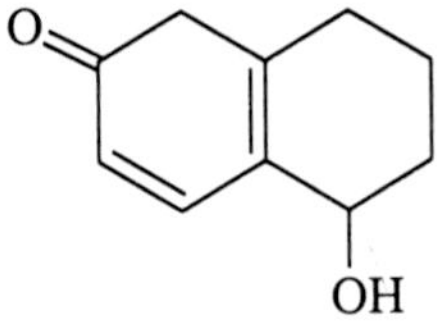

O

OH

This being an α,β-unsaturated cyclohexenone system, hence

Base value	215 nm
Ring residues γ (1)	18 nm
δ (2)	36 nm
One homoannular diene component	39 nm
One double bond extending conjugation	30 nm
Expected λ_{max} (EtOH)	338 nm

Example 5. Calculate λ_{max} for

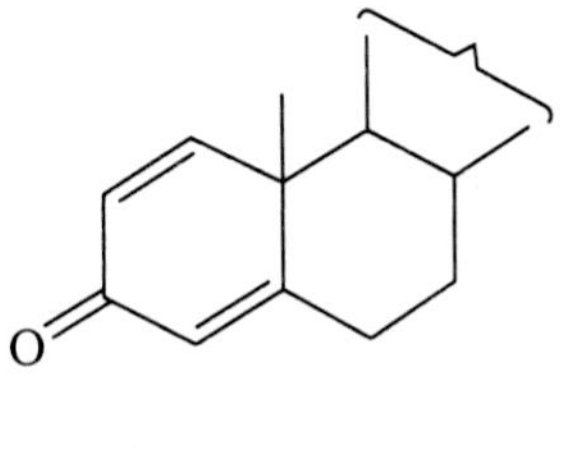

O

It is a six-membered cyclic α,β-unsaturated ketone in which the carbonyl group has α,β-unsaturation on either side, i.e. there is cross (branched) conjugation in this compound. In such cases, λ_{max} is calculated by considering most highly substituted conjugated system which gives higher value of λ_{max}. Thus,

Base value	215 nm
Two β-ring residues (2 × 12)	24 nm
One exocyclic carbon-carbon double bond	5 nm
Calculated λ_{max} (EtOH)	244 nm
Observed λ_{max} (EtOH)	245 nm

Example 6. Calculate the wavelength of the maximum UV absorption for the ethanolic solution of

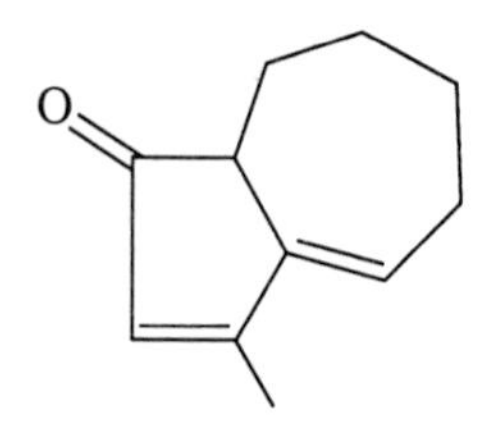

It is an α,β-unsaturated cyclopentenone system. Hence,

Base value	202 nm
One β-alkyl substituent	12 nm
One double bond extending conjugation	30 nm
One exocyclic carbon-carbon double bond	5 nm
One γ-ring residue	18 nm
One δ-ring residue	18 nm
Calculated λ_{max} (EtOH)	285 nm
Observed value (EtOH)	287 nm

Example 7. An organic compound can have one of the following structures:

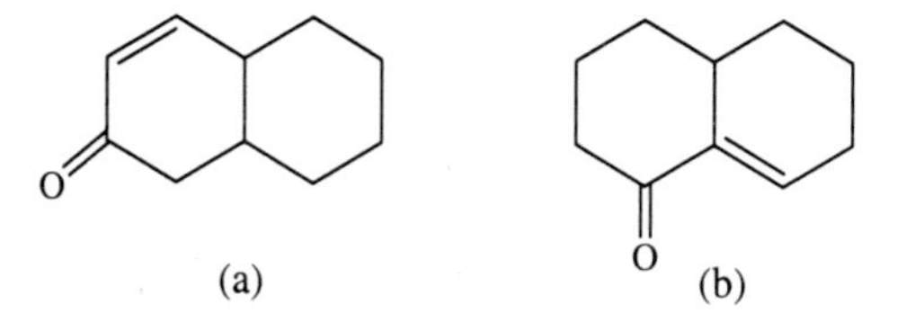

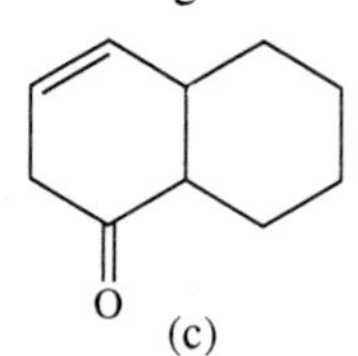

(a) (b) (c)

The λ_{max} of ethanolic solution of the compound is 242 nm. Which is the most likely structure of the compound? Explain your choice.

Let us calculate λ_{max} for each of the above structures.

The calculated λ_{max} (EtOH) for:

(a) 215 (base) + 12 (1 β-ring residue) = 227 nm.
(b) 215 (base) + 10(1 α-ring residue) + 12 (1 β-ring residue) + 5(1 exocyclic carbon-carbon double bond) = 242 nm.
(c) It has an unconjugated system. Hence its C=C will show $\lambda_{max} < 200$ nm due to $\pi \rightarrow \pi^*$ transition, and C=O will show two bands at ~180 and 285 nm due to $\pi \rightarrow \pi^*$ and $n \rightarrow \pi^*$ transitions, respectively.

Since the given λ_{max} of the compound is 242 nm, its most likely structure is (b) because the calculated λ_{max} for this structure is equal to the given value.

Example 8. The following α,β-unsaturated ketones have λ_{max} 236, 256 and 270 nm in ethanol. Explain which is which?

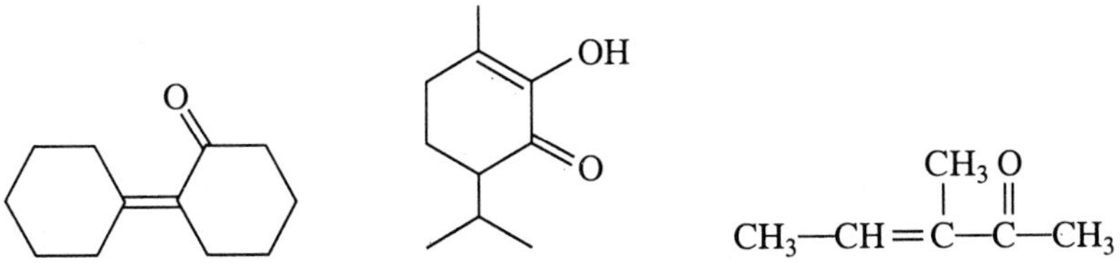

O

OH

O

CH_3 O

CH_3—CH=C—C—CH_3

(a) (b) (c)

The calculated λ_{max} for (a) is 215 (base) + 10 (1 α-ring residue) + 2 × 12 (2β-ring residues) + 10 (the same C=C exocyclic to two rings) = 259 nm (EtOH). Thus, (a) is the α,β-unsaturated ketone having λ_{max} 256 nm because this value agrees with the calculated λ_{max} for (a).

The calculated λ_{max} for (b) is 215 (base) + 35(1 α-OH) + 2 × 12 (2β-ring residues) = 274 nm (EtOH). Thus, (b) is the α,β-unsaturated ketone having λ_{max} 270 nm because this value agrees with the calculated λ_{max} for (b).

The calculated λ_{max} for (c) is 215 (base) + 10 (1 α-alkyl substituent) + 12 (1β-alkyl substituent) = 237 nm (EtOH). Thus, (c) is the α,β-unsaturated ketone having λ_{max} 236 nm because this value agrees with the calculated λ_{max} for (c).

Example 9. Predict the λ_{max} for the ethanolic solution of

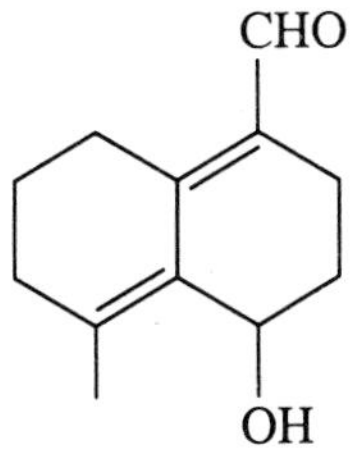

CHO

OH

This compound has an α,β-unsaturated aldehyde system. Thus,

Base value	207 nm
One double bond extending conjugation	30 nm
One α-ring residue	10 nm
One β-ring residue	12 nm
One γ-ring residue	18 nm
One δ-ring residue	18 nm
One δ-alkyl substituent	18 nm
Two carbon-carbon exocyclic double bonds (2 × 5)	10 nm
Predicted λ_{max} (EtOH)	323 nm

Example 10. Calculate λ_{max} for

The λ_{max} for the above compound can be calculated as:

Base value	215 nm
One double bond extending conjugation	30 nm
One α-ring residue	10 nm
One γ-alkyl substituent	18 nm
Two δ-ring residues (2 × 18)	36 nm
Two exocyclic C=C (2 × 5)	10 nm
Calculated λ_{max} (EtOH)	319 nm

Similar to that of conjugated dienes and trienes (Section 2.13), there are deviations from the calculated values of λ_{max} for α,β-unsaturated carbonyl compounds also due to distortion in the chromophore. For example, verbenone (VI) shows λ_{max} 253 nm but its calculated λ_{max} is 239 nm, i.e. there is an increment of 14 nm for strain.

Thus, in such bicyclic compounds, 14 nm should be added as the ring strain correction to the calculated values of λ_{max}.

(VI)
Verbenone

(VII)

Observed λ_{max} 243 nm, ε_{max} 1400
Calculated λ_{max} 249 nm

Similarly, in cyclic systems like (VII), the intensity of the *K*-band may be reduced to $<10^4$ due to steric hindrance which prevents coplanarity. If a carbonyl group is in a five-membered ring and the double bond is exocyclic to the five-membered ring, then the base value of 215 nm gives satisfactory results.

2.16 Dicarbonyl Compounds

Acyclic α-diketones exist in *s-trans* conformation and show the normal weak *R*-band and a weak *K*-band arising from the conjugation between carbonyl groups, e.g. biacetyl shows λ_{max} 275 nm (*R*-band) and ~450 nm (*K*-band). Cyclic α-diketones with α-hydrogen atom(s) exist almost exclusively in the enolic form. Thus, their absorption is related to α,β-unsaturated carbonyl compounds, e.g. diosphenol. In strong alkaline solution there is a bathochromic shift of 50 nm due to the formation of enolate ion because the greater the number of *n* electrons available for conjugation, lesser is the transition energy, i.e. longer is the wavelength of the maximum absorption. This enables the characterization of enolic structures like diosphenols (six-membered cyclic α-diketones).

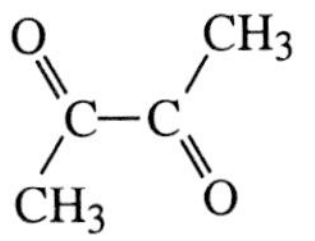

O CH3
C–C
CH3 O
s-trans-biacetyl
λ_{max} 450 nm

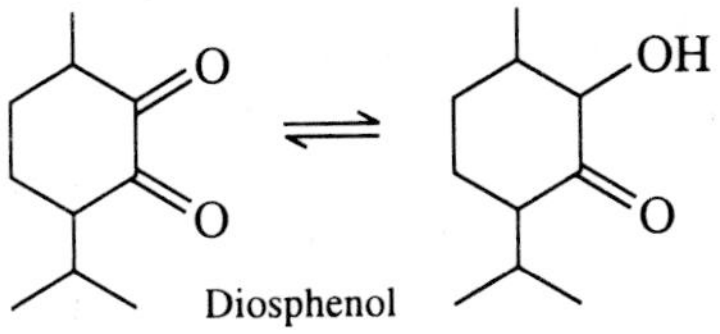

Diosphenol

Calculated λ_{max} (EtOH) = 215 (base) + 24 (2β-ring residues) + 35 (1 α-OH) = 274 nm

Observed λ_{max} (EtOH) = 270 nm

The absorption maxima of β-diketones depend on the concentration of the enol tautomer, i.e. the degree of enolization. For example, acetylacetone exists in enolic form up to 91–92% in solution in non-polar solvents or in the vapour phase. However, in this case the calculated λ_{max} 257 nm is not in good agreement with the observed value (λ_{max} 272 nm in isooctane).

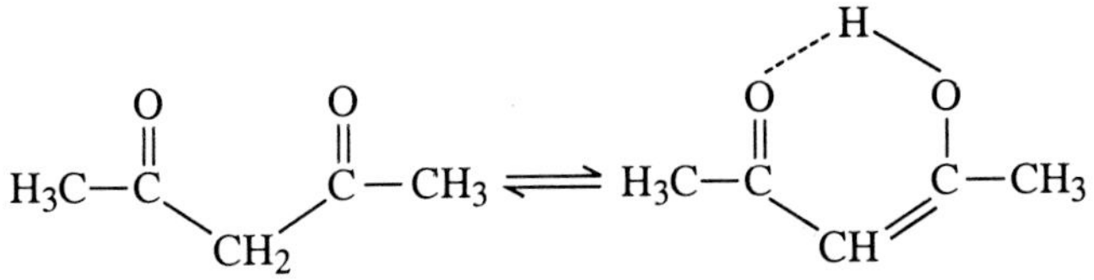

$H_3C-C(=O)-CH_2-C(=O)-CH_3 \rightleftharpoons H_3C-C(=O)-CH=C(OH)-CH_3$

λ_{max} (isooctane) = 272 nm, ε_{max} 12,000

λ_{max} (H_2O) = 274 nm, ε_{max} 2050

Similar to the cyclic α-diketones, cyclic β-diketones, like 1,3-cyclohexanedione exist almost exclusively in the enolic form even in polar solvents. The enolic forms show λ_{max} in the region 230–260 nm due to the $\pi \rightarrow \pi^*$ transition in the *s-trans* enone system, e.g. 1,3-cyclohexanedione shows λ_{max} (EtOH) 253 nm, ε_{max} 22,000 which is in good agreement with the calculated value (257 nm). This may be due to the difference in the configuration of enolic forms of acyclic β-diketones (*s-cis*) and that of the cyclic ones (*s-trans*). In these cases also, the formation of enolate ion in alkaline solution shifts λ_{max} into the 270–300 nm region.

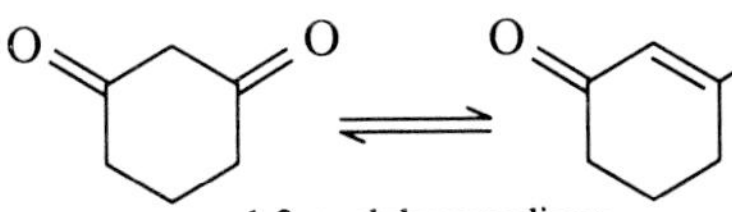

1,3-cyclohexanedione

Calculated λ_{max} (EtOH) : 215 (base) + 30 (1 α-OH) + 12 (1β-ring residue) = 257 nm

Observed λ_{max} (EtOH) = 253 nm

Quinones are α-, or vinylogons α-diketones. The spectrum of *p*-benzoquinone (VIII), is similar to that of a typical α,β-unsaturated ketone and shows λ_{max} 242 nm (strong *K*-band): 281 and 434 nm (weak *R*-bands). The weak $n \rightarrow \pi^*$ transitions (*R*-bands) stretching into the visible region are responsible for imparting colour to some simple quinones and α-diketones in the diketo form.

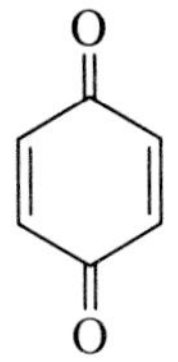

λ_{max} (hexane) = 242 nm, ε_{max} 24,000 *K*-bands

281 nm, ε_{max} 400 ⎤
434 nm, ε_{max} 20 ⎦ *R*-bands

(VIII)

2.17 α,β-Unsaturated Carboxylic Acids and Esters

The attachment of groups containing a lone pair of electrons to the carbonyl group, as in carboxylic acids, esters and amides, shifts the $n \rightarrow \pi^*$ transition to shorter wavelengths (200–220 nm) region with little effect on intensity. This hypsochromic shift is due to combined inductive and resonance effects. For example, compared to the *R*-band of acetaldehyde (λ_{max} 290 nm) that of acetic acid, ethyl acetate and acetamide appear at 204, 211 and 220 nm, respectively.

α,β-unsaturated acids, esters and amides display strong *K*-bands characteristic of the conjugated system. α,β-unsaturated acids and esters follow a trend similar to that of enones. Thus, we can calculate λ_{max} of their *K*-bands taking the base value 195 nm and using the data given in Table 2.4. As shown by the following examples, the calculated and observed values of λ_{max} match within ±5 nm.

Example 1. Calculate λ_{max} for

$$H_2C{=}C(CH_3){-}COOH$$

Base value	195 nm
One α-alkyl substituent	10 nm
Calculated λ_{max} (EtOH)	205 nm
Observed λ_{max} (EtOH)	210 nm

Example 2. Predict the λ_{max} for

(cyclohexylidene)=CH—COOH

Base value	195 nm
Two β-ring residues (2 × 12)	24 nm
One exocyclic C = C	5 nm
Predicted λ_{max} (EtOH)	224 nm
Observed λ_{max} (EtOH)	220 nm

2.18 Benzene and Its Derivatives

The UV spectrum of benzene (in hexane) shows three absorption bands, viz. at 184 nm, ε_{max} 60,000; at 204 nm, ε_{max} 7900 and a broad band with multiple peaks or fine structure in the region 230–270 nm; the λ_{max} of most intense peak in this band is 256 nm, ε_{max} 200. All these bands arise from $\pi \rightarrow \pi^*$ transitions. The most intense band (E_1-band) near 180 nm originates from an allowed transition, whereas the low intensity bands near 200 nm (E_2-band) and 260 nm (*B*-band) arise from forbidden transitions in the highly symmetrical benzene molecule.

The *B*-band of benzene and many of its homologues has fine structure, specially when the spectrum is recorded in the vapour phase or in non-polar solvents (Fig. 2.7(a)). The fine structure arises from sub-levels of vibrational absorption upon which the electronic absorption is superimposed. In polar solvents, due to interactions between solute and solvent molecules, the fine structure is either reduced or destroyed (Fig. 2.7(b)).

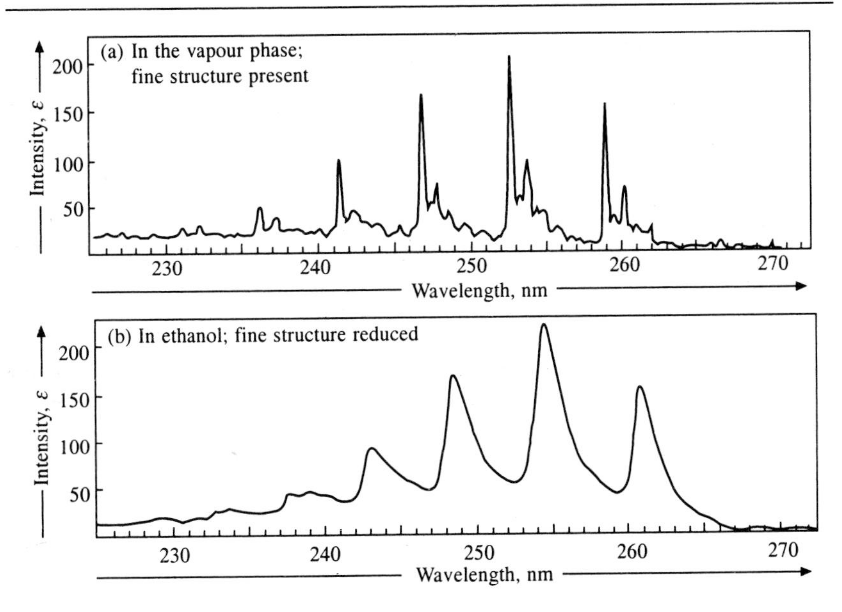

Fig. 2.7 UV spectrum of benzene showing *B*-band in: (a) vapour phase and (b) ethanol

There is a slight bathochromic shift of the *B*-band with a small increase in ε_{max} on introduction of an alkyl group into the benzene ring. This is due to hyperconjugation between the alkyl group and the π-electron system of the ring. The introduction of the second alkyl group into the *para* position is most effective in causing bathochromic shift. The *para* isomer has both the highest λ_{max} and ε_{max}, whereas the *ortho* isomer generally has the lowest λ_{max} and reduced ε_{max}. This effect is because of steric interactions between the *ortho* substituents which effectively reduce hyperconjugation.

CH_3–C_6H_5 ⟷ H^+ H–$\overset{-}{C}H$=C_6H_5 ⟷ H^+ CH_2=$C_6H_5^-$, etc.

B-band, λ_{max} = 261 nm, ε_{max} = 225

If substituents having non-bonding electrons such as OH, NH_2, OMe, etc. are present on the benzene ring, then because of $n \rightarrow \pi^*$, *E*- and *B*-bands are shifted to longer wavelengths with frequent increase in the intensity of the *B*-band (Table 2.6) and loss of its fine structure. The more extensive the conjugation, the less obvious is the vibrational fine structure of the *B*-band.

The spectral characteristics of phenols and aromatic amines change with the change of pH of the solution which causes protonation or deprotonation. For example, coversion of a phenol into the corresponding anion (alkaline solution of phenol) results in the shift of the E_2- and *B*-bands to longer wavelengths with an increase in ε_{max} (Table 2.6).

Table 2.6 Absorption characteristics of benzene and some of its monosubstituted derivatives Ph—*R*

R	λ_{max} (nm)	ε_{max}	λ_{max} (nm)	ε_{max}	Solvent
	E_2-band		*B*-band		
—H	204	7900	256	200	Hexane
—Me	206.5	7000	261	225	Water
$—NH_2$	230	8600	280	1430	Water
$—NH_3^+$	203	7500	254	160	Aq. acid
—NHAc	238	10,500	—	—	Water
—OH	210.5	6200	270	1450	Water
$—O^-$	235	9400	287	2600	Aq. alkali
—OMe	217	6400	269	1480	2% methanol
—OPh	255	11,000	272	2000	Cyclohexane
—SH	236	10,000	269	700	Hexane
—I	207	7000	257	700	Water
—Cl	209.5	7400	263.5	190	Water
—Br	210	7900	261	192	Water
	K-band		*B*-band		
$—CH{=}CH_2$	244	12,000	282	450	Ethanol
—C≡CH	236	12,500	278	650	Hexane
—CHO	244	15,000	280[a]	1500	Ethanol
—COMe	240	13,000	278[b]	1100	Ethanol
$—NO_2$	252	10,000	280[c]	1000	Hexane
—COOH	230	10,000	270	800	Water
$—COO^-$	224	8700	268	560	Aq. alkali
—CN	224	13,000	271	1000	Water
—COPh	252	20,000[d]	Submerged		Ethanol
—Ph	246	20,000	Submerged		Ethanol
—CH=CHPh (*cis*)	283	12,300	Submerged		Ethanol
—CH=CHPh (*trans*)	295	25,000	Submerged		Ethanol
—CH=CHCOOH (*trans*)	273	21,000	Submerged		Water

[a]*R*-band, λ_{max} 328 nm, ε_{max} 20; [b]*R*-band, λ_{max} 319 nm, ε_{max} 50;
[c]*R*-band, λ_{max} 330 nm, ε_{max} 125; [d]*R*-band, λ_{max} 325 nm, ε_{max} 180.

The larger the number of non-bonding electrons available for interaction (conjugation) with the π-electron system of the ring, smaller is the energy difference between the ground and excited states, i.e. the longer is the wavelength of absorption. Since an additional pair of non-bonding electrons in the anion is available for interaction with the π-electron system of the ring, its E_2- and *B*-band are shifted to longer wavelengths with an increase in ε_{max}. Thus, a suspected phenolic structure may be confirmed by comparison of the UV spectrum of the compound recorded in a neutral solution with that recorded in an alkaline solution (pH = 13).

Similarly, when aniline is converted into the anilinium cation (acidic solution of aniline), the pair of non-bonding electrons of the nitrogen atom of aniline is no longer available for interaction with the π-electrons of the ring. This makes the spectral data of the anilinium cation almost identical to that of benzene

(Table 2.6). Thus, a suspected aniline derivative may be confirmed by comparison of UV spectra recorded in neutral and acid solutions (pH = 1).

The interaction between the non-bonding electron pair(s) of the heteroatom attached to the benzene ring and the π-electrons of the ring is most effective when the π-orbital of the non-bonding electrons is parallel to the π orbitals of the ring. This arrangement is considerably disturbed due to twisting in sterically crowded molecules such as N,N-dimethylaniline resulting in a hypsochromic shift in the E_2-band accompanied by a marked reduction in ε_{max}. Thus, N,N-dimethylaniline shows λ_{max} 251 nm, ε_{max} 15,500, whereas 2-methyl-NN-dimethylaniline shows λ_{max} 248 nm, ε_{max} 6360.

If an unsaturated group (chromophore) is directly attached to the benzene ring, then because of π-π conjugation, a strong bathochromic shift of the *B*-band occurs with the appearance of an intense *K*-band (ε_{max} > 10,000) in the region 200–250 nm (Table 2.6).

In disubstituted benzenes, when an electron donating group and an electron attracting group (electronically complementary groups) are *para* to each other, there is a pronounced red shift and increase in intensity of the main absorption band (*K*-band) (Table 2.7) compared to the effect of either group separately (Table 2.6). This is due to the extension of the chromophore through resonance as shown below:

$H_2\ddot{N}-C_6H_4-N(=O)O^- \longleftrightarrow H_2\overset{+}{N}=C_6H_4=\overset{+}{N}(O^-)_2$

λ_{max} (EtOH) = 375 nm, ε_{max} = 16,000

When two groups are *ortho* or *meta* to each other or the groups situated *para* to each other are not complementary, the absorption spectrum usually has close resemblance with that of the separate, non-interacting chromophores. For example, *p*-dinitrobenzene has λ_{max} 260 nm, ε_{max} 13,000, whereas nitrobenzene has λ_{max} 252 nm, ε_{max} 10,000. On comparing the values for *ortho, meta*- and *para*- isomers with each other (Table 2.7) and with the values for the single substituent (Table 2.6), it is clear that the effect is most pronounced when complementary groups are *para* to each other.

Table 2.7 Absorption characteristics of some disubstituted benzene derivatives *R*—C_6H_4—*R′*

R	*R′*	Orientation	*K*-band		*B*-band	
			λ_{max} (nm)	ε_{max}	λ_{max} (nm)	ε_{max}
—NO_2	—NH_2	*o*-	283	5400	412	4500
—NO_2	—NH_2	*m*-	280	4800	358	1450
—NO_2	—NH_2	*p*-	381	13,500	Submerged	
—NO_2	—OH	*o*-	279	6600	351	3200
—NO_2	—OH	*m*-	274	6000	333	1960
—NO_2	—OH	*p*-	318	10,000	Submerged	
—OMe	—CHO	*o*-	253	11,000	319	4000
—OMe	—CHO	*m*-	252	8300	314	2800
—OMe	—CHO	*p*-	277	14,800	Submerged	

Similar to Woodward-Fieser rules, Scott formulated a set of rules for calculating the absorption maximum of the principal band of aromatic aldehydes, ketones, carboxylic acids and esters which are summarized in Table 2.8. In the absence of steric hindrance to coplanarity, the calculated values are usually within ±5 nm of the observed values.

Table 2.8 Rules for calculating λ_{max} of the principal band ($\pi \rightarrow \pi^*$ transition) of substituted benzene derivatives $R—C_6H_4—COG$

	Orientation	nm
Parent chromophore: R = H		
G = alkyl or ring residue; base value		246
G = H; base value		250
G = OH or *o*-alkyl; base value		230
Increment for each substituent:		
R = alkyl or ring residue	*o*-, *m*-	3
	p-	10
R = OH, OMe, *o*-alkyl	*o*-, *m*-	7
	p-	25
R = O^-	*o*-	11
	m-	20
	p-	78
R = Cl	*o*-, *m*-	0
	p-	10
R = Br	*o*-, *m*-	2
	p-	15
R = NH_2	*o*-, *m*-	13
	p-	58
R = NHAc	*o*-, *m*-	20
	p-	45
R = NHMe	*p*-	73
R = NMe_2	*o*-, *m*-	20
	p-	85
Calculated λ_{max} (EtOH) of $R—C_6H_4—COG$		Total nm

The following examples illustrate the application of these rules.

Example 1. Calculate λ_{max} for the ethanolic solution of *p*-chloroacetophenone.

$COCH_3$

Cl

In this compound G is an alkyl group. Thus,

Base value	246 nm
Para Cl	10 nm

Calculated λ_{max} (EtOH)	256 nm
Observed λ_{max} (EtOH)	254 nm

Example 2. Predict the absorption maximum for the ethanolic solution of 3,4-dihydroxyacetophenone.

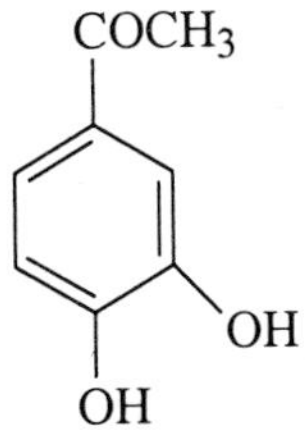

$COCH_3$

OH

OH

In this compound *G* is an alkyl group. Thus,

Base value	246 nm
m-OH	7 nm
p-OH	25 nm
Calculated λ_{max} (EtOH)	278 nm
Observed λ_{max} (EtOH)	281 nm

Example 3. Calculate the absorption maximum for *p*-bromobenzoic acid.

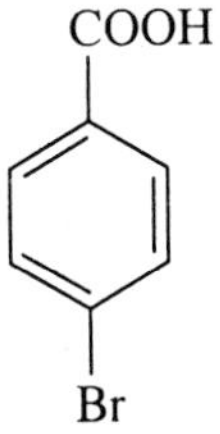

COOH

Br

In this case *G* is OH. Thus,

Base value	230 nm
p-Br	15 nm
Calculated λ_{max} (EtOH)	245 nm
Observed λ_{max} (EtOH)	245 nm

Example 4. Calculate λ_{max} for 6-methoxytetralone.

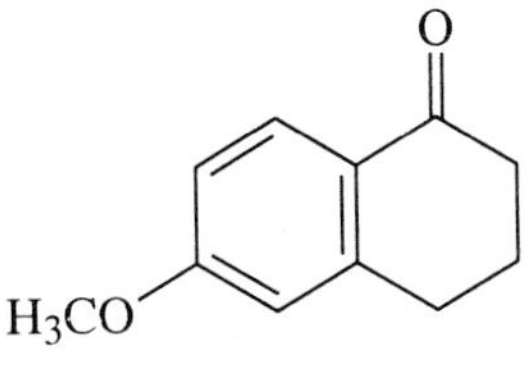

O

H_3CO

In this case *G* is a ring residue. Thus,

Base value	246 nm
o-ring residue	3 nm
p-methoxyl	25 nm
Calculated λ_{max} (EtOH)	274 nm
Observed λ_{max} (EtOH)	276 nm

Example 5. Calculate the absorption maximum for 3-carbethoxy-5-chloro-8-hydroxyl-4-methyltetralone.

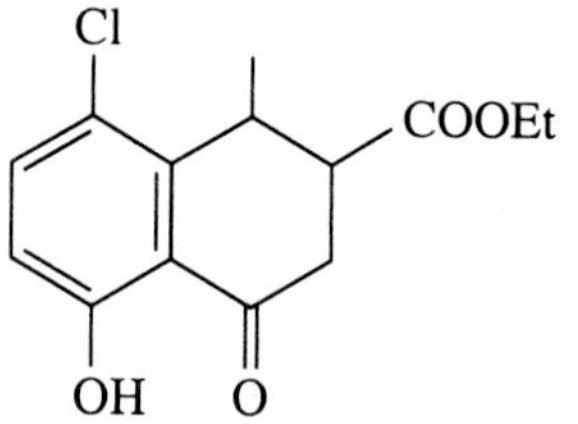

Cl COOEt OH O

In this case *G* is a ring residue. Thus,

Base value	246 nm
o-ring residue	3 nm
o-hydroxyl	7 nm
Calculated λ_{max} (EtOH)	256 nm
Observed λ_{max} (EtOH)	257 nm

Example 6. Predict the λ_{max} for the ethanolic solution of 3,4-dimethoxy-10-oxooctahydrophenanthrene.

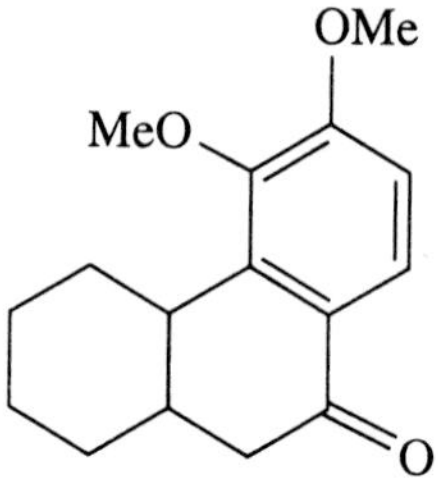

OMe MeO O

In this compound *G* is a ring residue. Thus,

Base value	246 nm
m-methoxyl	7 nm
p-methoxyl	25 nm
o-ring residue	3 nm
Predicted λ_{max} (EtOH)	281 nm
Observed λ_{max} (EtOH)	278 nm

2.19 Polynuclear Aromatic Compounds

Similar to that of benzene, the spectra of polynuclear aromatics are characterized by vibrational fine-structure. Biphenyl and its derivatives have two aromatic rings in conjugation. When the rings are coplanar, the conjugation is most effective overlap of the π-orbitals, but it is least effective when the rings are twisted up to

90°. In biphenyl, the angle of twist is small, hence conjugation between the rings is not affected. Consequently, biphenyl shows a very intense (ε_{max} 19,000) absorption band at 252 nm (*K*-band). In biphenyl derivatives with bulky substituents in the *ortho* positions, the rings are forced out of coplanarity because these molecules are more stable in twisted conformations than in the planar conformation which suffers steric strain due to bulky *ortho* substituents. This results in the loss of conjugation.

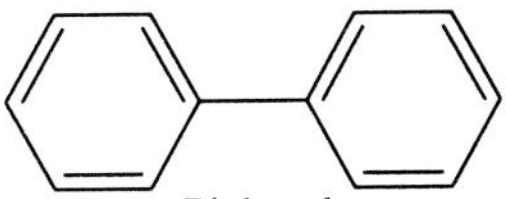

CH_3 CH_3 CH_3 CH_3

Biphenyl   2,2′-dimethylbiphenyl   *o*-xylene

K-band, λ_{max} 252 nm, ε_{max} 19,000

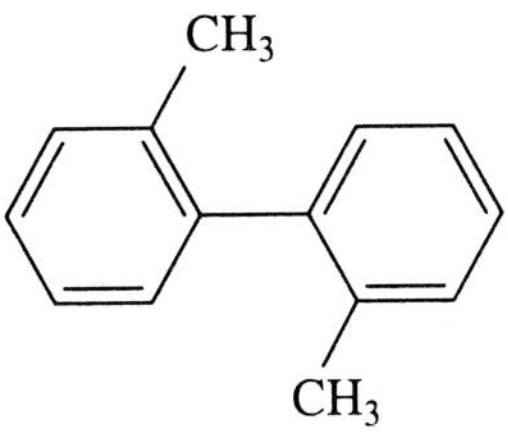

B-band, λ_{max} 270 nm, ε_{max} 800

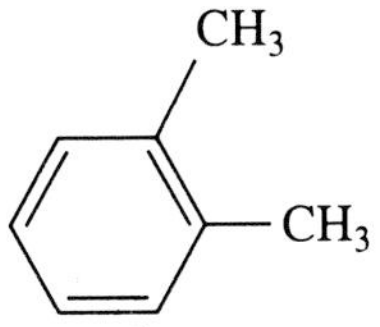

B-band, λ_{max} = 262 nm, ε_{max} = 270

For example, the loss of conjugation due to forcing the rings out of coplanarity is reflected in the UV spectral data of 2,2′-biphenyl which are similar to those of *o*-xylene but different from that of the biphenyl which shows a very intense *K*-band.

We have seen that coplanarity is required for the most effective conjugation which lowers the transition energy resulting in a bathochromic shift accompanied by increased intensity. For example, *trans*-stilbene absorbs at a longer wavelength (295 nm) with a greater intensity (ε_{max} 25,000) compared to the corresponding band (283 nm) in *cis*-stilbene because in the latter there is the destruction of coplanarity by steric interference resulting in the loss of conjugation.

H C=C H   H C=C H

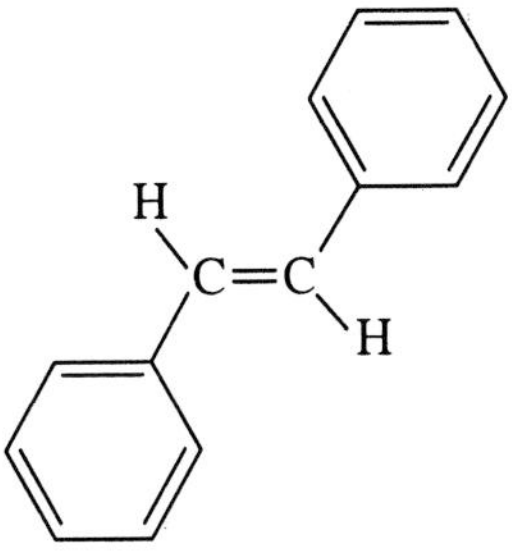

trans-stilbene
λ_{max}(EtOH) = 295 nm, ε_{max} = 25,000

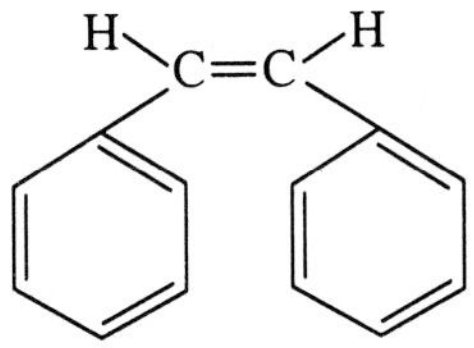

cis-stilbene
λ_{max}(EtOH) = 283 nm, ε_{max} = 12,300

There are two common series of fused-ring aromatic compounds, viz. the linear series such as anthracene and the angular series like phenanthrene. The spectra of the linear series of compounds retain the vibrational fine structure as well as the other absorption bands typical of the benzene ring. The spectra of the angular series of compounds are relatively complicated. In both series, as the number of fused rings increases, the absorption is shifted to progressively longer wavelengths and reaches the visible region. For example, benzene absorbs at 256 nm, ε_{max} 200; naphthalene absorbs at 312 nm, ε_{max} 289, whereas pentacene absorbs in the visible region at 580 nm, ε_{max} 12,600 and is blue.

2.20 Non-benzenoid Aromatic Compounds

Spectra of non-benzenoid aromatics are considerably similar to that of benzenoid aromatics. Actually, the UV spectroscopy is one of the methods for determining whether a particular compound has aromatic character, e.g. tropolone and its derivatives absorb in the region 220–250 nm, ε_{max} ~30,000 and 340–375 nm, ε_{max} ~8000; the latter absorption has vibrational fine structure typical of aromatic systems. The UV spectra of azulene and its derivatives are complicated and consist of a number of intense bands in the UV region (up to 360 nm) and a number of relatively weak bands in the visible region (between 500 and 700 nm). As a consequence of the absorption in the visible region, azulene and most of its derivatives are blue.

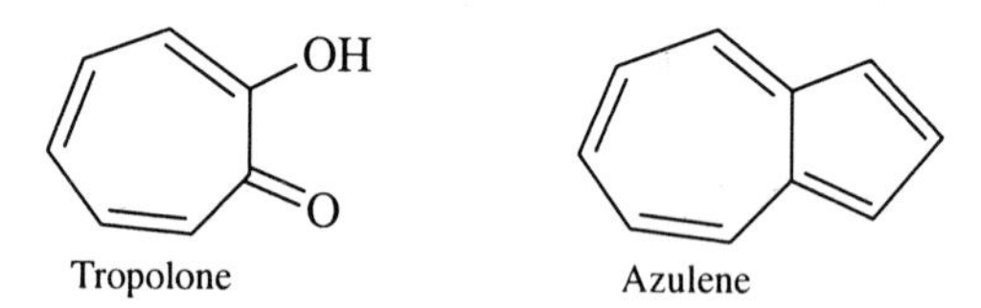

2.21 Heteroaromatic Compounds

In general, the spectra of heteroaromatics are almost similar to that of their corresponding hydrocarbons and their long wavelength band has fine structure analogous to the *B*-band of benzene. Thus, the absorptions of five-membered heteroaromatics are compared to that of cyclopentadiene which absorbs near 200 nm, ε_{max} 10,000 and near 238 nm, ε_{max} 3400. For example, pyrrole shows comparable absorptions at 211 nm, ε_{max} 15,000 and at 240 nm, ε_{max} 300; the latter absorption band has fine structure analogous to the *B*-band of benzene. The presence of an auxochromic or chromophoric substituent on the five-membered heteroaromatic ring causes a bathochromic shift and increased intensity of the bands of the parent molecule.

Similarly, the spectrum of a six-membered aromatic heterocycle, e.g. pyridine, is comparable with that of benzene; the only difference being that the *B*-band of pyridine (λ_{max} 270 nm, ε_{max} 450 at pH > 7) is relatively more intense and has less distinct fine structure. The transition ($\pi \rightarrow \pi^*$) resulting in the *B*-band is allowed for pyridine but forbidden for the more symmetrical benzene molecule. Pyridine shows another absorption band at 257 nm, ε_{max} 2750 (pH > 7). An increase in the solvent polarity produces a hyperchromic effect on the *B*-band of pyridine and its homologues, whereas it has little or no effect on the position or intensity of the *B*-band of benzene. This effect is due to hydrogen bonding through the lone pair of electrons of the nitrogen atom which makes pyridine molecules far less symmetrical than benzene molecules. The effect of substituents on absorption characteristics of pyridine is not clearly defined. The spectra of diazines are similar to that of pyridine.

In general, spectra of simple heteroaromatics resemble spectra of benzenoid aromatics in the sense that they consist of an absorption band of relatively higher intensity (ε_{max} 5000-15,000) at short wavelength (λ_{max} 190-240 nm) and a fine structure band of lower intensity (ε_{max} 1-400) at longer wavelength (λ_{max} 240-300 nm).

2.22 Applications of Ultraviolet and Visible Spectroscopy

A few functional groups (chromophores) may be detected by the ultraviolet and visible (electronic) spectroscopy, but it is especially useful for detecting the presence and elucidating the nature of conjugated systems including aromatic rings. In the application of the electronic spectroscopy for structural analysis, only the region above 200 nm is really useful and the region below 200 nm is hardly useful for this purpose. Some important applications of UV and visible spectroscopy to organic chemistry are summarized as follows.

(i) Detection of a Functional Group (Chromophore)

The presence or absence of a particular chromophore may be indicated by the presence or absence of an absorption band in the expected wavelength region. For example, the presence of a low intensity band in the region 270–300 nm indicates the presence of an aldehydic or ketonic carbonyl group. If the spectrum is transparent above 200 nm, it shows the absence of

(a) an aldehydic or ketonic carbonyl group.
(b) a conjugated-system.
(c) an aromatic ring.
(d) a bromine or iodine atom in the molecule.

However, an unconjugated C=C bond or some other atoms or groups may be present in the molecule if it does not absorb above 200 nm. Thus, no definite conclusion can be drawn regarding the structure of the molecule if it absorbs below 200 nm.

(ii) Detection of Conjugation and Elucidation of Its Nature

Compounds containing a conjugated system including aromatics are characterized by their absorptions above 200 nm. We have noted that the longer the conjugated system, the longer is the wavelength of absorption, usually accompanied by the increased intensity. Also, substitutions on a conjugated system generally cause bathochromic and hyperchromic effects. Thus, we can elucidate the nature of conjugation by comparing the values of λ_{max} and ε_{max} for the compound under study with that of a probable analogous compound.

UV spectral data for a variety of compounds are available for comparison, e.g. acyclic conjugated dienes show intense *K*-bands in region 215–230 nm (ε_{max} ~21,000). Similarly, α,β-unsaturated aldehydes and ketones show an intense *K*-band in the 215–250 nm region with ε_{max} usually 10,000–20,000, and a weak *R*-band in the region 310–330 nm (ε_{max} 10–100). Aromatic compounds are characterized by an intense band (E_2-band) near 200 nm (ε_{max} ~8000), and a weak *B*-band with fine structure near 260 nm (ε_{max} ~200). In all the above cases, additional conjugation as well as substitution generally causes bathochromic and hyperchromic effects.

A particular conjugated system may be recognized in molecules of widely varying complexities because a large portion of such molecules may be transparent in the UV region. This results in a spectrum similar to that of a much simpler molecule. For example, an α,β-unsaturated ketone moiety is easily recognized in a complex steroidal molecule 4-cholesten-3-one by resemblance of its spectrum with that of simpler mesityl oxide molecule (Fig. 2.8).

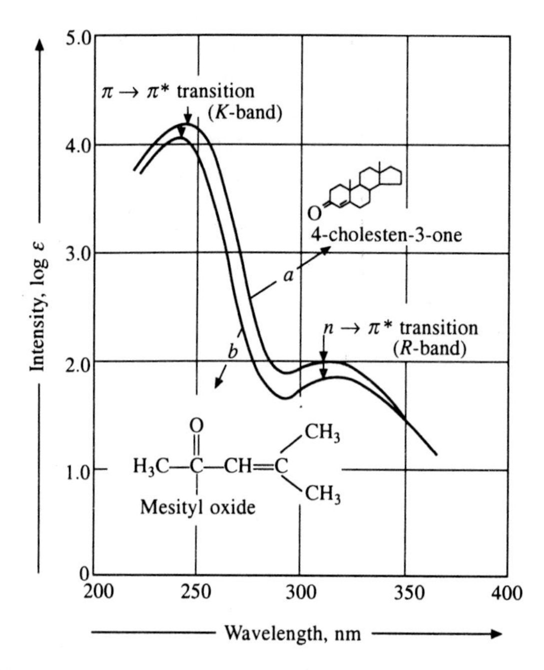

Fig. 2.8 UV spectra of (a) 4-cholesten-3-one and (b) mesityl oxide

(iii) Study of Extent of Conjugation

The values of λ_{max} and ε_{max} increase as the number of conjugated multiple bonds increases, thus the extent of conjugation can be estimated. It has been found that the absorption occurs in the visible region if a polyene has eight or more conjugated double bonds. A polyene with such a sufficient conjugation becomes coloured, e.g. β-carotene (orange) and lycopene (red) having eleven conjugated double bonds are coloured and absorb in the visible region at 450 nm, ε_{max} 14×10^4 and 474 nm, ε_{max} 18.6×10^4, respectively.

(iv) Distinction Between Conjugated and Unconjugated Compounds

In general, electronic spectroscopy can distinguish isomeric conjugated and unconjugated compounds. For example, isomeric dienes 2,4-hexadiene and 2,5-hexadiene can be readily differentiated because the former being a conjugated diene will absorb above 200 nm (227 nm), whereas the latter being an unconjugated diene will absorb below 200 nm (~170 nm). Similarly, an α,β-unsaturated ketone can be readily distinguished from its β,γ-isomer because the former having conjugated system will show both the $\pi \rightarrow \pi^*$ and $n \rightarrow \pi^*$ transition bands at longer wavelengths compared to that of the latter which is an unconjugated compound.

(v) Study of Strain

In molecules like 2-substituted biphenyls, there is steric strain which forces the rings out of coplanarity resulting in the loss of conjugation. This causes hypsochromic and hypochromic effects which are measures of steric strain in such molecules, i.e. the larger are these effects, the greater will be the steric strain.

(vi) Determination of Configurations of Geometrical Isomers

It is possible when there is loss of coplanarity of one isomer due to steric hindrance resulting in the loss of conjugation accompanied by hypsochromic and hypochromic effects. Obviously, this isomer is the *cis*-isomer in which groups are closer to each other to cause steric strain and to force the groups out of coplanarity. Thus, *cis*-isomers absorb at shorter wavelengths and have lower intensity than the *trans*-isomers, e.g. *cis*-stilbene shows λ_{max} 283 nm, ε_{max} 12,300, whereas *trans*-stilbene shows λ_{max} 295 nm, ε_{max} 25,000 (Section 2.18).

(vii) Study of Tautomerism

UV spectroscopy can be used for identifying the predominant (stable) tautomer. For example, 2-hydroxypyridine (IX, R=H) and 2-pyridone (X, R=H equilibrium has been shown to lie far to the right, i.e. 2-pyridone predominates, because the UV spectrum of the solution resembles with that of a solution of N-methyl-2-pyridone (X, R = Me) but not with that of 2-methoxypyridine (IX, R = Me).

(viii) Confirmation of Suspected Phenols and Aromatic Amines

The spectral characteristic of phenols and aromatic amines change with the change of pH of the solution. Thus, suspected phenols and aniline derivatives may be confirmed by comparison of UV spectra recorded in neutral and alkaline or acid solutions (for details, see Section 2.18).

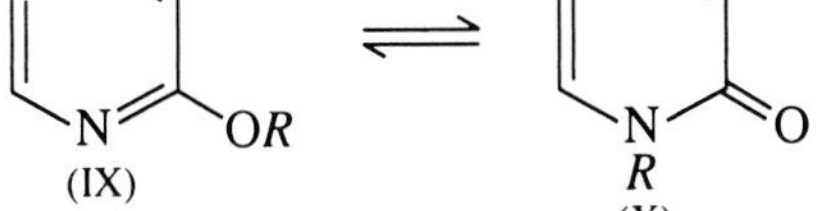

N OR ⇌ N O
R

(IX)
R = Me
λ_{max} < 205 nm (ε_{max} > 5300)
269 nm (ε_{max} = 3230)

(X)
R = Me
λ_{max} = 226 nm (ε_{max} = 6100)
297 nm (ε_{max} = 5700)
R = H
λ_{max} = 224 nm (ε_{max} = 7230)
293 nm (ε_{max} = 5900)

(ix) Study of Structural Features in Different Solvents

In certain cases, the structure of a compound changes with the change in the solvent. For example, chloral hydrate shows an absorption maximum at 290 nm (*R*-band) but this band disappears when the spectrum is recorded in aqueous solution. This shows that the compound has a carbonyl group in hexane solution and its structure is $CCl_3CHO \cdot H_2O$, whereas in aqueous medium, it changes to $Cl_3CCH(OH)_2$.

PROBLEMS

1. What is electronic spectroscopy? Discuss various types of electronic transitions giving at least one example in each case.
2. Discuss the origin of UV-visible spectra and arrange the following electronic excitations in order of their decreasing energy:

$$\pi \to \pi^*, \quad n \to \pi^*, \quad \sigma \to \sigma^*, \quad n \to \sigma^*$$

3. Why are absorption bands formed in UV spectrum instead of sharp peaks?
4. Explain the following:
   (a) Effect of increasing solvent polarity on $n \to \pi^*$, $n \to \sigma^*$ and $\pi \to \pi^*$ transitions.
   (b) Transition probability.
5. (a) Arrange the expected electronic transitions for 2-pentanone in order of their increasing energy.
   (b) What is effect of hydrogen bonding on UV absorptions?
6. Write notes on:
   (a) *K*- and *R*-bands
   (b) Chromophores and auxochromes
   (c) Bathochromic and hypsochromic shifts
   (d) Hyperchromic and hypochromic effects
7. Explain how the presence of two conjugated chromophores in a molecule shifts both the λ_{max} and ε_{max} to higher values.
8. Compounds *A*, *B* and *C* have the same molecular formula C_5H_8 and on hydrogenation, all yield *n*-pentane. Their UV spectra show the values of λ_{max} as: *A*, 176 nm; *B*, 212 nm and *C*, 215 nm (1-pentene has λ_{max} 178 nm). What are the likely structures of *A*, *B* and *C*?
9. Using Woodward-Fieser rules, calculate the λ_{max} of UV absorption for the following compounds:

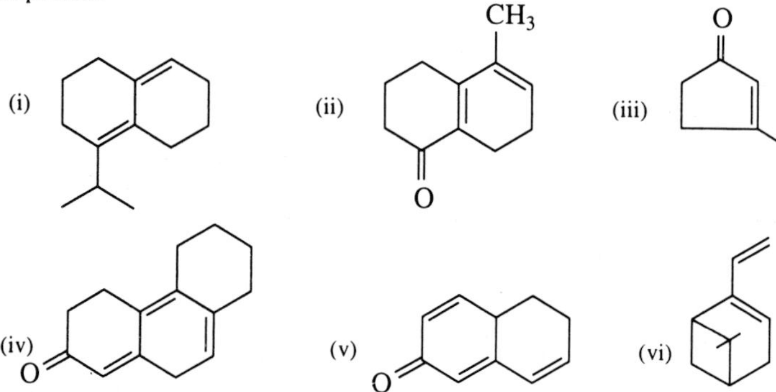

(Hint. For (vi), 15 nm should be added to the calculated value as the ring strain correction.)

10. Discuss the following spectral data (nm):

(i) PhOH (210) (ii) PhOMe (217) (iii) PhO^- (235)

11. Explain why

(a) PhMe, Ph_2CH_2 and Ph_3CH have similar UV spectra (λ_{max} ~262 nm).

(b) 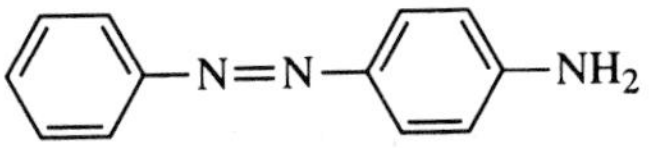 C_6H_5–N=N–C_6H_4–NH_2 has yellow colour, whereas

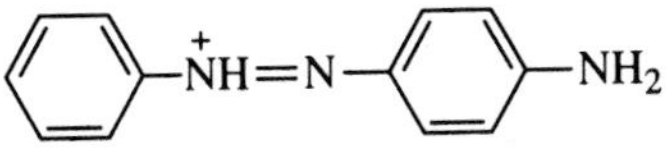

C_6H_5–$\overset{+}{N}H$=N–C_6H_4–NH_2 has violet colour.

(c) The UV spectrum of anilinium cation resembles that of benzene.

(d) The ethanolic solution of phenol shows λ_{max} 210 nm but on addition of dilute NaOH to the solution, the λ_{max} is shifted to 235 nm.

12. Calculate λ_{max} for the following benzene derivatives:

(i) 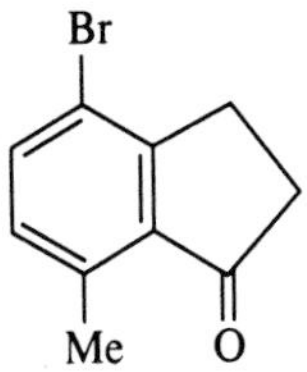 Br, Me, O (ii) 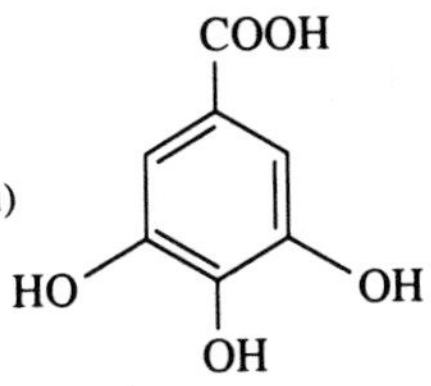 COOH, HO, OH, OH (iii) 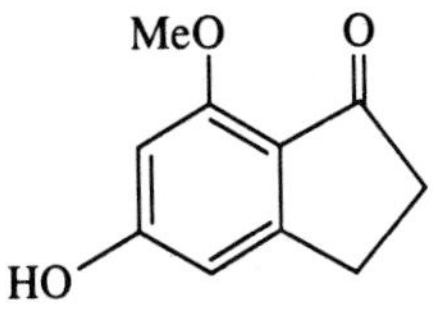 MeO, O, HO

13. The λ_{max} of acetone shifts in different solvents: 279 nm (hexane), 272 nm (ethanol) and 264.5 (water). Explain.

14. Explain why the *cis* isomer of the following compound is colourless, whereas its *trans*-isomer is coloured:

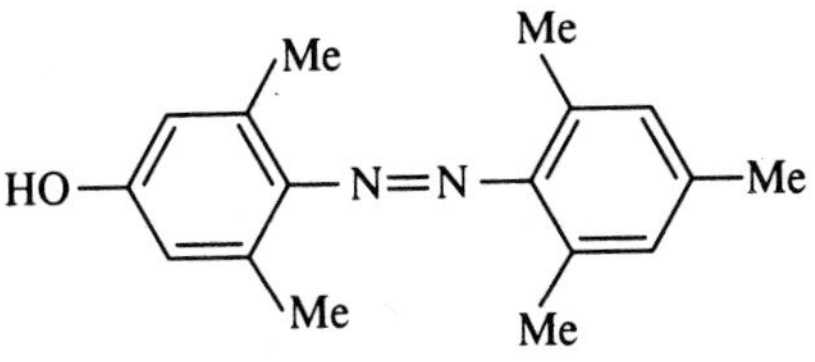

HO, Me, Me, N=N, Me, Me, Me

15. Arrange the following compounds in order of their increasing wavelength of UV absorption maxima:

(a) Ethylene (b) Naphthalene (c) Anthracene (d) 1,3-Butadiene

16. How will you distinguish between the following pairs of compounds by UV spectroscopy:

(a) 1,3-pentadiene and 1,4-pentadiene
(b) Benzene and anthracene
(c) 1,3-hexadiene and 1,3-cyclohexadiene

17. The following α,β-unsaturated ketones have λ_{max} at 254 nm (ε_{max} 9550), 259 nm (ε_{max} 10,790) and 315 nm (ε_{max} 7000) in ethanol. Explain which is which?

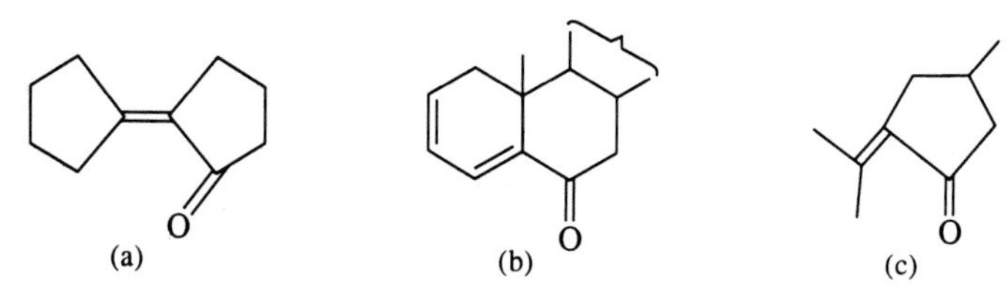

(Hint. For (a) and (c), base value 215 nm should be used.)

18. Discuss the structural features which may cause a bathochromic or a hypsochromic effect in an organic compound.

19. An organic compound can have one of the following structures:

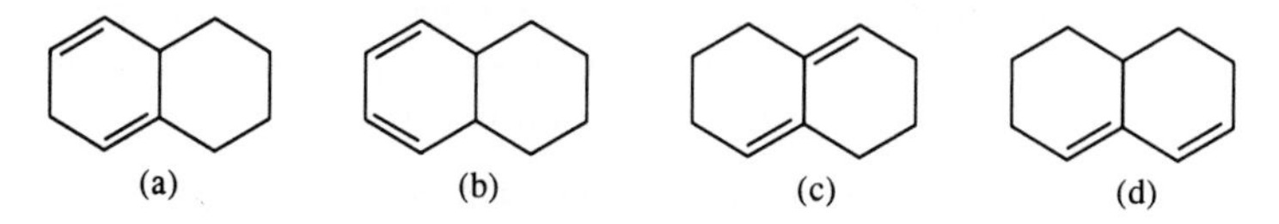

The observed λ_{max} of the compound is 235 nm. Which is the most likely structure of the compound. Explain your choice.

20. Two isomeric compounds *A* and *B* with molecular formula C_6H_8 decolourize bromine solution in CCl_4 and also alkaline $KMnO_4$. On catalytic hydrogenation, both yield cyclohexane (C_6H_{12}). *A* shows an absorption maximum at 256 nm and *B* is transparent above 200 nm. Assign structures to *A* and *B*.

21. Which of the following is the preferred conformaton of 2-chlorocyclohexanone having the λ_{max} 293 nm (cyclohexanone has λ_{max} 282 nm).

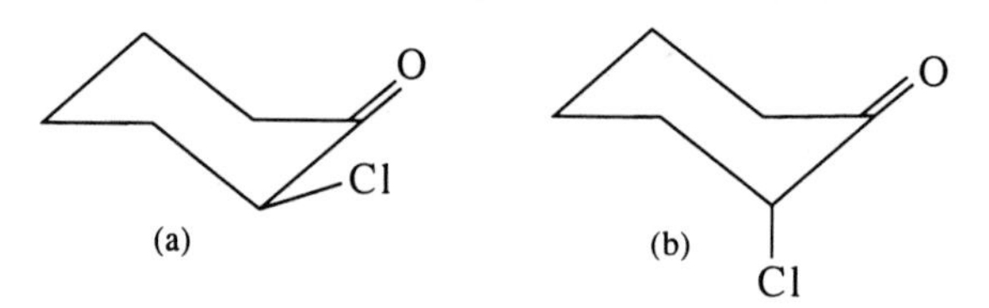

(Hint. The presence of a polar group, e.g., a halogen, in the axial position raises the λ_{max} by 10–30 nm and that in equational position lowers it by 4–10 nm compared to the parent ketone.)

22. A very useful antifungal compound griseofulvin shows two significant absorption bands at λ_{max} 290 nm and 252 nm. Correlate these values to the calculated value of λ_{max} for this compound.

MeO O OMe

O

MeO O

Cl

Griseofulvin

23. The following medium ring compound seems to have unconjugated C = O group but it shows λ_{max} 238 nm, ε_{max} 2535. Explain.

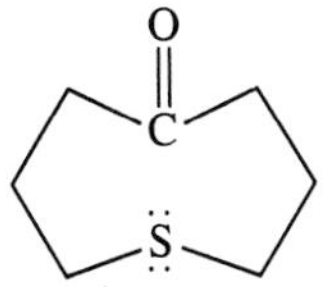

(Hint. Here, effective overlap (conjugation) of the π-orbital of the C = O group and the p (n)-orbitals of sulphur atom results in a transannular conjugation.)

24. (a) Which of the following do not absorb in the near UV region: Benzene, 2-propanol, acrolein, ethanol, methyl iodide, heptene, water, acetaldehyde, dioxane and benzoic acid.

(b) Expain why

(i) Methyl iodide has both the λ_{max} and ε_{max} higher than that of methyl chloride.

(ii) No $n \rightarrow \sigma^*$ transition is observed in protonated triethylamine.

(iii) The region below 200 nm is called vacuum UV region.

(iv) Dimethylamine has higher value of λ_{max} in hexane solution than that in water.

25. Applying Woodward-Fieser rules, calculate the λ_{max} of the following trienone:

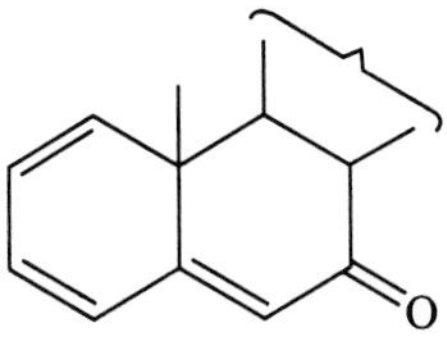

How do you correlate the calculated value with the observed values of λ_{max} which are 230 nm (ε_{max} 18,000), 278 nm (ε_{max} 3720) and 348 nm (ε_{max} 11,000). (Hint. Simple conjugated polyenes show several subsidiary peaks but the longest wavelength peak is in good agreement with the calculated value of λ_{max}. Similar is the case with the above trienone; the calculated λ_{max} 349 nm is in excellent agreement with the observed longest wavelength peak, i.e. 348 nm.)

26. 4-cholesten-3-one gives an enol acetate which shows λ_{max} 238 nm. Suggest the structure for the enol acetate.

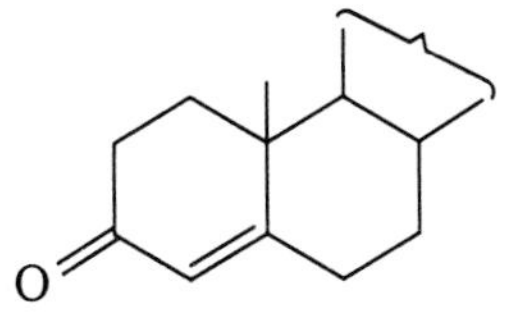

O ⟶ the enol acetate, Obs. λ_{max} = 238 nm

(Hint. The following two enol acetates are possible:

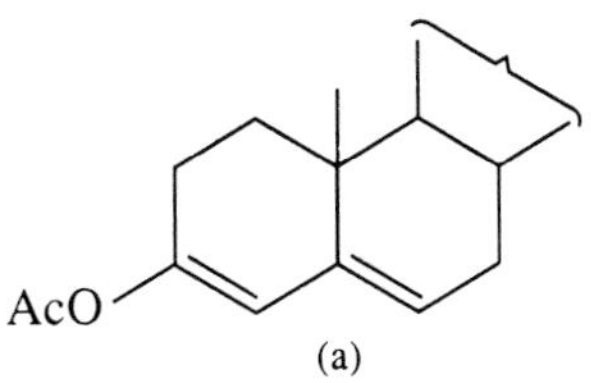

AcO (a)

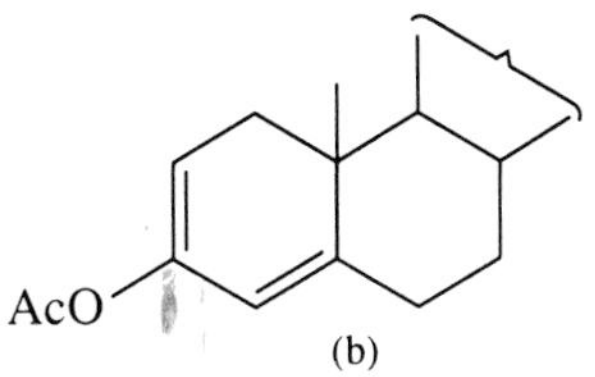

AcO (b)

Calculate their λ_{max} and by comparing with that of the observed value and assign the structure.)

27. Explain the increasing order (a) → (c) of the λ_{max} (EtOH) for the following pyrrole derivatives:

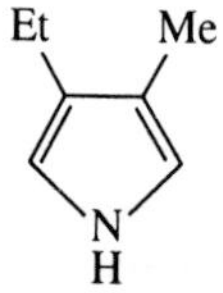

(a) λ_{max} 203 nm (ε_{max} 5670)

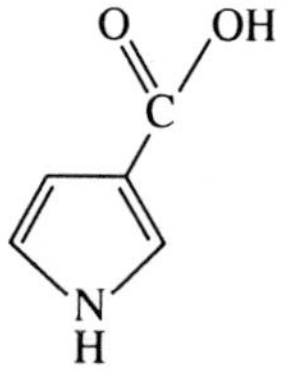

(b) λ_{max} 245 nm (ε_{max} 4800)

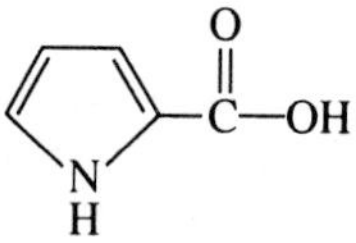

(c) λ_{max} 262 nm (ε_{max} 12,000)

(Hint. In (b) and (c), the conjugation present from the lone pair of nitrogen through the pyrrole ring to the electron attracting —COOH group increases the length of the chromophore resulting in the higher λ_{max}. This extension of conjugation is greatest in (c) and least in (a) compared with disubstituted benzene in Section 2.18.)

28. Benzene is colourless and *p*-benzoquinone is yellow in colour; which of these will undergo more easy electronic excitation?

29. (a) Write the electronic transitions responsible for the following bands in the UV spectrum of acrolein:

(i) λ_{max} 210 nm (ε_{max} 11,500)
(ii) λ_{max} 315 nm (ε_{max} 14)

(b) What are the exceptions to Woodward-Fieser rules for calculating λ_{max} in dienes and trienes?

30. Calculate λ_{max} for the following compounds:

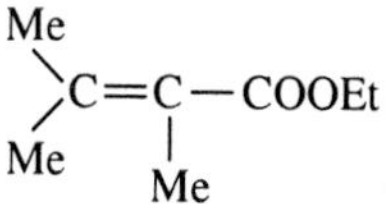

(i) 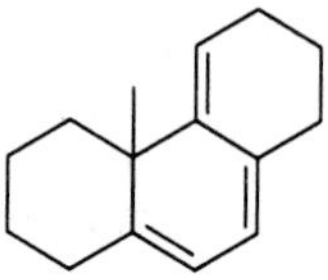 (ii) 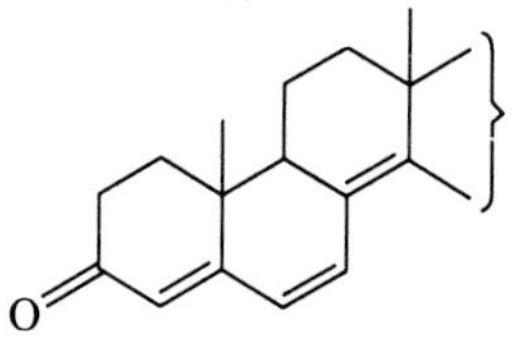 (iii) 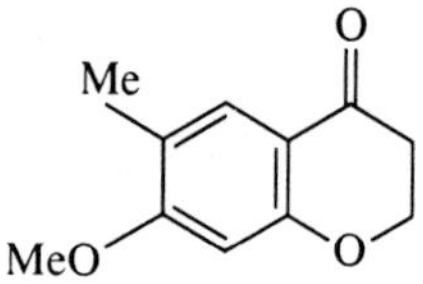 (iv)

References

1. A.I. Scott, Interpretation of Ultraviolet Spectra of Natural Products, Pergamon Press, New York, 1964.

2. A. Knowles and C. Burgess (Eds.), Practical Absorption Specroscopy, Chapman and Hall, London, 1984.
3. B.M. Trost, Problems in Spectroscopy: Organic Structure Determination by NMR, UV and Visible and Mass Spectra, Benjamin, 1967.
4. C.N.R. Rao, Ultraviolet and Visible Spectroscopy, 2nd Edition, Plenum, New York, 1967.
5. E.S. Stern and T.C.J. Timmons, Electronic Absorption Spectroscopy in Organic Chemistry, St. Martin's Press, New York, 1971.
6. H.H. Jaffe' and M. Orchin, Theory and Applications of Ultraviolet Spectroscopy, Wiley, New York, 1962.
7. J.C.D. Brand and G. Eglinton, Applications of Spectroscopy to Organic Chemistry, London, 1965.
8. J.P. Phillips, Spectra Structure Correlation, Academic Press, New York, 1964.
9. R. Denny and R. Sinclair, Visible and Ultraviolet Spectroscopy, Wiley, Chichester, 1987 (ACOL Text).
10. R.A. Friedel and M. Orchin, Ultraviolet Spectra of Aromatic Compounds, Wiley, New York, 1951, Revised Edition, 1958.
11. R.M. Silverstein, G.C. Bassler and T.C. Morrill, Spectrometric Identification of Organic Compounds, 3rd Edition, Wiley, New York, 1974.
12. S. Sternhell and J.R. Kalman, Organic Structures from Spectra, Wiley, Chichester, 1986.

3

Infrared (IR) Spectroscopy

3.1 Introduction

Infrared spectroscopy deals with the recording of the absorption of radiations in the infrared region of the electromagnetic spectrum. The position of a given infrared absorption is expressed in terms of wavelength in micron μ or more commonly in terms of wavenumber $\bar{\nu}$ (cm^{-1}) since it is directly proportional to energy. Note that wavenumbers are often called frequencies, although strictly it is incorrect. However, it is not a serious error as long as we keep in mind that $\bar{\nu} = \frac{1}{\lambda}$ and $\nu = \frac{c}{\lambda}$. The ordinary infrared region 2.5-15 μ (4000-667 cm^{-1}) is of greatest practical use to organic chemists. The region 0.8-2.5 μ (12,500-4000 cm^{-1}) is called the near infrared and the region 15-200 μ (667-50 cm^{-1}) the far infrared. The absorption of infrared radiation by a molecule occurs due to quantized vibrational and rotational energy changes when it is subjected to infrared irradiation. Thus, IR spectra are often called *vibrational-rotational spectra.*

Unlike UV spectra which have relatively few absorption bands, IR spectra have a large number of absorption bands and therefore provide plenty of structural information about a molecule. Different bands observed in an IR spectrum correspond to various functional groups and bonds present in the molecule. Thus, IR spectroscopy is most widely used for the detection of functional groups and identification of organic compounds.

3.2 Instrumentation

Most IR spectrophotometers are double-beam instruments consisting of the following main parts:

(i) Radiation source
(ii) Sample and reference cells
(iii) Attenuator and comb (photometer)
(iv) Monochromator
(v) Detector and amplifier
(vi) Recorder

(i) Radiation Source

Infrared radiation is usually produced by electrically heating a Nernst filament (mainly composed of oxides of zirconium, thorium and cerium) or a globar (rod

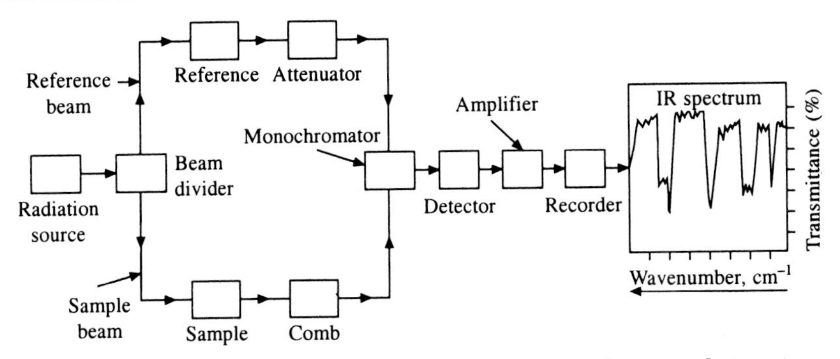

Fig. 3.1 Schematic diagram of a double-beam infrared spectrophotometer

of silicon carbide) to 1000–1800°C. The infrared radiation of succesively increasing wavelength is used. The radiation from the source is divided into sample and reference beams of equal intensity by beam divider.

(ii) Sample and Reference Cells

Reference and sample beams pass through the reference cell and sample cell, respectively. Glass and quartz cannot be used as windows of cells and optical prisms, etc. because they absorb strongly in most of the IR region. Thus, alkali metal halides such as NaCl, NaBr, KCl and KBr are most commonly used as these are transparent to most of the IR region.

(iii) Attenuator and Comb (Photometer)

The reference beam passes through the attenuator and the sample beam through the comb. Then the two beams can be alternately reflected out of the optical system and to the entrance slit of the monochromator with the help of several mirrors. Thus, the photometer combines the reference and sample beams into a single beam of alternating segments. The comb allows balancing of the two beams.

(iv) Monochromator

The combined beam passes through the prism or grating of the monochromator which disperses the beam into various frequencies. Since the prism or grating rotates slowly, it sends individual frequency bands to the detector, thus allowing a scan of frequency bands. Gratings that give better resolutions than prisms consist of a series of parallel and straight thin lines on a smooth reflecting surface; the spacing between lines is of the order of few angstrom (Å) depending on the desired wavelength range.

(v) Detector and Amplifier

The detector is a thermocouple which measures radiant energy by means of its heating effect that produces current. Due to difference in the intensity of the two beams falling on the detector, an alternating current starts flowing from the detector to the amplifier where it is amplified and relayed to the recorder.

(vi) Recorder

It records IR spectra as a plot of wavelengths λ or wavenumbers $\bar{\nu}$ of absorbed radiations against the intensity of absorption in terms of transmittance T or absorbance A. Presently we use the wavenumber unit as it is directly proportional to energy

$$T = \frac{I}{I_0}$$

$$T\,\% \text{ (percent transmittance)} = \frac{I}{I_0} \times 100$$

$$A = \log_{10} \frac{1}{T}$$

where I_0 is the intensity of the incident radiation and I the intensity of the radiation emerging from the sample.

IR absorption intensities are rarely described quantitatively, but they are described as *s* (strong), *m* (medium), *w* (weak) or *υ* (variable). In IR spectroscopy, the magnitude of the molar extinction coefficient ε varies from near zero to around 2000. Here, ε is proportional to the square of the change in the dipole moment of the molecule that the particular absorption causes.

Wavenumbers are commonly calibrated by using one of the several peaks, e.g. 2850, 1603 or 906 cm^{-1} of a polystyrene film. It is common practice to record an IR spectrum along with polystyrene band (usually at 1603 cm^{-1}) on it.

At present, FT-IR (Fourier transform infrared) spectrophotometers have become common. The FT-IR instrument gives same information as a simple IR spectrophotometer but the former is much efficient, as it is quick, has high sensitivity and requires very small quantity of the sample.

3.3 Sample Handling

Infrared spectra of compounds may be recorded in the vapour phase, as pure liquids, in solution and in the solid state. The sample should be dry because water absorbs near ~3710 cm^{-1} and near ~1630 cm^{-1}.

(i) In Vapour Phase

The vapour or gas is introduced into a special cell which is usually about 10 cm long and the walls of its both the ends are normally made of NaCl which is transparent to IR radiation. The vapour phase technique is limited because of the too low vapour pressure of most organic compounds to produce a useful absorption spectrum.

(ii) As a Liquid Film

A drop of neat liquid is placed between two flat plates of NaCl to give a thin film. Thick samples of neat liquids usually absorb too strongly to give satisfactory spectrum. This is the simplest of all sampling techniques.

(iii) In solution

Usually, a 1–5% solution of the compound is introduced into a special cell of

0.1-1 mm thickness and made of NaCl. In order to cancel out solvent absorptions, a compensating cell containing pure solvent is placed in the reference beam. The chosen solvent must not absorb in the region of interest. If the entire spectrum is of interest, then it should be recorded in different solvents in order to see all the bands clearly. For example, carbon tetrachloride and carbon disulphide may be used because the former shows little absorption above 1333 cm^{-1} and the latter below 1333 cm^{-1} has little absorption. The most commonly used solvents in IR spectroscopy are carbon disulphide, carbon tetrachloride and chloroform. The selected solvent should be inert to the sample. For example, carbon disulphide should not be used as solvent for primary or secondary amines.

(iv) In Solid State

(a) *As a mull or paste:* About 2–5 mg of a solid is finely ground in a agat mortar with one or two drops of the mulling agent. The mull is examined as a thin film between two flat plates of NaCl. The most commonly used mulling agent is nujol (a high-boiling petroleum oil). When C—H bands interfere with the spectrum, another mulling agent, hexachlorobutadiene, may be used.

(b) *As a pressed disc:* A solid about 0.5–1 mg is intimately mixed with about 100 mg of dry and powdered KBr. The mixture is pressed with special dies, under a pressure of 10,000–15,000 pounds per square inch, to give a transparent disc (pellet). The use of KBr eliminates the problem of bands due to the mulling agent and gives better spectra.

(c) *As deposited films:* The spectra of solids can also be recorded by depositing a glassy thin film of the compound on an alkali metal halide (NaCl or KBr) disc. The deposition takes place by putting a drop of the solution of the sample (in a volatile solvent) on the disc and then evaporating the solvent. Resins, plastics and waxy materials give satisfactory spectra by this technique.

In general, a dilute solution in a non-polar solvent gives the best, i.e. least distorted, IR spectrum because of far less intermolecular interactions. However, in the solid state or neat liquids, especially when polar groups capable of hydrogen bonding are present in the molecule, there are marked changes in spectral characteristics due to intermolecular interactions.

3.4 Theory (Origin) of Infrared Spectroscopy

IR absorption spectra originate from transitions in vibrational and rotational energy levels within a molecule. On absorption of IR radiation, vibrational and rotational energies of the molecule are increased. When a molecule absorbs IR radiation below 100 cm^{-1}, the absorbed radiation causes transitions in its rotational energy levels. Since these energy levels are quantized, a molecular rotational spectrum consists of discrete lines.

When a molecule absorbs IR radiation in the range 100–10,000 cm^{-1}, the absorbed radiation causes transitions in its vibrational energy levels. These energy levels are also quantized, but vibratonal spectra appear as bands rather than discrete lines. The energy differences between various rotational energy levels of a molecule are far less than that between its vibrational energy levels. Thus, a single transition in vibrational energy levels is accompanied by a large number of transitions in rotational energy levels and so the vibrational spectra appear as vibrational-rotational bands instead of discrete lines. Organic chemists are mainly concerned with these vibrational-rotational bands, especially with those occurring in the region 4000–667 cm^{-1} (2.5–15 μ).

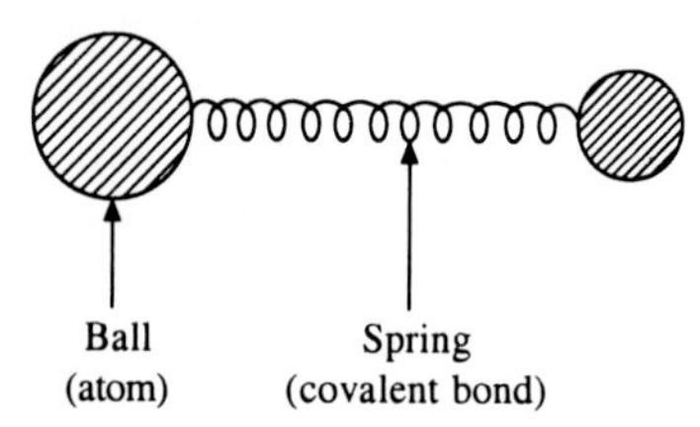

Various atoms in a molecule may be regarded as balls of different masses and the covalent bonds between them as weightless tiny springs holding such balls together. Atoms in a molecule are not still but they vibrate. The two types (modes) of fundamental molecular vibrations known are: (a) stretching and (b) bending vibrations (deformations).

(i) Stretching Vibrations

In stretching vibrations, the distance between two atoms increases or decreases, but the atoms remain in the same bond axis. Stretching vibrations are of two types:

(a) Symmetrical stretching. In this mode of vibration, the movement of atoms with respect to the common (or central) atom is simultaneously in the same direction along the same bond axis (Fig. 3.2(a)).

(b) Asymmetrical stretching. In this vibration, one atom approaches the common atom while the other departs from it (Fig. 3.2(b)).

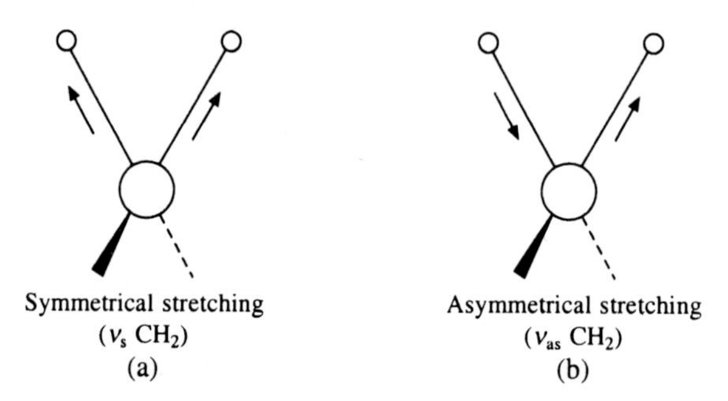

Fig. 3.2 Stretching vibrations of a CH_2 group (ν CH_2)

(ii) Bending Vibrations (Deformations)

In such vibrations, the positions of the atoms change with respect to their original bond axes. Bending vibrations are of four types:

(a) Scissoring. In this mode of vibration, the movement of atoms is in the opposite direction with change in their bond axes as well as in the bond angle they form with the central atom (Fig. 3.3(a)).

(b) Rocking. In this vibration, the movement of atoms takes place in the same direction with change in their bond axes (Fig. 3.3(b)).

Scissoring and rocking are in-plane bendings.

(c) Wagging. In this vibration, two atoms simultaneously move above and below the plane with respect to the common atom (Fig. 3.3(c)).

(d) Twisting. In this mode of vibration, one of the atom moves up and the other moves down the plane with respect to the common atom (Fig. 3.3(d)).

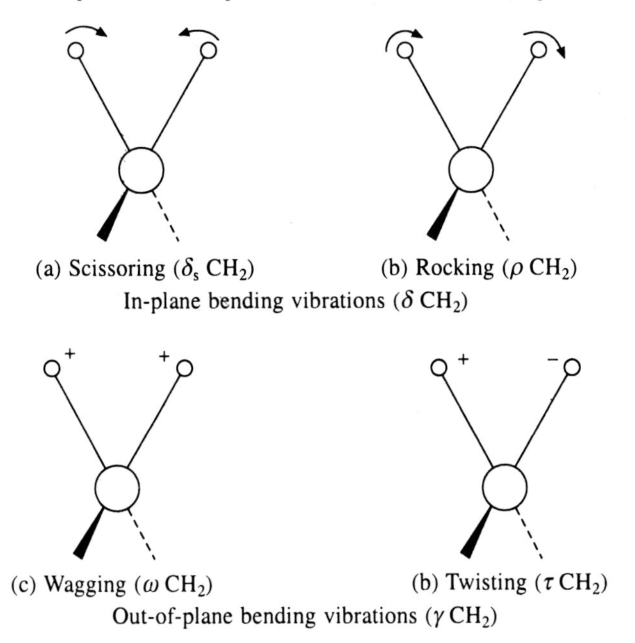

Fig. 3.3 Bending vibrations (deformations) of a CH_2 group (+ and – signs indicate movements perpendicular to the plane of the paper)

Infrared radiation is absorbed when the oscillating dipole moment, due to a molecular vibration, interacts with the oscillating electric field of the infrared beam. This interaction occurs and hence, an absorption band appears only when a molecular vibration produces a change in the dipole moment of the molecule. Otherwise, the vibration is said to be *infrared inactive* and it will show no absorption band in the infrared spectrum. Usually, larger the change in dipole moment, the higher is intensity of absorption. It is not necessary for a molecule to have a permanent dipole moment for IR absorption.

It may be noted that various stretching and bending vibrations of a bond have a definite and quantized frequency of their own. When infrared radiation of the same frequency is incident on the molecule, there is absorption of energy resulting in the increase of the amplitude of that particular absorption; now the molecule is in the excited state. Such an absorption gives rise to a peak in the IR spectrum. From the excited state, the molecule returns to the ground state by release of extra energy through rotational, collision and translational processes resulting in the increase of temperature of the sample under investigation.

We know that lesser energy is required to bend a spring than that required for stretching it. Analogously, bending vibrations of a bond require lesser energy than its stretching vibrations. Thus, the absorption bands resulting from bending vibrations appear at lower wavenumbers those that resulting from stretching vibrations of the same bond. Stretching vibrations usually cause peaks of high intensity. A typical IR spectrum showing various mode of vibrations is given in Fig. 3.4.

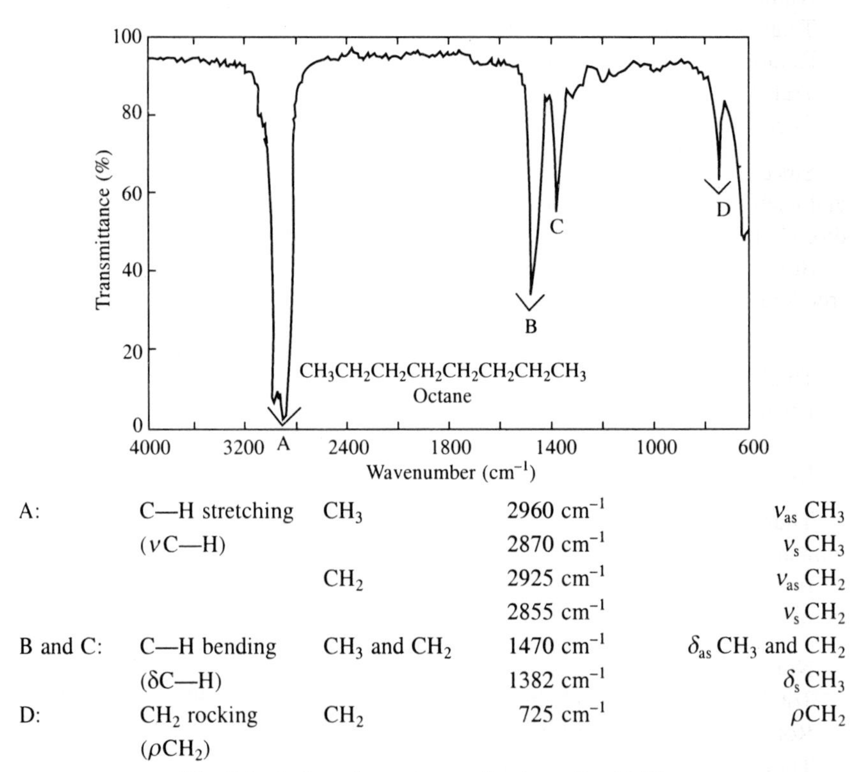

A:	C—H stretching (νC—H)	CH_3	2960 cm^{-1}	ν_{as} CH_3
			2870 cm^{-1}	ν_s CH_3
		CH_2	2925 cm^{-1}	ν_{as} CH_2
			2855 cm^{-1}	ν_s CH_2
B and C:	C—H bending (δC—H)	CH_3 and CH_2	1470 cm^{-1}	δ_{as} CH_3 and CH_2
			1382 cm^{-1}	δ_s CH_3
D:	CH_2 rocking (ρCH_2)	CH_2	725 cm^{-1}	ρCH_2

Fig. 3.4 Infrared spectrum of octane, liquid film

3.5 Number of Fundamental Vibrations

The IR spectra of polyatomic molecules may exhibit more than one vibrational absorption bands. The number of these bands corresponds to the number of fundamental vibrations in the molecule which can be calculated from the degrees of freedom of the molecule. The degrees of freedom of a molecule are equal to the total degrees of freedom of its individual atoms. Each atom has three degrees of freedom corresponding to the three Cartesian coordinates (x, y and z) necessary to describe its position relative to other atoms in the molecule. Therefore, a molecule having n atoms will have $3n$ degrees of freedom. In case of a non-linear molecule, three of the degrees of freedom describe rotation and three describe translation. Thus, the remaining $(3n - 3 - 3) = 3n - 6$ degrees of freedom are its vibrational degrees of freedom or fundamental vibrations, because

$$\text{Total degrees of freedom } (3n) = \text{Translational} + \text{Rotational} + \text{Vibrational degrees of freedom}$$

In case of a linear molecule, only two degrees of freedom describe rotation (because rotation about its axis of linearity does not change the positions of the atom) and three describe translation. Thus, the remaining $(3n - 2 - 3) = 3n - 5$ degrees of freedom are vibrational degrees of freedom or fundamental vibrations.

The number of vibrational degrees of freedom for the linear carbon dioxide molecule can be calculated as follows:

Number of atoms $(n) = 3$
Total degrees of freedom $(3n) = 3 \times 3 = 9$
Rotational degrees of freedom = 2
Translational degrees of freedom = 3
Therefore, vibrational degrees of freedom = 9 – 2 – 3 = 4

Since each vibrational degree of freedom corresponds to a fundamental vibration and each fundamental vibration corresponds to an absorption band, for carbon dioxide molecule there should be four theoretical fundamental bands.

Similarly, for a non-linear molecule ethane (C_2H_6), the vibrational degrees of freedom can be calculated as:

Number of atoms $(n) = 8$
Total degrees of freedom $(3n) = 3 \times 8 = 24$
Rotational degrees of freedom = 3
Translational degrees of freedom = 3
Hence, vibrational degrees of freedom = 24 – 3 – 3 = 18

Thus, theoretically there should be 18 absorption bands in the IR spectrum of ethane.

In case of benzene (C_6H_6), the number of vibrational degrees of freedom can be calculated as follows:

Number of atoms $(n) = 12$
Total degrees of freedom $(3n) = 3 \times 12 = 36$
Rotational degrees of freedom = 3
Translational degrees of freedom = 3
Therefore, vibrational degrees of freedom = 36 – 3 – 3 = 30

Thus, theoretically, there should be 30 fundamental vibrational bands in the IR spectrum of benzene.

It has been observed that in actual IR spectra, the theoretical number of fundamental bands are seldom obtained because there are certain factors which decrease, whereas certain factors increase the number of bands. The following factors decrease the theoretical number of fundamental bands:

(i) The frequencies of fundamental vibrations which fall outside of the region 4000–667 cm^{-1}.
(ii) Fundamental bands which are so weak that they are not observed.
(iii) Fundamental bands which are so close that they coalesce.
(iv) The occurrence of a degenerate band from several absorptions of the same frequency in highly symmetrical molecules such as carbon dioxide.

(v) Certain fundamental vibrational bands which do not appear in the infrared spectrum due to lack of the required change in dipole-moment of the molecule, e.g. in carbon dioxide molecule.

The carbon dioxide molecule is linear and has four fundamental vibrations $(3 \times 3) - 5 = 4$ (Fig. 3.5). Thus, four theoretical fundamental bands are expected but actually it shows only two. The symmetrical stretching vibration in carbon dioxide is IR inactive because it produces no change in the dipole moment of the molecule. The two bending vibrations are equivalent and absorb at the same wavenumber (667.3 cm^{-1}). Thus, the IR spectrum of carbon dioxide shows only two fundamental absorption bands, one at 2350 cm^{-1} due to asymmetrical stretching vibration, and the other at 667.3 cm^{-1} due to the two bending vibrations.

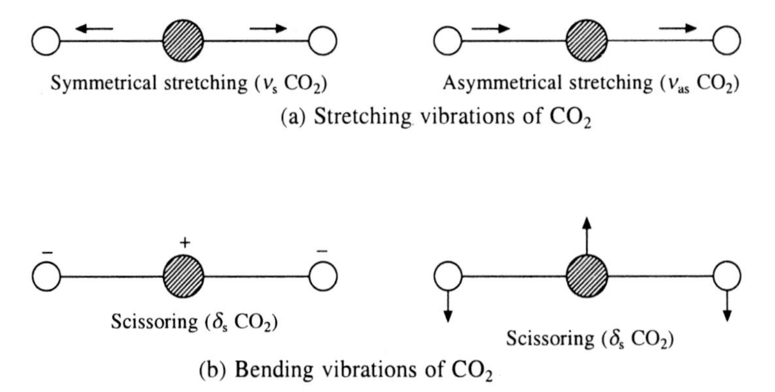

Fig. 3.5 Fundamental vibrations of the linear carbon dioxide molecule (+ and – signs indicate movements perpendicular to the plane of paper)

In a molecule having a centre of symmetry, the vibrations which are symmetrical about the centre of symmetry are IR inactive. For example, the symmetrical stretching vibrations of carbon dioxide and the symmetrical C=C stretching vibrations of ethylene do not result in an absorption band in the IR spectra. It should be noted that ethylene shows bands for C—H stretching vibrations. Similarly, symmetrical diatomic molecules such as H_2, N_2, O_2 and Cl_2 do not absorb IR radiation, but unsymmetrical diatomic molecules like carbon monoxide and iodine chloride (I—Cl) absorb IR radiation. Fundamental vibrations which are IR inactive may be Raman active and may give rise to observable bands in the Raman spectra.

The appearance of the following types of additional (non-fundamental) bands increases the number of bands as compared to that expected from the theoretical number of fundamental vibrations. All these bands have one-tenth to one-hundredth intensity of the fundamental bands.

(i) Overtone Bands

These may arise if a molecule is excited, e.g. from its first vibrational energy level to the third vibrational energy level; the energy required is almost twice of that required for the excitation to second vibrational energy level. In this way, if there are two fundamental bands at x and y cm^{-1}, then the overtone bands can be expected, e.g. at $2x$, $2y$, $3x$ and $3y$ cm^{-1}. The intensity of the overtone decreases as the order of the overtone increases, e.g. the second overtone ($3x$ or $3y$) is less intense than the first overtone ($2x$ or $2y$). Consequently, second and higher overtones are rarely observed, whereas first overtones are observed only for strong bands.

(ii) Combination Bands

If there are two fundamental bands at x and y cm^{-1}, then the combination bands can be expected, e.g. at $(x + y)$, $(x + 2y)$ and $(2x + y)$ cm^{-1}.

(iii) Difference Bands

If there are two fundamental bands at x and y cm^{-1}, then the difference bands can be expected, e.g. at $(x - y)$, $(x - 2y)$ and $(2x - y)$ cm^{-1}.

3.6 Calculation of Vibrational Frequencies

The stretching vibrations of two bonded atoms may be regarded as the vibration of two balls connected by a spring, a situation for which Hooke's law applies. Thus, an approximate value for the stretching vibrational frequency, i.e., the stretching absorption frequency of a bond, can be calculated by Hooke's law

$$\bar{\nu} = \frac{1}{2\pi c}\sqrt{\frac{f}{\mu}} \qquad (3.1)$$

where $\mu = \dfrac{m_1 m_2}{m_1 + m_2}$ is the reduced mass, m_1 and m_2 are the masses (g) of the atoms linked to the particular bond, f the force constant of the bond in dynes/cm, c the velocity of light (2.998×10^{10} cm/sec) and $\bar{\nu}$ the stretching vibrational frequency of the bond in cm^{-1} (wavenumber).

The force constant of a bond is related to its bond strength. The value of force constant is approximately 5×10^5 dynes/cm for single bonds and approximately twice and thrice of this value for double and triple bonds, respectively.

The force constant is a characteristic property of a bond and like other physical constants, it is another physical constant. The frequencies of IR absorptions are commonly used to calculate the force constants of various bonds.

Applying Eq. (3.1) (Hooke's law), the frequency of the C—H stretching vibration can be calculated as follows:

The atomic mass of carbon is 12 and that of hydrogen is 1. Thus,

$$\bar{\nu} = \frac{1}{2\pi \times 2.998 \times 10^{10}\ \text{cm sec}^{-1}} \times \sqrt{\frac{5 \times 10^5\ \text{dynes cm}^{-1}\left(\frac{12}{6.023} + \frac{1}{6.023}\right) \times 10^{-23}\ \text{g}}{\left(\frac{12}{6.023} \times 10^{-23}\ \text{g}\right)\left(\frac{1}{6.023} \times 10^{-23}\ \text{g}\right)}}$$

Therefore, $\bar{\nu} = 3032$ cm^{-1}.

Actually, C—H stretching vibrations of CH_3 and CH_2 groups are generally observed in the range 2850–3000 cm^{-1}.

The value of stretching vibrational frequency depends on the bond strength and the reduced mass. Obviously, the stretching vibrational frequency of a bond increases with an increase in the bond strength or decrease in the reduced mass of the bonded atoms. Thus, due to the greater bond strength (value of f) of C=C and C=O bonds, they are expected to absorb at higher stretching frequencies than that of C—C and C—O, respectively. Similarly, O—H stretching absorption appears at a higher frequency than that of C—C because of the lower value of reduced mass for O—H compared to that for C—C. For the same reason, we would predict O—H stretching vibration to be of higher frequency than O—D stretching vibration.

On the basis of reduced mass, we can expect that F—H bond should absorb at a lower stretching frequency than the C—H bond. Actually, F—H absorbs at a higher frequency (4138 cm^{-1}) than the C—H group (3040 cm^{-1}). This can be explained because there is an increase in the force constant when we proceed from left to right in the first two periods of the periodic table and this increase is more effective than the mass increase which tends to lower the absorption frequency. Thus, the relative contributions of bond strengths and atomic masses must be considered while applying the Hooke's law for predicting the vibrational frequencies of bond stretchings.

3.7 Factors Affecting Vibrational Frequencies

It should be noted that any factor which affects the force constant of a bond will affect its stretching frequency. There are various interrelated factors which shift the vibratonal frequencies from their expected values. For this reason, the values of vibrational frequencies of the bonds calculated by the application of Hooke's law are not exactly equal to their observed values. The force constant of a bond changes with the electronic and steric effects of the other groups present in the molecule, and so the vibrational frequencies are shifted from their normal values. Also, frequency shifts may occur when the IR spectrum of the same compound is recorded in different states, viz. solid, liquid or vapour. Usually, a substance absorbs at higher frequency in the vapour state than that in the liquid or solid state. Following are some important factors which affect the vibrational frequencies of bonds.

(i) Coupled Vibrations

For an isolated C—H bond, only one stretching vibrational frequency is expected but a methylene group shows two absorptions corresponding to symmetrical and asymmetrical stretchings (Fig. 3.2(a) and (b)). This is because there is mechanical coupling or interaction between the C—H stretching vibrations in the CH_2 group. In all such cases, asymmetric stretching vibrations occur at higher frequencies or wavenumbers than the symmetric stretching vibrations. The C—H coupled vibrations if CH_2 groups are of different frequencies than that of CH_3 groups.

Coupling accounts for the two N—H stretching bands in the region 3077–

3497 cm^{-1} in the spectra of primary amines and primary amides. A strong vibrational coupling is present in carboxylic acid anhydrides in which symmetrical and asymmetrical stretching vibrations appear in the region 1720–1825 cm^{-1} and are separated by about 60 cm^{-1}. Here, two carbonyl groups are coupled through the oxygen. The interaction is very effective probably because of the partial double-bond character in the carbonyl oxygen bonds due to resonance which also keeps the system planar for effective coupling. The asymmetrical stretching band in acyclic anhydrides is more intense, whereas the symmetrical stretching band is more intense in cyclic anhydrides. This characteristic can be used for differentiating acyclic and cyclic anhydrides.

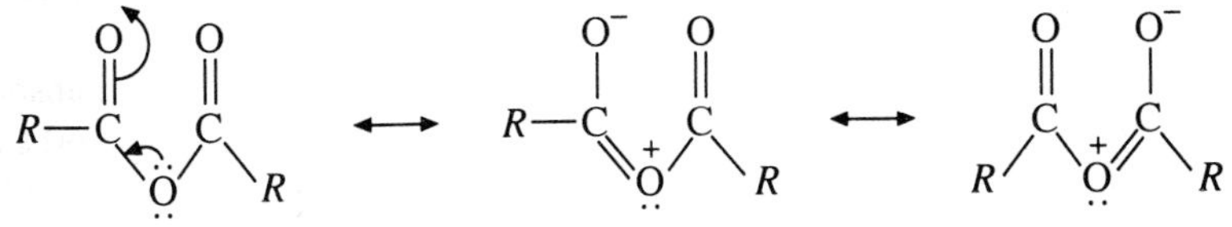

Resonating structures of an acid anhydride

Following are the requirements for effective coupling interaction:

(i) Interaction is greatest when the individual absorption frequencies of the coupled groups are nearly the same.
(ii) Strong coupling between stretching vibrations occur when the groups have a common atom between them.
(iii) Coupling is negligible when groups are separated by one or more carbon atoms and the vibrations are mutually perpendicular.
(iv) Coupling between bending vibrations can occur when there is a common bond, e.g. there is coupling between the bending vibrations of the adjacent C—H bonds in aromatic rings.
(v) If the stretching bond forms one side of the changing angle, then coupling between stretching and bending vibrations can take place.

(ii) Fermi Resonance

We have discussed above the coupling between two fundamental vibrational modes. The coupling between fundamental vibrations and overtone or combination-tone vibrations of very similar frequency is known as *Fermi resonance* (after Enrico Fermi who first observed it in the case of CO_2). In Fermi resonance, a molecule transfers its energy from fundamental vibrational level to overtone or combination tone level and back. According to quantum mechanics, the resonance pushes the two levels apart and mixes their character, consequently each level has partly fundamental and partly overtone or combination tone character. Thus, Fermi resonance give rise to a pair of transitions of almost equal intensity, and so the resulting absorption bands usually appear as a doublet.

For example (as seen in Section 3.5), the symmetrical stretching vibration of CO_2 is IR inactive. For this stretching, Raman spectrum shows a strong band at 1337 cm^{-1}. The two bending vibrations (Fig. 3.4) are equivalent and absorb at the same frequency, of 667.3 cm^{-1}. The first overtone of this is $2 \times 667.3 =$ 1334.6 cm^{-1}. Since it is very close to the frequency of the fundamental vibration (symmetrical stretching), Fermi resonance occurs. Thus, there is mixing of 1337 cm^{-1} (fundamental) and 1334.6 cm^{-1} (overtone) levels in accordance with Fermi resonance to give two bands at 1285.5 cm^{-1} and at 1388.3 cm^{-1} with the intensity ratio 1:0.9, respectively. Some other examples of Fermi resonance are:

The absorption of aldehydic C—H appears as a doublet at ~2820 and ~2720 cm^{-1} due to the interaction of C—H stretching vibration (fundamental) and the first overtone of C—H bending.

Appearanace of a doublet in case of certain lactones and cycloketones can be explained with the help of Fermi resonance. Cyclopentanone shows carbonyl absorption at 1746 and 1750 cm^{-1}. This splitting is due to Fermi resonance with an overtone or combination band of an α-methylene group and the carbonyl band.

Fermi resonance is a common phenomenon in IR and Raman spectroscopy. It is observed only when the interacting groups are so located in the molecule that there is considerable mechanical coupling and that the fundamental and overtone or combination-tone vibrational levels are of the same symmetry species.

(iii) Hydrogen Bonding

Hydrogen bonding remarkably lowers the stretching frequencies of both the groups involved in it, and also changes the shape and intensity of the absorption bands. Usually, absorption bands become more intense and broad on hydrogen bonding. The stronger the hydrogen bond, lower is the O—H stretching frequency. Thus, the value of O—H stretching frequency is a test for hydrogen bonding as well as a measure of the strength of hydrogen bonds.

Non-hydrogen-bonded (free) O—H group of alcohols and phenols show sharp and strong absorption bands in the region 3590–3650 cm^{-1}. Sharp, non-hydrogen-bonded O—H bands are observed only in the vapour phase, in very dilute solution, in non-polar solvents or when hydrogen bonding is prevented by steric hindrance. Pure samples and concentrated solutions of alcohols and phenols show broad O—H stretching bands in the region about 3200–3600 cm^{-1} due to intermolecular hydrogen bonding. The N—H stretching frequencies of amines are also affected by hydrogen bonding in the same way as that of the hydroxyl group but frequency shifts for amines are lesser than that for hydroxyl compounds. This is because nitrogen is less electronegative than oxygen and so the hydrogen bonding in amines is weaker than that in hydroxy compounds. For example, non-hydrogen-bonded primary amines exhibit two bands, one near 3400 cm^{-1} and the other near 3500 cm^{-1} due to symmetrical and asymmetrical N—H stretching modes, respectively. In pure amines, these bands respectively appear in the range 3250–3330 cm^{-1} and 3330–3400 cm^{-1} due to intermolecular hydrogen bonding.

In liquid or solid state, and in concentrated solutions, carboxylic acids exist as dimers due to strong intermolecular hydrogen bonding.

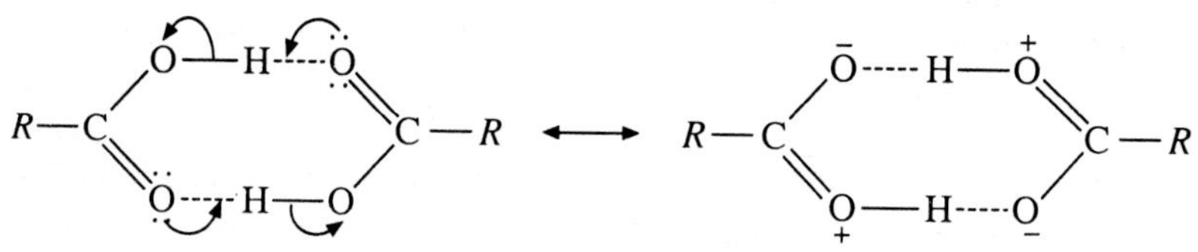

The strong hydrogen bonding between C=O and OH groups lowers their stretching frequency. Because of more polar O—H bond, carboxylic acids form stronger hydrogen bond than alcohols. Carboxylic acid dimers show very broad, intense O—H stretching absorption in the region 2500–3000 cm^{-1}, whereas non-hydrogen-bonded carboxylic acids (monomers) show the O—H absorption near 3550 cm^{-1}. Hydrogen bonding weakens the C=O bond, i.e. its force constant is reduced, resulting in absorption at a lower frequency than the monomer. Thus, C=O stretching bands of monomers of saturated carboxylic acids appear at 1760 cm^{-1}, whereas that of dimers appear at about 1710 cm^{-1}.

In many cases, some O—H groups are hydrogen bonded and some are free. Hence, bands due to both may be observed.

Why does hydrogen bonding lower the absorption frequency?

Let us take the example of hydroxy compounds. On hydrogen bonding, the original O—H bond is lengthened (weakened) due to electrostatic attraction between the hydrogen atoms of one molecule and oxygen atom of the other, and thus the force constant of the O—H bond is reduced resulting in a decrease in its stretching frequency.

$$R{-}O{-}\overset{\delta+}{H}\cdots\overset{\delta-}{O}(H)R$$

Weakened bond   Electrostatic attraction

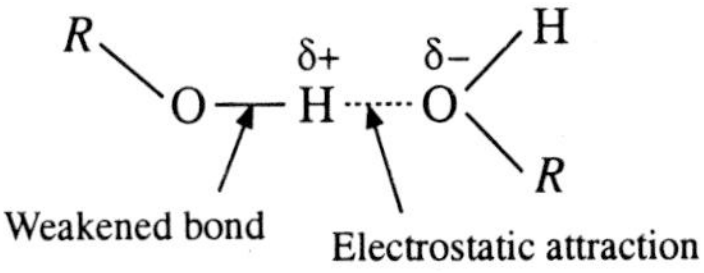

Further, the hydrogen bond may be regarded as a resonance hybrid of the following resonating structures, consequently the O—H bond is weakened and its stretching frequency is lowered.

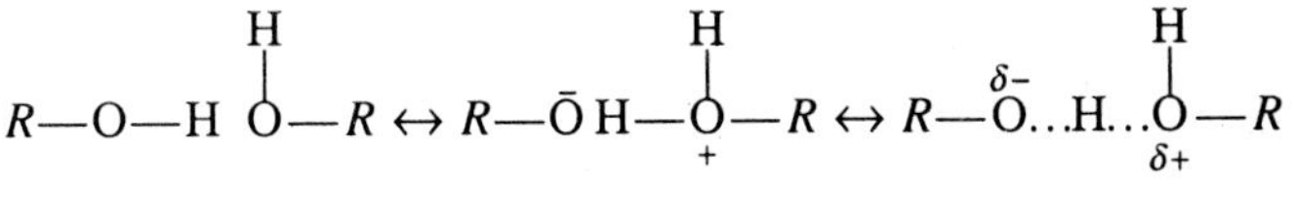

$$R{-}O{-}H\ \ \overset{H}{\underset{}{O}}{-}R \leftrightarrow R{-}\bar{O}\ H{-}\overset{H}{\underset{+}{O}}{-}R \leftrightarrow R{-}\overset{\delta-}{O}\ldots H\ldots\overset{H}{\underset{\delta+}{O}}{-}R$$

Distinction between inter- and intramolecular hydrogen bondings

In very dilute solutions (in non-polar solvents), intermolecular distances are too large to form intermolecular hydrogen bonds. On the other hand, in pure liquids, solids and concentrated solutions, the molecules are closer to form intermolecular hydrogen bond. Thus, intermolecular hydrogen bonding is concentration dependent. On dilution with non-polar solvents, intermolecular hydrogen bonds are broken. Hence, there is decrease in intensity or disappearance of the hydrogen-bonded O—H stretching band and increase in intensity or appearance of free O—H stretching absorption. Thus, in very dilute solutions (in non-polar solvents, concentration $< \sim 0.01$ M), the O—H stretching frequency is shifted to a higher frequency in case of intermolecular hydrogen bonding.

Intramolecular hydrogen bonding is within the same molecule, hence it is not affected by change in intermolecular distances. Thus, intramolecular hydrogen bonds are unaffected by dilution, and so the absorption band is also unaffected. Intramolecular hydrogen bonding (chelation) is very strong in enols and compounds like methyl salicylate etc., due to resonance stabilization of the chelate ring. For example, in enols, the O—H group involved in chelation shows broad absorption band in the range 2500–3200 cm^{-1}, whereas the C=O stretching band in enolic form occurs at 1630 cm^{-1} and that in the *keto* form at 1725 cm^{-1}. From the relative intensities of the two bands, it is possible to determine ratio of the *keto* and *enol* forms.

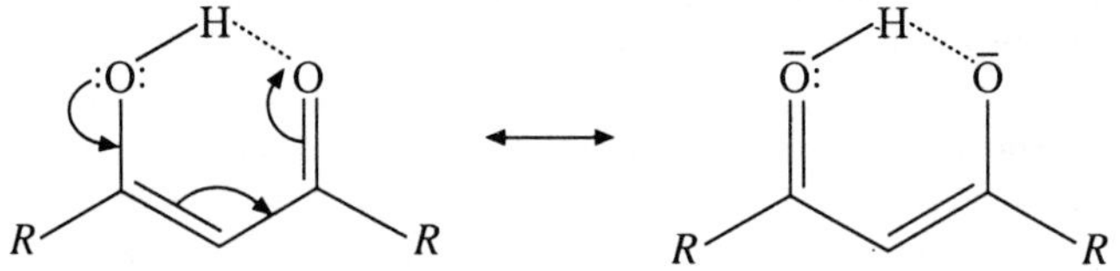

enol of $RCOCH_2COR$

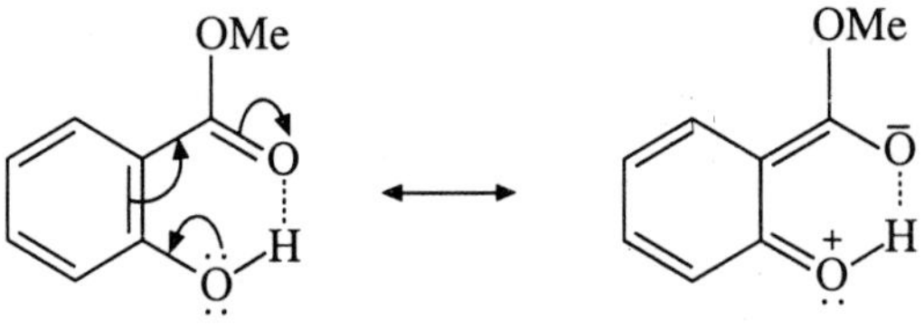

Methyl salicylate

Due to interactions of π electrons of Lewis bases (such as alkenes and benzene) with acidic hydrogen, lengthening and hence weakening of O—H bonds has also been observed. For example, the O—H stretching frequency of phenols is lowered by 40–100 cm^{-1} when the spectrum is recorded in benzene solution as compared to that recorded in CCl_4 solution.

(iv) Inductive and Mesomeric (Resonance) Effects

Absorption frequencies of a particular group are affected by electronic effects, especially inductive and mesomeric (resonance) effects of the nearby groups. These effects are generally interrelated and their individual contribution can only be estimated approximately. Strength (force constants) of a particular bond is changed by these effects and hence its stretching frequency is also changed with respect to the normal values. Both of these effects may operate in the same direction or opposite to each other where one dominates the other.

A carbonyl compound may be considered as a resonance hybrid of the following structures:

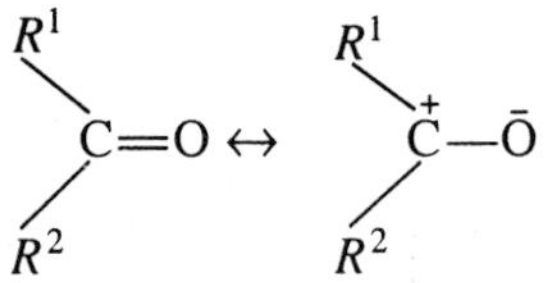

$$R^1R^2C{=}O \leftrightarrow R^1R^2\overset{+}{C}{-}\bar{O}$$

(I) (II)

The stretching frequency of a carbonyl group decreases with increasing number of alkyl groups attached to it. This is due to *+I* effect of alkyl groups which favours structure (II) and lengthens (weakens) the carbon-oxygen double bond, and hence its force constant is decreased resulting in the lowering of the C=O stretching frequency. For example, HCHO, CH_3CHO and CH_3COCH_3 show $\nu_{C=O}$ absorption at 1750, 1730 and 1720 cm^{-1}, respectively. It should be noted why aldehydes absorb at higher frequency than ketones. Similarly, when a group with $-I$ effect is attached to a C=O group, it favours structure (I) and its stretching frequency is increased due to increase in the bond order (force constant) of the carbon-oxygen double bond. For example, CH_3COCH_3, CH_3COCF_3 and CF_3COCF_3 show $\nu_{C=O}$ bands at 1720, 1769 and 1801 cm^{-1}, respectively.

Amides show $\nu_{C=O}$ band at a lower frequency than that of esters. Due to lesser electronegativity of nitrogen than oxygen, its lone pair of electrons are more readily involved in resonance than that of oxygen. Thus, the carbon-oxygen double bond character is more reduced by resonance in amides than that in esters resulting in a much lower $\nu_{C=O}$ frequency of amides than that of esters. For example, benzamide shows $\nu_{C=O}$ band at 1663 cm^{-1} and methyl benzoate at 1735 cm^{-1}.

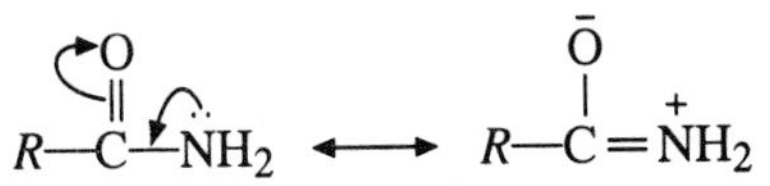

$$R-\overset{\overset{O}{\|}}{C}-\ddot{N}H_2 \longleftrightarrow R-\overset{\overset{\bar{O}}{|}}{C}=\overset{+}{N}H_2$$

An amide

$$R-\overset{\overset{O}{\|}}{C}-\ddot{O}-R \longleftrightarrow R-\overset{\overset{\bar{O}}{|}}{C}=\overset{+}{\ddot{O}}-R$$

An ester

Conjugation of a carbonyl group with an olefinic double bond or an aromatic ring lowers the stretching frequency of the C=O groups by about 30 cm^{-1}. This is because the double bond character of the C=O group is reduced by mesomeric effect.

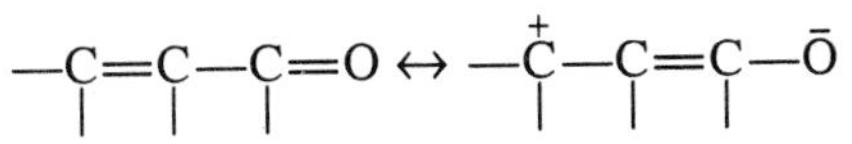

$$-\underset{|}{C}=\underset{|}{C}-\underset{|}{C}=O \leftrightarrow -\overset{+}{\underset{|}{C}}-\underset{|}{C}=\underset{|}{C}-\bar{O}$$

For example, the stretching frequency of C=O group in acetone is 1720 cm^{-1}, whereas acetophenone shows $\nu_{C=O}$ band at 1697 cm^{-1} due to its conjugation with the aromatic ring. In this case, $-I$ effect of the sp^2-carbon of the ring is dominated by the $+M$ effect of the ring.

3.8 Characteristic Absorptions in Common Classes of Compounds

Various groups present in various classes of organic compounds are responsible for IR absorptions in different frequency regions. Most of these group absorptions are characteristic of a particular class of compounds. These characteristic group frequencies (given in Table 3.1) along with effects of structural environments on them are discussed as follows.

(i) Alkanes and Alkane Residues

The most characteristic absorptions in alkanes and alkane residues ($-CH_3$,

—CH_2— and >CH—) are C—H stretching and bending absorptions occurring in the regions 2840–3000 cm^{-1} and 1340–1485 cm^{-1}, respectively (Table 3.1). The C—H stretching commonly appears as two bands—one for symmetrical and the other for asymmetrical stretching. Due to their inert nature, the absorption positions of alkane residues are hardly affected by their chemical environment (Fig. 3.4).

Absorptions due to $\nu_{C—C}$ vibrations occur in the range 800–1200 cm^{-1} and are generally weak, hence practically not useful for identification. This low frequency range compared to C—H vibrations is due to differences in atomic masses of C—H and C—C groups. Since most organic compounds possess alkane residues, their absorption bands are of little value in structure determination. Of course, the absence of an absorption due to an alkane residue is an evidence for the absence of such a part structure in the molecules. When alkane residues are attached to a carbonyl group, a heteroatom (e.g. N or O) or an aromatic ring, their stretching bands appear below 3000 cm^{-1} (Table 3.1).

(ii) Cycloalkanes (Cycloparaffins)

The C—H stretching frequencies in cycloalkanes increase with increasing ring strain. However, in strainless rings such as cyclohexane they are almost same as in alkanes (Table 3.1). For example, CH_2 and CH groups in monoalkyl cyclopropanes absorb in the region 2985–3060 cm^{-1}, whereas that in cyclohexanes around 2850–2930 cm^{-1}. The frequency of CH_2 absorbs at 1468 cm^{-1}, whereas that of cyclopropane and cyclohexane at 1442 cm^{-1} and 1452 cm^{-1}, respectively.

(iii) Alkenes

Unconjugated C=C bonds have stretching frequency in the region 1620–1680 cm^{-1} (Table 3.1). More substituted double bonds absorb at higher frequencies compared to less substituted C=C bonds. The larger the change in dipole moment during the vibration, stronger is the absorption. Thus, C=C stretching absorption is very weak when the double bond is more or less symmetrically substituted. For the same reason, the intensity of $\nu_{C=C}$ bands of *trans* alkenes is lesser than that of *cis* alkenes but the absorption frequencies of these bands are higher for *trans* alkenes than that for *cis* alkenes (Table 3.1).

Conjugation of the C=C double bond with a double bond or an aromatic ring decreases the frequency of the C=C bond. The IR spectrum of an asymmetrical conjugated diene like 1,3-pentadiene shows two stretching bands, one near 1600 cm^{-1} and the other near 1650 cm^{-1}. The asymmetrical alkene like 1,3-butadiene shows only one band near 1600 cm^{-1} due to asymmetrical stretching. The symmetrical stretching of such molecules is inactive in the infrared. Cumulative double bonds (C=C=C) show strong absorption near 1900–2000 cm^{-1}.

Olefinic =C—H stretching appears in the region 3000–3100 cm^{-1}. Terminal double bonds absorb more strongly than internal double bonds. The most characteristic vibrational modes of olefins are the out-of-plane bending vibrations which appear in the region 650–1000 cm^{-1} (Table 3.1). These are the strongest bands in the spectra of olefins. A pair of intense bands in this region (one near 910 cm^{-1} and the other near 990 cm^{-1}) shows the presence of vinyl (*R*—CH=CH_2)

double bond, whereas only one band near 890 cm^{-1} shows the presence of vinylidene ($R_2C{=}CH_2$) type double bond. Bands due to out-of-plane bending of olefinic C—H of *trans* alkenes are relatively stronger and appear at higher frequencies than that of *cis* alkenes.

(iv) Alkynes

Alkynes show a C≡C stretching band with variable intensity in the region 2100–2260 cm^{-1}. Due to symmetry, weak or no C≡C stretching band is observed in the spectra of acetylene and symmetrically substituted alkynes. Because of symmetry reasons, a terminal C≡C gives stronger stretching band than an internal C≡C (pseudosymmetry). The ≡C—H stretching band of alkynes occurs around 3300 cm^{-1}. This band is stronger and narrower than the hydrogen-bonded O—H or N—H stretching bands occurring in the same region, hence can be distinguished from them. The absorption due to ≡C—H bending of alkynes appears as a strong and broad band in the region 610–645 cm^{-1} (Table 3.1).

(v) Halogen Compounds

As the electronegativity of halogen increases, symmetrical and asymmetrical stretching frequencies of C—H bond in CH_3—X decrease, whereas its bending vibrations are shifted to higher frequency. The bending vibrations of the C—H bond in CH_3—*X* appear in the region 1250–1500 cm^{-1}, in CH_3F being at the highest frequency and in CH_3I at the lowest. The bending vibrations of the C—*X* bond in halogen compounds are strong and appear in the regions: C—F, 1000–1400; C—Cl, 600–800; C—Br, 500–750 and C—I, ~500 cm^{-1}, i.e. the frequency of bending absorption decreases with decreasing electronegativity of the halogen (Table 3.1).

(vi) Aromatic Hydrocarbons

Aromatic C—H stretching bands occur in the region 3000–3100 cm^{-1}. Out-of-plane bending bands of the ring C—H appear in the region 675–900 cm^{-1}. In-plane C—H bending bands appear in the region 1000–1300 cm^{-1}. In aromatic compounds, the most characteristic medium bands due to C—C stretching appear around 1600, 1580 (when ring is further conjugated), 1500 and 1450 cm^{-1}. The absence of these bands indicates that the compound is not aromatic. Monosubstituted benzenes show characteristic, strong bands in the regions 690–710 and 730–770 cm^{-1}. *Ortho*-disubstituted benzenes show a strong band in the region 735–770 cm^{-1}, but no band in the region 690–710 cm^{-1}. The *meta*-disubstituted benzenes show two strong bands in the regions 680–725 and 750–810 cm^{-1}, whereas *para*-disubstituted benzenes exhibit only one band in the region 800–860 cm^{-1} (Table 3.1). Fig. 3.6 shows IR spectrum of a typical aromatic compound.

(vii) Alcohols and Phenols

Alcohols and phenols show characteristic bands arising from O—H stretching and C—O stretching in the regions 3200–3650 and 1000–1260 cm^{-1}, respectively (Table 3.1). In Section 3.7(iii), we have already discussed the excellent hydrogen bonding property of alcohols and phenols. As we have seen, the non-hydrogen-

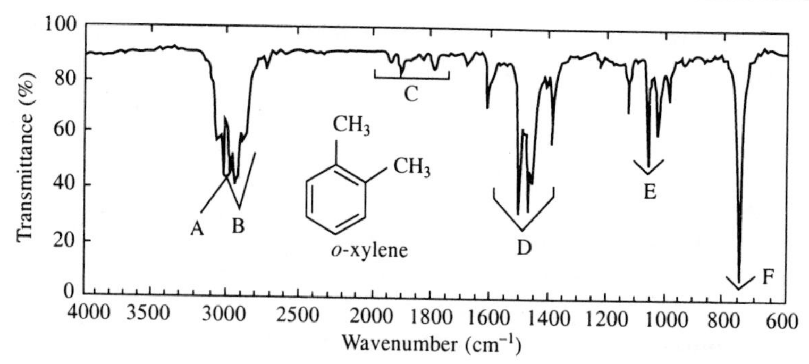

A: Aromatic C—H stretching, 3008 cm^{-1}
B: Methyl C—H stretching, 2965, 2938, 2918 and 2875 cm^{-1}
C: Overtone or combination bands, 1667–2000 cm^{-1}
D: C$\overset{...}{=}$C ring stretching, 1605, 1495 and 1466 cm^{-1}
E: In-plane C—H bending, 1052 and 1022 cm^{-1}
F: Out-of-plane C—H bending, 742 cm^{-1}

Fig. 3.6 IR spectrum of *o*-xylene, neat

bonded hydroxyl groups of alcohols and phenols strongly absorb in the region 3590–3650 cm^{-1}, whereas their free O—H stretching bands appear in the region 3200–3600 cm^{-1}. We should note that the frequency of O—H stretching band depends on concentration, nature of the solvent and temperature because these factors affect the hydrogen bonding which affects the absorption frequency.

IR spectra of alcohols and phenols exhibit a strong band in their IR spectra due to their C—O stretching vibrations in the region 1000–1260 cm^{-1}. Primary, secondary and tertiary alcohols can be recognized with the help of their O—H stretching and C—O stretching bands (Table 3.1). The O—H in-plane bending vibrations occur in the region 1330–1420 cm^{-1}. Alcohols and phenols in the liquid state show a broad absorption band in the region 650–769 cm^{-1} because of out-of-plane bending of hydrogen-bonded O—H group. Figs. 3.7 and 3.8 show IR spectra of a typical alcohol and a phenol, respectively.

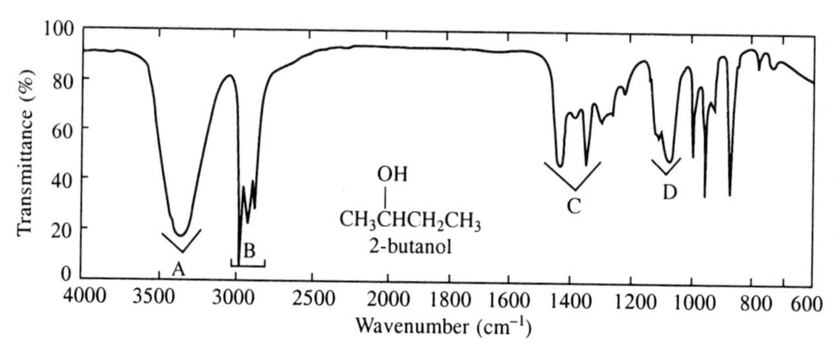

A: O—H stretching, intermolecularly hydrogen bonded, 3330 cm^{-1}
B: C—H stretching (Fig. 3.4), ~2900 cm^{-1}
C: C—H bending (Fig. 3.4), ~1460 and ~1360 cm^{-1}
D: C—O stretching, ~1100 cm^{-1}

Fig. 3.7 IR specturm of 2-butanol, neat

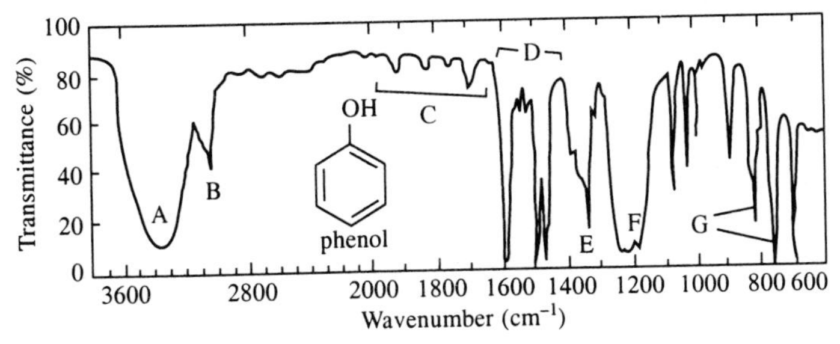

A: Broad, intermolecularly hydrogen bonded, O—H stretching 3333 cm^{-1}

B: Aromatic C—H stretching, 3045 cm^{-1}

C: Overtone or combination bands (Fig. 3.6), 1667–2000 cm^{-1}

D: C ═ C ring stretching, 1580, 1495 and 1468 cm^{-1}

E: In-plane O—H bending, 1359 cm^{-1}

F: C—O stretching, 1223 cm^{-1}

G: Out-of-plane C—H bending, 805 and 745 cm^{-1}

Fig. 3.8 IR spectrum of phenol, melt

The axial or equatorial position occupied by the O—H group can be detected with the help of IR spectroscopy. The stretching absorption frequency of the axial O—H group is higher than that of the equatorial. This is due to the 1,3-diaxial interaction (i.e. the interaction between the axial O—H group and the axial hydrogen at position-3 with respect to this O—H group) which hinders the stretching vibration of the O—H group and thus its frequency is raised.

In 2,6-disubstituted phenols intermolecular hydrogen bonding is prevented by steric hindrance. Hence, no band due to hydrogen-bonded O—H stretching is observed in the region 3200–3600 cm^{-1}. Thus, such phenols show a strong band near 3640 cm^{-1}. In case of 2-substituted phenols, some O—H groups are hydrogen bonded and some are free, hence bands due to both are observed, i.e. one band around 3640 cm^{-1} and the other near 3200–3600 cm^{-1}. For example, 2,6-di-*t*-butyl phenol shows only one O—H stretching band at 3642 cm^{-1}, whereas 2-*t*-butyl phenol shows two O—H stretching bands, one at 3608 cm^{-1} and the other at 3643 cm^{-1}.

(viii) Ethers and Epoxides

Ethers exhibit characteristic C—O—C stretching bands (Table 3.1). The atomic masses of C and O are fairly comparable. Thus, their force constants are close, and so C—O—C and C—C—C stretchings have almost the same frequency. Since vibrations of C—O—C result in greater dipole moment changes than that of C—C—C due to greater electronegativity of oxygen, more intense IR bands are observed for ethers. Aliphatic ethers show a characteristic, strong band in the region 1060–1150 cm^{-1} because of asymmetrical C—O—C stretching. The symmetrical stretching band is usually weak and clearly appears in the Raman spectrum.

Epoxides (cyclic ethers) show two stretching bands, one near 1250 cm^{-1} owing to symmetrical stretching (ring breathing) and the other in the range

810–950 cm^{-1} due to asymmetrical ring stretching. Larger ring ethers absorb closer to the usual values for acyclic ethers.

(ix) Aldehydes and Ketones

Aldehydes, ketones, carboxylic acids, carboxylic esters, lactones, acid halides, anhydrides, amides and lactams show a strong C═O stretching band generally in the region 1630–1870 cm^{-1}. This band has relatively constant position, high intensity and has far less interfering bands in its region making it one of the most readily recognizable bands in IR spectra. The C═O stretching frequency depends on inductive effect, mesomeric effect, hydrogen bonding, field effect, steric effect and ring strain. These factors either increase or decrease the stretching frequency of both the aldehydic and ketonic C═O groups compared to the normal value 1715 cm^{-1} (stretching absorption frequency of a neat sample of a saturated aliphatic ketone). Non-polar solvents increase, whereas polar solvents decrease the frequency of absorption but the overall range of solvent effect does not exceed 25 cm^{-1}. Figs. 3.9 and 3.10 show IR spectra of a typical aldehyde and a ketone, respectively.

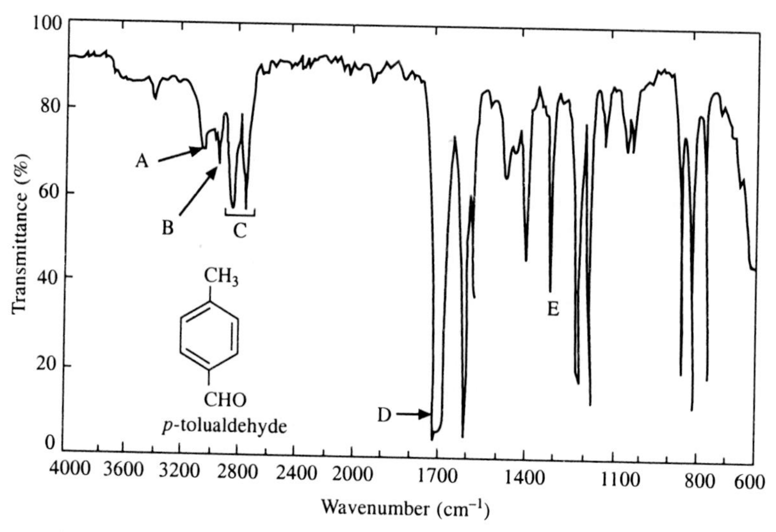

A: Aromatic C—H stretching, near 3060 cm^{-1} (Fig. 3.6)
B: Methyl C—H stretching, near 2950 cm^{-1} (Fig. 3.6)
C: Aldehydic C—H stretching, 2825 and 2717 cm^{-1}; doublet due to Fermi resonance with overtone of band at 1389 cm^{-1} (E)
D: Conjugated aldehydic C═O stretching, 1700 cm^{-1}
E: Aldehydic C—H bending, 1389 cm^{-1}

Fig. 3.9 IR spectrum of *p*-tolualdehyde, neat

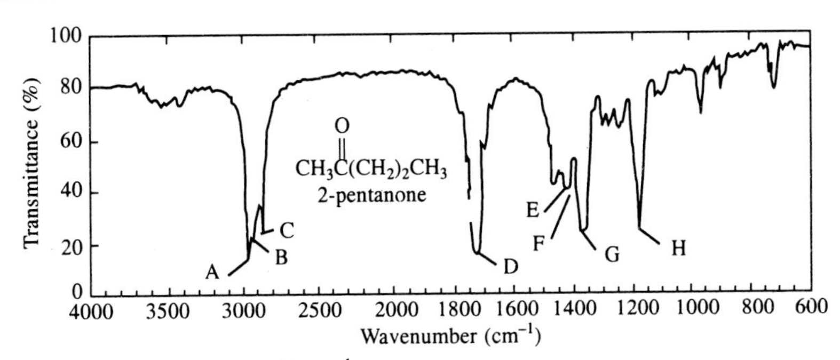

A: ν_{as}, methyl, 2955 cm^{-1}
B: ν_{as}, methylene, 2930 cm^{-1}
C: ν_{s}, methyl, 2866 cm^{-1}
D: Normal C═O stretching, 1725 cm^{-1}
E: δ_{as}, CH_3, ~1430 cm^{-1}
F: δ_{s}, CH_2, ~1430 cm^{-1}
G: δ_{s}, CH_3 of CH_3CO unit, 1370 cm^{-1}
H: C—CO—C stretching and bending, 1172 cm^{-1}

Fig. 3.10 IR spectrum of 2-pentanone, neat

As discussed in Section 3.7(iv), a carbonyl compound may be considered as a resonance hybrid of the following structures:

$$R^1R^2C{=}O \longleftrightarrow R^1R^2\overset{+}{C}{-}\overset{-}{O}$$

(I) (II)

An aldehyde or a ketone

The $-I$ effect of R^1 and R^2 groups favour structure (I), whereas $+I$ effect of these groups favour structure (II). Thus, in the former the length of C═O bond is reduced resulting in the increase of its force constant and the frequency of absorption, whereas in the latter C═O bond is lengthened resulting in the decrease of its absorption frequency. Because of the larger $+I$ effect operating in ketones, their C═O stretching frequency is lower than that of aldehydes. As seen in Section 3.7(iv), conjugation of an aldehydic or ketonic carbonyl group with an olefinic double bond or an aromatic ring lowers the $\nu_{C=O}$ frequency due to mesomeric effect which reduces the double bond character (i.e. force constant) of the C═O group (Table 3.1).

Intermolecular hydrogen bonding between a C═O group and a hydroxylic solvent causes a slight decrease in the absorption frequency of the carbonyl group. The effect of intermolecular and intramolecular hydrogen bondings on carbonyl absorption has already been discussed in Section 3.7(iii). Similar is the case with aldehydes and ketones.

In the absence of steric hindrance, a conjugated system tends to remain in a planar conformation because it allows mesomerism, i.e. stabilization. Steric effects reduce the effect of conjugation by reducing the coplanarity of the conjugated system. α,β-unsaturated ketones may exist in *s-cis* and *s-trans* conformations and then will show absorption for each of these forms. For example, benzalacetone is present in both the *s-cis* and *s-trans* form in carbon disulphide and shows absorption for each of these forms. Because of steric inhibition of mesomerism, the *s-cis* form has high value for C=O stretching frequency (1699 cm^{-1}) than that for the *s-trans* form (1674 cm^{-1}) where the double bond character of the C=O group is reduced by mesomerism causing the lowering of the absorption frequency.

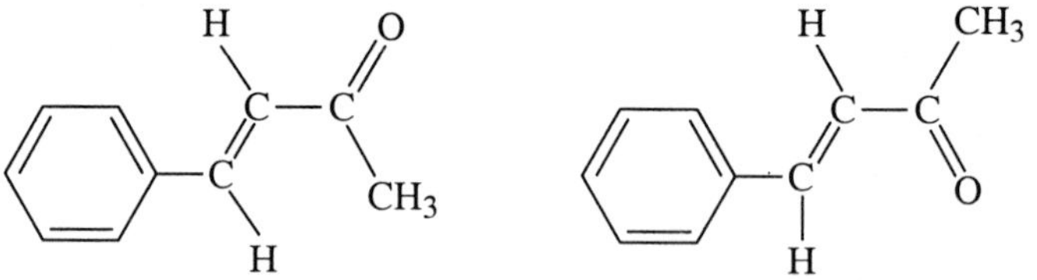

s-trans $\nu_{C=O}$ 1674 cm^{-1} *s-cis* $\nu_{C=O}$ 1699 cm^{-1}

Benzalacetone

In α-halocyclohexanone, the equatorial halogen atom is near the carbonyl group and its negative field repels the non-bonding electrons of the carbonyl oxygen (field effect*). Thus, the force constant of the C=O bond is increased shifting its stretching band to a higher frequency. When α-halogen is in axial position, it is distant from the C=O group and so there is no field effect resulting in the shift of the absorption band.

The carbonyl stretching frequency in cyclic ketones having ring strain is shifted to a higher value. The C—CO—C bond angle in strained rings is reduced below the normal value of 120° (acyclic and six-membered cyclic ketones have the normal C—CO—C angle of 120°). This leads to an increase in *s* character in the sp^2 orbital of carbon involved in the C=O bond. Hence, the C=O bond is shortened (strengthened) resulting in an increase in the $\nu_{C=O}$ frequency. This increase in the *s* character of the outside sp^2 orbital is there because it gives more *p* character to the sp^2 orbitals of the ring bonds which relieves some of the strain, as the preferred bond angle of *p* orbitals is 90°. In ketones where C—CO—C angle is greater than the normal angle (120°), an opposite effect operates and they have lower $\nu_{C=O}$ frequency. For example, in di-*t*-butyl ketone, where the C—CO—C angle is pushed outward above 120°, has very low $\nu_{C=O}$ frequency (1698 cm^{-1}).

In another explanation, it has been suggested that the C=O stretching is influenced by the adjacent C—C stretching. In strained rings, the interaction with C—C bond stretching increases the energy required to cause C=O stretching and thus increases its frequency.

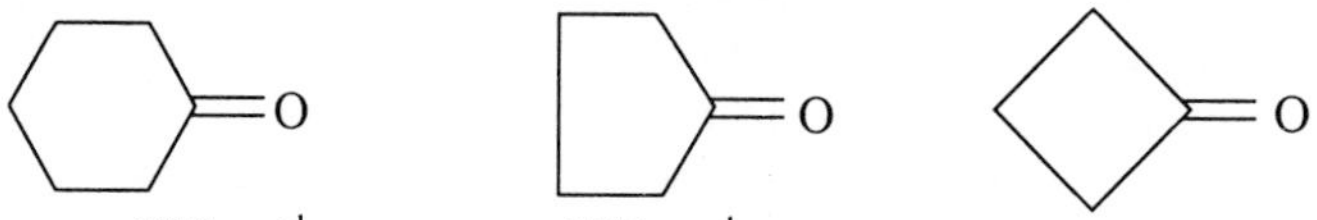

$\nu_{C=O}$ 1715 cm^{-1} 1745 cm^{-1} 1780 cm^{-1}

Increasing strain →

*Field effect is the electrostatic interaction of two charged centres within the same molecule through space or through solvent.

Quinones having both C=O groups in the same ring show characteristic absorption bands due to C=O and C=C stretchings in the region 1660–1690 cm^{-1} and near 1600 cm^{-1}, respectively. Quinones with extended conjugation in which the carbonyl groups are present in different rings absorb in the region 1635–1655 cm^{-1} (Table 3.1).

Most of the aldehydes exhibit aldehydic $\nu_{C—H}$ absorption as a doublet in the region 2700–2900 cm^{-1} (Fig. 3.9). The appearance of doublet is due to Fermi resonance between the fundamental aldehydic C—H stretch and the first overtone of bending vibration which usually appears around 1390 cm^{-1}. The aldehydes whose C—H bending is shifted considerably from 1390 cm^{-1} show only one $\nu_{C—H}$ band (Table 3.1).

(x) Carboxylic Acids and Carboxylate Anions

Carboxyl group (—COOH) is the functional group which can be most easily detected by IR spectroscopy. This is because the $\nu_{C=O}$ absorption occurs in the region (1655–1740 cm^{-1}) having far less interfering bands and in addition, it shows a broad $\nu_{O—H}$ band in the region 2500–3000 cm^{-1} (Table 3.1). Due to strong hydrogen bonding, carboxylic acids exist as dimers (Section 3.7(iii)) in the liquid or solid state and in concentrated CCl_4 solution (> 0.01 M). Carboxylic acid dimers show very broad, intense $\nu_{O—H}$ absorption in the region 2500–3000 cm^{-1}, whereas that of the monomer appears at ~3550 cm^{-1}. The $\nu_{C=O}$ absorption of the dimer of saturated aliphatic acid appears at ~1710 cm^{-1}, whereas that of the monomer at ~1760 cm^{-1}. The carboxylic acid dimer has centre of symmetry. Thus, only asymmetrical C=O stretching vibration absorbs in the infrared. The lower $\nu_{C=O}$ absorption frequency of carboxylic acid dimers is due to hydrogen bonding and resonance which lengthen (weaken) the C=O bond.

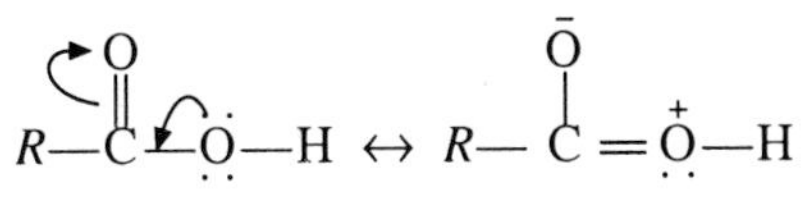

$$R—C(=O)—\ddot{O}—H \leftrightarrow R—C(—\bar{O})=\overset{+}{\ddot{O}}—H$$

Resonating structures of a carboxylic acid

The α, β-unsaturated acids and aryl acids show $\nu_{C=O}$ absorption at more lower frequency because of the mesomeric effect which decreases the double bond character of the C=O bond and hence the force constant is decreased. Fig. 3.11 shows IR spectrum of a typical carboxylic acid.

The carboxylate anion is a resonance hybrid of the following equivalent resonating structures:

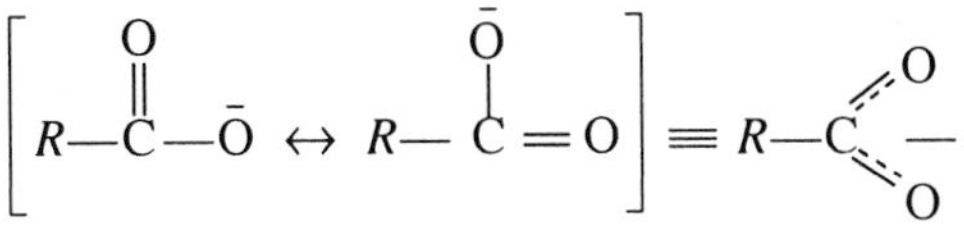

$$\left[R—C(=O)—\bar{O} \leftrightarrow R—C(—\bar{O})=O\right] \equiv R—C\overset{\cdot\cdot O}{\underset{\cdot\cdot O}{}}{}^{-}$$

Resonating structures of the carboxylate anion

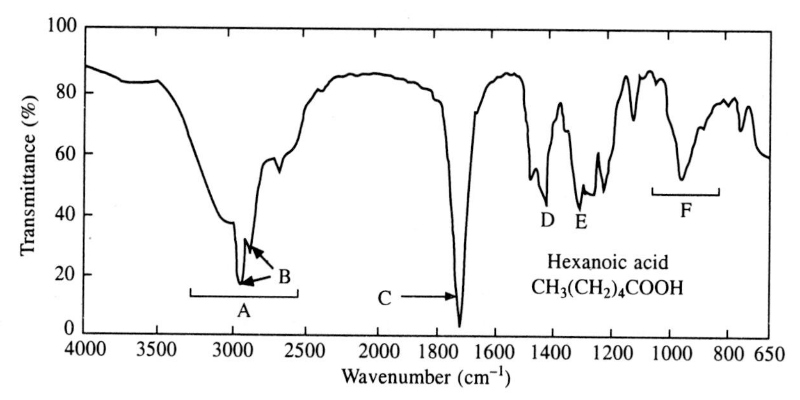

A: Broad O—H stretching 2500–3300 cm^{-1}
B: C—H stretching (superimposed upon O—H stretching) 2950, 2920 and 2850 cm^{-1}
C: Normal, dimeric carboxylic C=O stretching, 1715 cm^{-1}
D: O—H in-plane bending, 1408 cm^{-1}
E: C—O stretching, 1280 cm^{-1}
F: O—H out-of-plane bending, 930 cm^{-1}

Fig. 3.11 IR spectrum of hexanoic acid, neat

We know that resonance is much more important between the exactly equivalent structures as in the carboxylate anion than between non-equivalent structures as in the carboxylic acid. Thus, the decrease in the double bond character of the C=O group by resonance is greater in the carboxylate anion than that in the carboxylic acid. Thus, the $\nu_{C=O}$ absorption occurs at a lower wavenumber for a carboxylate anion than that for a carboxylic acid. The carboxylate anion gives two bands, a strong asymmetrical stretching band near 1550–1650 cm^{-1} and a weaker symmetrical stretching band near 1300–1400 cm^{-1} (Table 3.1). A suspected carboxylic acid structure may be confirmed by comparison of IR spectrum of the compound recorded in a neutral solvent (or as a neat sample) with that recorded in the presence of a base such as triethyl amine. The carboxylate ion thus formed also shows an additional ammonium band in the region 2200–2700 cm^{-1}.

(xi) Esters and Lactones

Esters and lactones (cyclic esters) have two characteristic, strong bands due to C=O and C—O stretchings. Saturated aliphatic esters (except formates) show $\nu_{C=O}$ absorption bands in the region 1735–1750 cm^{-1}. The $\nu_{C=O}$ bands of formates, α,β-unsaturated and aryl esters appear in the relatively lower region 1715–1730 cm^{-1} because of mesomeric effect. The force constant of the C=O bond in esters is increased due to the $-I$ effect of the adjacent oxygen. Thus, $\nu_{C=O}$ band of esters appear at a higher wavenumber than that of a ketone. The stretching frequency of the ester C=O group is influenced by various factors almost in the same way as that of ketones. The $\nu_{C=O}$ bands of esters occur in the region 1000–1300 cm^{-1} (Table 3.1). Fig. 3.12 shows IR spectrum of a typical ester.

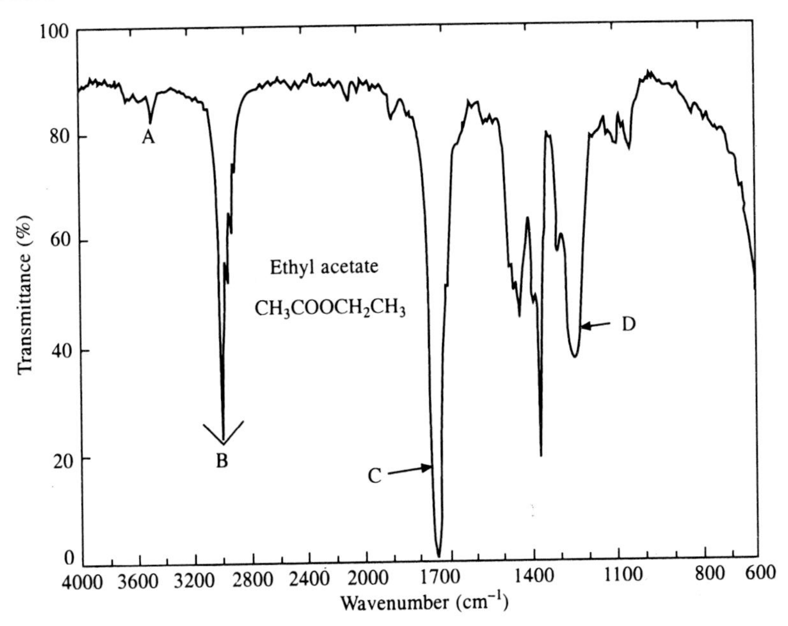

A: Overtone band of C=O stretching, frequency twice that of C=O stretching, 3478 cm^{-1}
B: Methyl and methylene stretching bands, around 2900 cm^{-1}
C: Normal ester C=O stretching, 1740 cm^{-1}
D: C—O—C stretching, 1259 cm^{-1}

Fig. 3.12 IR spectrum of ethyl acetate, neat

When a C=C or an aromatic ring is attached to the oxygen of the C—O group of an ester or lactone, there is a marked increase in the carbonyl frequency along with a decrease in the C—O frequency. This is because of the mesomeric effect as shown below which increases the force constant of the C=O bond and decreases that of the C—O bond

$$R-\overset{\overset{\displaystyle O}{\|}}{C}-\ddot{O}-C=C- \longleftrightarrow R-\overset{\overset{\displaystyle O}{\|}}{C}-\overset{+}{O}=C-\overset{-}{C}-$$

Saturated δ-lactones (six-membered ring) show the carbonyl absorption in the same region as straight-chain, unconjugated esters. Similar to that of ketones (Section 3.8(ix)), as ring strain in lactones increases, the carbonyl stretching frequency also increases.

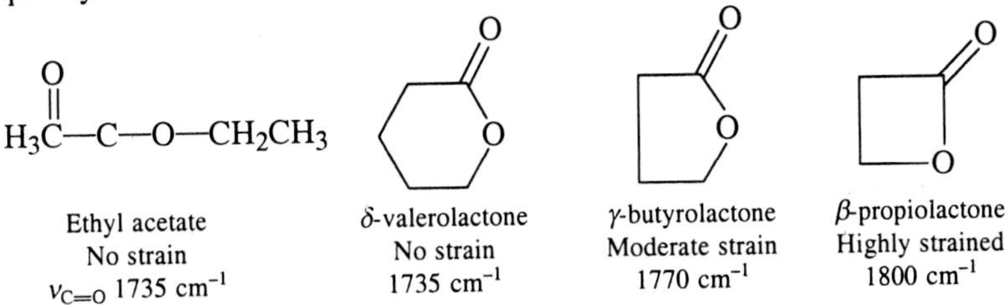

Unsaturated α to the C=O group decreases the C=O absorption frequency through mesomerism. Unsaturated α to the —O— group increases the $\nu_{C=O}$ frequency and decreases the $\nu_{C—O}$ frequency through resonance in the same way as shown for acyclic esters, for example

Conjugation of C=C with C=O
$\nu_{C=O}$ 1720 cm^{-1}

Conjugation of C=C with —O—
$\nu_{C=O}$ 1760 cm^{-1}

(xii) Acid Halides

Acid halides show strong C=O absorption band in the carbonyl stretching region (Table 3.1). Unconjugated acid chlorides show $\nu_{C=O}$ band in the region 1790–1815 cm^{-1}. Since the resonance reduces the force constant of the C=O bond, conjugated acid halides absorb at slightly lower frequency. For example, aromatic acid chlorides absorb strongly in the region 1750–1800 cm^{-1}. The decreasing order of $\nu_{C=O}$ frequency of various acid halides is as follows:

acid fluorides > acid chlorides > acid bromides > acid iodides

(xiii) Acid anhydrides

Acid anhydrides show two bands in the carbonyl region—one due to symmetrical and the other due to asymmetrical stretching of the C=O group (Table 3.1). Saturated acyclic anhydrides show two $\nu_{C=O}$ bands in the region 1710–1825 cm^{-1} (Fig. 3.13) due to strong vibrational coupling of the two carbonyl groups through the oxygen as discussed in Section 3.7(a). Aryl and α,β-unsaturated anhydrides show $\nu_{C=O}$ bands in relatively lower regions due to resonance which decreases the force constant of the C=O bond. In acyclic anhydrides, the high

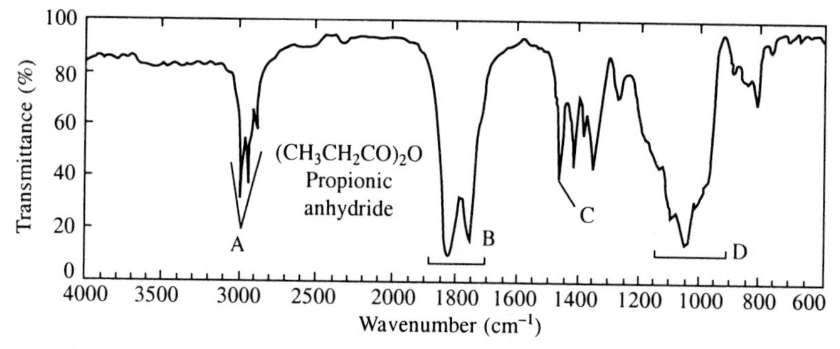

A: C—H stretching, 2990, 2950 and 2880 cm^{-1}

B: Asymmetrical and symmetrical C=O coupled, 1825 and 1758 cm^{-1}, respectively

C: δ_s CH_2 (scissoring), 1465 cm^{-1}

D: C—CO—O—CO—C stretching, 1040 cm^{-1}

Fig. 3.13 IR spectrum of propionic anhydride, neat

frequency $\nu_{C=O}$ band is more intense, whereas the low frequency $\nu_{C=O}$ band is more intense in cyclic anhydrides. This characteristic can be used for differentiating acyclic and cyclic anhydrides. In cyclic anhydrides, both the $\nu_{C=O}$ bands shift towards higher frequencies than that of acyclic anhydrides. Cyclic anhydrides with five-membered rings show $\nu_{C=O}$ absorption at higher wavenumbers than acyclic anhydrides due to ring strain (Table 3.1). All types of anhydrides exhibit one or two strong $\nu_{C=O}$ bands in the region 1050–1300 cm^{-1}.

(xiv) Amides, Lactams and Imides

We have already discussed in Section 3.7(d) that amides show $\nu_{C=O}$ band at a lower frequency than that of esters due to lesser electronegativity of nitrogen than oxygen. All amides show a $\nu_{C=O}$ band in the region 1630–1700 cm^{-1} which is known as the amide I band (Table 3.1). The absorption frequency of this band decreases on hydrogen bonding. Primary amides show two ν_{N-H} bands due to symmetrical and asymmetrical stretchings, whereas secondary amides and lactams (cyclic amides) show only one ν_{N-H} band (Table 3.1). The frequency of ν_{N-H} absorption bands is also reduced by hydrogen bonding though to a lesser degree than that of ν_{O-H}. Fig. 3.14 shows IR spectrum of a typical amide.

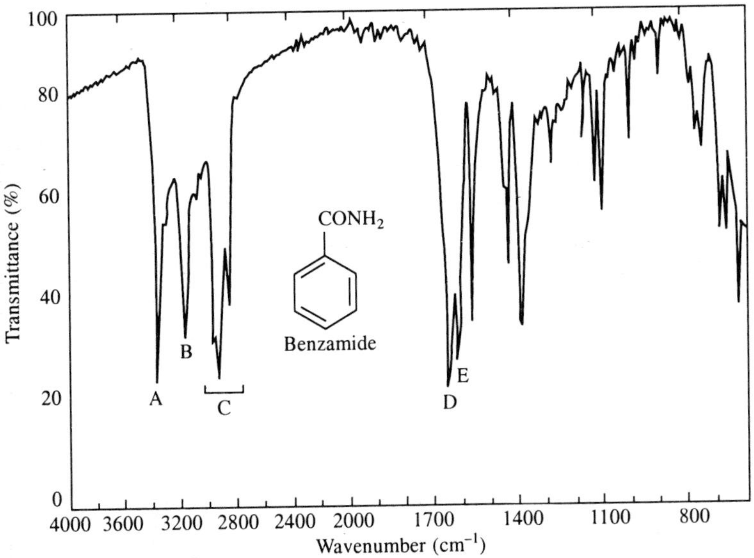

N—H stretching, coupled, primary amide, hydrogen bonded:

A: Asymmetric, 3370 cm^{-1}

B: Symmetric, 3170 cm^{-1}

C: Nujol and aromatic ν_{C-H} bands, 2860-3030 cm^{-1}

D: C=O stretching, amide I band, 1660 cm^{-1}

E: N—H bending, amide II band, 1631 cm^{-1}

Fig. 3.14 IR spectrum of benzamide, as a nujol mull

In dilute solutions, primary and secondary amides and a few lactams exhibit a band or bands (amide II band) mainly due to NH_2 or NH bending in the region

1510–1650 cm^{-1}. Amide I band is generally more intense than amide II band. In solid state, amide I and II may overlap.

The frequency of $\nu_{C=O}$ absorption in lactams depends on the ring size. Similar to that of ketones and lactones (Sections 3.8(i) and 3.8(k)), as the ring strain in lactams increases, the carbonyl stretching frequency also increases. Fusion of the lactam ring with another ring generally increases $\nu_{C=O}$ frequency by 20–50 cm^{-1}. For example, in penicillin this absorption occurs at 1770 cm^{-1}.

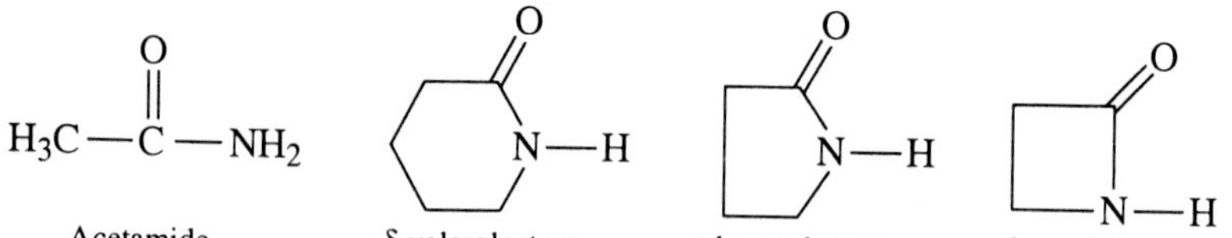

H_3C—C(=O)—NH_2

Acetamide	δ-valerolactam	γ-butyrolactam	β-propiolactam
No strain	No strain	Moderate strain	Highly strained
$\nu_{C=O}$ 1680 cm^{-1}	1670 cm^{-1}	1700 cm^{-1}	1745 cm^{-1}

In cyclic imides, the $\nu_{C=O}$ frequency decreases with increasing ring size, whereas it is shifted to a higher value with α,β-unsaturation. For example, succinimide shows two strong C=O bands at ~1770 and 1700 cm^{-1}, whereas benzimide shows at ~1730 and 1670 cm^{-1}.

Summary of carbonyl absorptions

Effects of structural variations on the C=O stretching frequencies may be summarized as follows:

(i) The greater the electronegativity of the group X in the compound $RCOX$, the higher is the frequency.

(ii) α,β-unsaturation causes a lowering of frequency by 15–40 cm^{-1}. Amides are exceptions, where little shift is observed and that too usually to a higher frequency.

(iii) Additional conjugation has relatively little effect.

(iv) Hydrogen bonding to a carbonyl group shifts its frequency to a lower value by 40–60 cm^{-1}. Carboxylic acids, amides, enolized β-keto compounds and *o*-hydroxy and *o*-aminophenyl carbonyl compounds show this effect. In the solid state, the carbonyl stretching frequency of all the carbonyl compounds is slightly lowered compared to that in dilute solutions.

(v) Ring strain in cyclic compounds causes large shift towards higher frequency. This provides a reliable test for distinguishing clearly between four, five and larger membered ring ketones, lactones and lactams. Six-membered ring and larger membered ring ketones, lactones and lactams show the normal frequency found for the corresponding open-chain compounds.

(vi) When more than one structural influences are operating on a particular carbonyl group, the net effect is usually almost additive.

(xv) Amines and their Salts

Amines show characteristic ν_{N-H} absorption in the region 3300–3500 cm^{-1}. The position of absorption is affected by hydrogen bonding. In dilute solution, primary amines exhibit two sharp bands due to symmetrical and asymmetrical stretchings, secondary amines give only one weak band, whereas tertiary amines give no band due to free N—H stretching in the region 3300–3500 cm^{-1} (Table 3.1). These bands are shifted to lower wavenumbers due to hydrogen bonding. Aromatic primary amines show $\nu_{N—H}$ bands at slightly higher wavenumbers. Fig. 3.15 shows a typical IR spectrum of an amine.

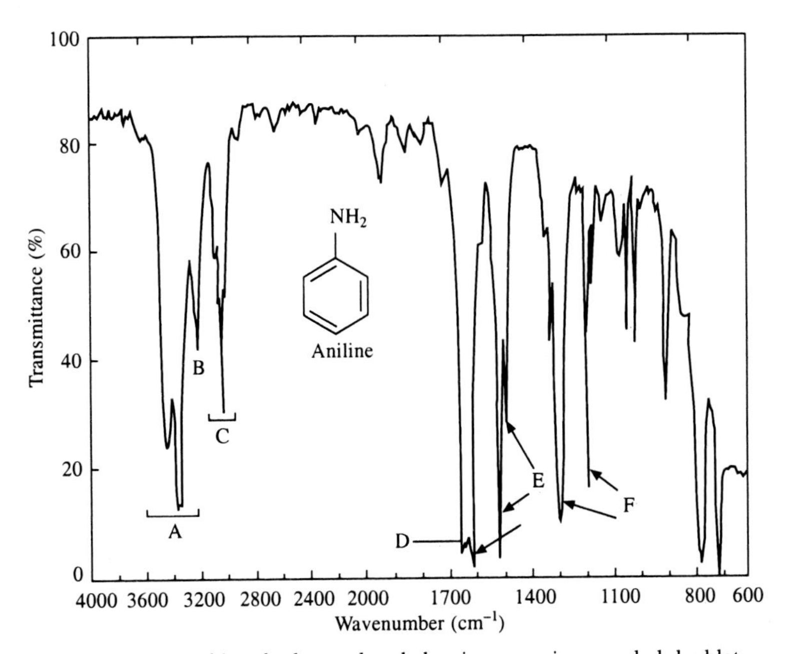

A: N—H stretching, hydrogen bonded, primary amine coupled doublet asymmetrical 3440 cm^{-1}, symmetrical 3350 cm^{-1}
B: Fermi resonance band, 3200 cm^{-1}; with overtone of band at 1620 cm^{-1} (E)
C: Aromatic C—H stretch, around 3050 cm^{-1}
D: N—H bending, 1620 cm^{-1}
E: C⩦C ring stretching, 1600, 1500 and 1468 cm^{-1}
F: C—N stretching, 1280 and 1175 cm^{-1}

Fig. 3.15 IR spectrum of aniline, neat

The N—H bending vibration of primary amines appears in the region 1590–1650 cm^{-1}. This band moves to slightly higher frequency on hydrogen bonding. The N—H bending band is seldom detectable in the IR spectra of aliphatic secondary amines.

Salts of primary amines show strong, broad band due to N^+H_3 symmetrical and asymmetrical stretchings in the region 2700–3000 cm^{-1}, whereas salts of secondary and tertiary amines absorb in the region 2250–2700 cm^{-1}. In addition, bending N—H vibrations occur in amine salts in the region 1500–1600 cm^{-1}. All the amines and amine salts show bands due to C—N vibrations in the fingerprint region.

(xvi) Amino Acids and their Salts

Amino acids exist in the following three forms as:

(a) Zwitter ions ($H_3\overset{+}{N}—\underset{|}{\overset{|}{C}}—CO\bar{O}$)

These show characteristic absorptions due to N^+H_3 and COO^- groups. A broad, strong N^+H_3 stretching band is observed in the region 3030–3130 cm^{-1}. The COO^- group absorbs strongly in the region 1590–1600 cm^{-1} and more weakly near 1400 cm^{-1} due to C=O asymmetrical and symmetrical stretchings, respectively.

(b) Hydrochlorides (or other acid) salts ($\bar{Cl}H_3\overset{+}{N}—\underset{|}{\overset{|}{C}}—COOH$)

In these, the $\nu_{C=O}$ bands are shifted to higher frequencies. Hydrochlorides of α-amino acids show strong carbonyl absorption in the range 1730–1755 cm^{-1}. Superimposed O—H and N^+H_3 stretching bands appear as a broad strong band in the region 2380–3333 cm^{-1}.

(c) Sodium (or other cation) salts ($H_2N—\underset{|}{\overset{|}{C}}—CO\bar{O}\overset{+}{Na}$)

These salts exhibit the normal $\nu_{N—H}$ band in the range 3200–3400 cm^{-1} similar to other amines. The characteristic COO^- stretching bands appear near 1590–1600 cm^{-1} and near 1400 cm^{-1}.

In addition, all the three forms of amino acids show weak, asymmetrical bending band near 1590–1610 cm^{-1} and a relatively strong symmetrical bending band near 1481–1550 cm^{-1} due to N^+H_3 (or NH_2) bendings (Table 3.1).

(xvii) Nitro Compounds and Nitrites

Nitro compounds are considered to be a resonance hybrid of the following structures:

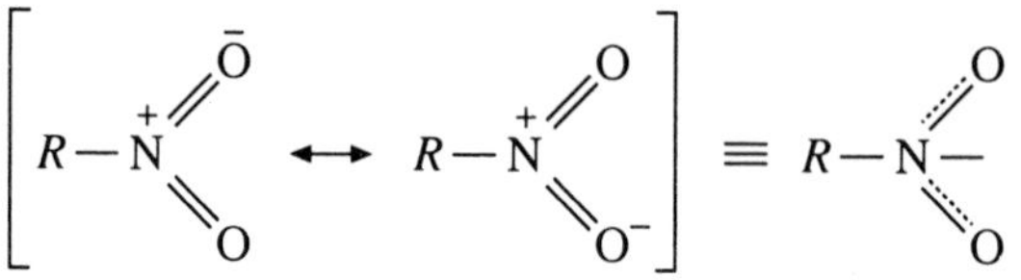

$$\left[R—\overset{+}{N}(=O)(—\bar{O}) \longleftrightarrow R—\overset{+}{N}(=O)(—O^-) \right] \equiv R—N\overset{\cdot\cdot\cdot O}{\underset{\cdot\cdot\cdot O}{}}—$$

Reasoning structures of a nitro compound

Nitroalkanes exhibit two bands near 1550 and 1370 cm^{-1} due to asymmetrical and symmetrical stretching of the N=O group, respectively. Conjugation lowers the frequency of both the bands causing absorption in the regions near 1500–1550 cm^{-1} and near 1290–1360 cm^{-1} (Table 3.1). Aromatic nitro compounds absorb around the same frequencies as conjugated aliphatic nitro compounds. Following is the order of $\nu_{N=O}$ frequency of nitro compounds:

primary > secondary > tertiary

Nitrites (R—O—N=O) show two strong $\nu_{C=O}$ bands at higher frequencies than that of the nitro compounds. They exhibit one band near 1650–1680 cm^{-1} and the other near 1611–1625 cm^{-1}. The nitrite absorption bands are one of the strongest bands observed in IR spectra.

(xviii) Nitriles and Related Compounds

Nitriles show weak to medium $\nu_{C\equiv N}$ band in the region 2200–2260 cm^{-1}. Conjugated nitriles show absorption band in the lower end of this wavenumber region due to resonance which decreases the triple bond character of the C≡N group. Occasionally, the $\nu_{C\equiv N}$ band is very weak or absent, e.g. some cyanohydrins show no $\nu_{C\equiv N}$ absorption.

$$R-\overset{|}{C}=\overset{|}{C}-C\equiv N \longleftrightarrow R-\overset{+}{\underset{|}{C}}-\underset{|}{C}=C=\overset{-}{N}$$

Reasonating structures of conjugated nitrile

Isocyanides (isonitriles), cyanates, isocyanates, thiocyanates isothiocyanates, azides and carbodiimides show absorption band in the region 1990–2275 cm^{-1} (Table 3.1) due to the C≡N or cumulated double bond stretching. Schiff bases, oximes and thiazoles, etc. exhibit $\nu_{C=N}$ absorption in the region 1470–1690cm^{-1}.

The N=N stretching absorption of symmetrical diazo compounds is either very weak or inactive in IR but appears in the Raman spectrum near ~1575 cm^{-1}. The bands are weak due to the non-polar nature of the bonds. Diazonium salts absorption band due to —N^{+}≡N is near 2260 cm^{-1}.

(xix) Sulphur and Phosphorus Compounds

For characteristic absorption frequencies of sulphur or phosphorus containing groups, see at the end of Table 3.1.

Table 3.1 Characteristic infrared absorption frequencies

Type	Group	Absorption frequency (cm^{-1})	Intensity*	Assignment and remarks
Alkanes	—CH_3, >CH_2, —>CH	2840–3000	$m \to s$	C—H stretch; two or three bands
	—>C—C<—	800–1200	w	C—C stretch; of little value
	—CH_3	1430–1470 1370–1380	m s	Asym. C—H bending Sym. C—H bending; very characteristic
	>CH_2	1445–1485	m	Asym. C—H bending
	—>CH	~1340	w	C—H bending
	—>C—C<—	below 500	w	C—C bending; outside the ordinary IR region
Cycloalkanes	—CH_2—	2840–3050	m	Asym. and sym. C—H stretch; two bands

(*Contd.*)

Type	Group	Absorption frequency in cm^{-1}	Intensity*	Assignment and remarks
	$—CH_2—$ (cyclopropane)	3040–3060	*m*	Asym. C—H stretch
		2975–2985	*m*	Sym. C—H stretch
		1015–1045	*m*	Skeletal vibration
Alkenes	$>C=C<$	1620–1680	*v*	
	$>C=CH_2$	1640–1660	*m*	
	$HC=CH$ (*cis*)	1655–1660	*m*	C=C stretch; in dienes, trienes etc., 1650 (*s*) and 1600 (*s*); on conjugation with aromatic ring, ~1625 (*m*); in α,β-unsaturated carbonyl compounds, 1590–1640 (*s*)
	$HC=CH$ (*trans*)	1670–1675	*w*	
	$>C=C<^{H}$	3000–3100	*m*	C—H stretch; almost the same position in the *cis* and *trans* isomers
	$>C=CH_2$	885–895	*s*	
	$RCH = CH_2$	985–995	*s*	
		900–940	*s*	
	$HC=CH$ (*cis*)	675–730	*m*	C—H out-of-plane bending
	$HC=CH$ (*trans*)	960–970	*s*	
	$>C=C<^{H}$ (trisubstituted)	790–840	*m*	
Alkynes	—C≡C—	2100–2260	*v*	C≡C stretch
	—C≡C—H (terminal)	2100–2140	*s*	C≡C stretch

Type	Group	Absorption frequency in cm^{-1}	Intensity*	Assignment and remarks
	—C≡C— (non-terminal)	2190–2260	*v*	C≡C stretch; intensity of band decreases as symmetry of molecule increases; may be absent if the molecule is symmetrical about the triple bond
	—C≡C—H	3300–3310	*s*	C—H stretch
		615–645	*s*	C—H bending; outside the ordinary IR region
Halogen compounds	CH_3—*X* (*X* = F, Cl, Br, I)	near 3000	*s*	Asym. and sym. C—H stretch
	>C—F	1000–1400	*s*	C—F stretch
	>C—Cl	600–800	*s*	C—Cl stretch
	>C—Br	500–750	*s*	C—Br stretch
	>C—I	~500	*s*	C—I stretch
	CH_3—*X* (*X* = F, Cl, Br, I)	1441–1471	*v*	C—H asym. bending
		1255–1475	*v*	C—H sym. bending
Aromatic compounds	Aromatic C—H	3000–3100	*m*	C—H stretch
	Aromatic C—C	1600±5	*v*	C=C skeletal stretch
		1580±5	*m*	Skeletal stretch; present when ring is further conjugated
		1500 ± 25	*v*	Skeletal stretch
		1450 ± 10	*m*	Skeletal stretch
	Monosubstituted benzenes	730–770	*s*	C—H out-of-plane bending
		690–710	*s*	
	o-disubstituted benzenes	735–770	*s*	
	m-disubstituted benzenes	750–810	*s*	
		680–725	*m*→*s*	
	p-disubstituted benzenes	800–860	*s*	
	Aromatic C—H	1000–1300	*w*	In-plane C—H bending; usually difficult to assign because they are weak and C—O and other C—H bands occur in the same region

(Contd.)

Type	Group	Absorption frequency in cm^{-1}	Intensity*	Assignment and remarks
Alcohols and phenols	O—H	3590–3650	*v*	Free O—H stretch
		3200–3600	*v*	Intermolecular hydrogen bonded O—H stretch
		2500–3200	*s*	Intramolecular hydrogen bonded O—H stretch
	C—O	1000–1260	*m*→*s*	C—O stretch
		~1050	*s*	C—O stretch primary alcohols
		~1100	*s*	C—O stretch secondary alcohols
		~1150	*s*	C—O stretch tertiary alcohols
		~1200	*s*	C—O stretch phenols
	O—H	1330–1420	*s*	In-plane O—H bending
		650–769	*s*	Out-of-plane O—H bending
Ethers and epoxides	—O—CH_3	2810–2850	*m*	C—H stretch
	—O—CH_2—	~3050	*m*	Epoxide C—H stretch
	—O—CH_2—O—	2770–2790	*m*	C—H stretch
	>C—O—C<	1060–1150	*s*	C—O—C asym. stretch
	>C=C(—)—O—	1020–1075	*s*	C—O—C sym. stretch
	Aryl—O—	1200—1275	*s*	C—O—C asym. stretch
	>C—C< (epoxide, bridged by O)	1240–1260	*s*	C—O—C sym. stretch
		810–950	*m*→*s*	C—O—C asym. stretch; larger ring ethers absorb closer to the usual values for acyclic ethers
Aldehydes	*R*—C(H)=O	1720–1740	*s*	C=O stretch; saturated, aliphatic
		1695–1715	*s*	C=O stretch; aromatic
		1680–1705	*s*	C=O stretch; *α,β*-unsaturated, aliphatic
		1660–1680	*s*	C=O stretch; *α,β,γ,δ*-unsaturated, aliphatic
		1645–1670	*s*	C=O stretch; *β*-keto aldehyde in enol form; lowering caused by intramolecular hydrogen bonding
		2700–2900	*w*	C—H stretch; usually two

Type	Group	Absorption frequency in cm^{-1}	Intensity*	Assignment and remarks
				bands, one near 2720 cm^{-1} due to Fermi resonance
Ketones	$R(R')C=O$	1705–1725	*s*	C=O stretch; saturated, acyclic
		1680–1700	*s*	C=O stretch; aryl
		1660–1670	*s*	C=O stretch; diaryl
		1665–1685	*s*	C=O stretch; α,β-unsaturated, and α, β,α', β' unsaturated, acyclic
		1705–1725	*s*	C=O stretch; six- and larger-membered ring
		1740–1750	*s*	C=O stretch; five-membered ring
		~1775	*s*	C=O stretch; four-membered ring
		1685–1705	*s*	C=O stretch; cyclopropyl
		1665–1685	*s*	C=O stretch; α,β-unsaturated cyclic, six- and larger-membered ring
		1708–1725	*s*	C=O stretch; five-membered ring
		1540–1640	*s*	C=O stretch; β-diketone, enolic
	o-amino- and *o*-hydroxy-aryl ketones	1635–1655	*s*	C=O stretch
	1,4- and 1,2-benzoquinones	1660–1690	*s*	C=O stretch
Carboxylic acids	—COOH	1700–1725	*s*	C=O stretch; saturated, aliphatic
		1690–1715	*s*	C=O stretch; α,β-unsaturated, aliphatic
		1680–1700	*s*	C=O stretch; aryl
		1720-1740	*s*	C=O stretch; α-halo
		1655–1665	*s*	C=O stretch; intramolecularly hydrogen-bonded
		2500–3000	*m*	O—H stretch; hydrogen-bonded, free O—H stretch, 3550 cm^{-1}

(Contd.)

Type	Group	Absorption frequency in cm^{-1}	Intensity*	Assignment and remarks
Carboxylate anions	—COO^-	1550–1650	*s*	C=O asym. stretch
		1300–1400	*s*	C=O sym. stretch
Esters	*R*—COO*R′*	1735–1750	*s*	C=O stretch; saturated acyclic
		1715–1730	*s*	C=O stretch; $R = \alpha,\beta$-unsaturated or aryl group
		1750–1800	*s*	C=O stretch; $R' = \alpha,\beta$-unsaturated or aryl group
		1650	*s*	C=O stretch; β-ketoesters, enolic
		1000–1300	*s*	C—O—C sym. and asym. stretch; in all types of esters
Lactones (cyclic esters)	O=C(–O–)(CH_2)$_n$ (ring)	1735–1750	*s*	C=O stretch; δ-lactone and larger rings, $n \geq 4$
		1760–1780	*s*	C=O stretch; γ-lactone, $n = 3$
		1820–1840	*s*	C=O stretch; β-lactone, $n = 2$
		1717–1730	*s*	C=O stretch; α,β-unsaturated δ-lactone
		1740–1770	*s*	C=O stretch; α,β-unsaturated γ-lactone
		~1800	*s*	C=O stretch; β,γ-unsaturated γ-lactone
Acid halides	—COCl	1790–1815	*s*	C=O stretch; fluorides higher, bromides and iodides successively lower
		1750–1800	*s*	C=O stretch; aryl and α,β-unsaturated
Acid anhydrides	—CO—O—CO—	1800–1850 1740–1790	*s* *s*	C=O stretch; two bands
		1780–1830 1710–1770	*s* *s*	C=O stretch; two bands, aryl and α,β-unsaturated, acyclic
		1820–1870 1750–1800	*s* *s*	C=O stretch; two bands, saturated five-membered ring
		1780–1830 1710–1770	*s* *s*	C=O stretch; two bands, aryl and α,β-unsaturated
		1050–1300	*s*	C—O stretching; one or two bands in all types of anhydrides

Type	Group	Absorption frequency in cm^{-1}	Intensity*	Assignment and remarks
Amides	*R*—$CONH_2$ (primary)	~1690	*s*	Amide I, C=O stretch; free, i.e. in dil. solution
		~1650	*s*	C=O stretch; associated, in solid state and conc. solution
	R—CONHR (secondary)	1670–1700	*s*	C=O stretch; free, i.e. in dil. solution
		1630–1680	*s*	C=O stretch; associated, in solid state
	R $CONR_2$ (tertiary)	1630–1670	*s*	C=O stretch; free and associated
	R—$CONH_2$	~3500 ~3400	*m* *m*	N—H stretching; free, two bands
		3350 3180	*m* *m*	N—H stretching; associated, two bands
	R—CONH*R*	3430	*m*	N—H stretch; free, one band
		3140–3320	*m*	N—H stretch; associated, one band
	R $CONH_2$	1590–1620	*s*	Amide II, N—H bending; dil. solution
	R CONH*R*	1510–1550	*s*	Amide II, N—H bending; dil. solution
Lactams (cyclic amides)	H–N(–C=O)–$(CH_2)_n$ (ring)	~1680	*s*	C=O stretch; $n \geq 4$ δ-lactam, dil. solution
		~1700	*s*	C=O stretch; $n = 3$, γ-lactam, dil. solution
		1730–1760	*s*	β-lactam, C=O stretch; $n = 2$, dil. solution
		~3400	*m*	N—H stretch; free lactam, associated, 3070–3175 cm^{-1}
Amines and their salts	*R*NH_2	3300–3500	*m*	N—H stretch; free, two bands due to sym. and asym. stretch
	R_2NH	3310–3550	*m*	N—H stretch; free, one band
	$R_2C{=}NH$ (imine)	3300–3400	*m*	N—H stretch; one band
	$R\overset{+}{N}H_3$	~3000	*s*	$\overset{+}{N}H_3$ sym. and asym. stretch

(Contd.)

Type	Group	Absorption frequency in cm^{-1}	Intensity*	Assignment and remarks
	$R_2\overset{+}{N}H_2$ and $R_3\overset{+}{N}H$	2250–2700	*s*	$\overset{+}{N}H_2$ and $\overset{+}{N}H$ stretch
	$>N-CH_3$	2780–2820	*m*	C—H stretch
	$R\,NH_2$	1590–1650	$m \rightarrow s$	N—H in-plane bending
	R_2NH	1550–1650	*w*	N—H in-plane bending
	Amine salts	1575–1600	*s*	$\overset{+}{N}H_3$ and $\overset{+}{N}H_2$ bending
		~1500	*s*	$\overset{+}{N}H_3$ sym. bending
Amino acids and their salts	$H_3\overset{+}{N}-\overset{\vert}{\underset{\vert}{C}}-CO\bar{O}$	1590–1600	*s*	C=O asym. stretch
		~1400	*w*	C=O sym. stretch
		3030–3130	*s*	N—H stretch
		1730–1755	*s*	C=O stretch
	$Cl\overset{+}{N}H_3-\overset{\vert}{\underset{\vert}{C}}-COOH$	2380–3333	*s*	O—H and $\overset{+}{N}H_3$ stretch; superimposed
	$H_3N-\overset{\vert}{\underset{\vert}{C}}-CO\bar{O}\overset{+}{N}a$	1590–1600	*s*	C=O asym. stretch
		~1400	*w*	C=O sym. stretch
		3200–3400	*s*	N—H stretch
	All amino acids and their salts	1590–1610	*w*	NH_2 or $\overset{+}{N}H_3$ asym. bending
		1481–1550	*s*	NH_2 or $\overset{+}{N}H_3$ sym. bending
Nitro compounds and nitrites	$R—NO_2$ (aliphatic)	1550–1570	*s*	Asym. N=O stretch
		1370–1380	*s*	Sym. N=O stretch
	$Ar-NO_2$ (aromatic)	1500–1570	*s*	Asym. N=O stretch
		1300–1370	*s*	Sym. N=O stretch
	—O—N=O	1610–1680	*s*	N=O stretch; two bands
Nitriles and related compounds	*R*—C≡N (nitriles)	2200–2260	*v*	C≡N stretch
	R—N≡C (isonitriles)	2070–2220	*m*	N≡C stretch
	R—S—C≡N (thiocyanates)	2140–2175	*s*	C≡N stretch
	R—N=C=N—*R* (carbodiimides)	2130–2155	*s*	N = C = N stretch
	R—N=C=O (isocyanates)	2240–2275	*m*	N = C = O stretch

Type	Group	Absorption frequency in cm^{-1}	Intensity*	Assignment and remarks
	R—N=C=S (isothiocyanates)	1990–2140	*s*	N=C=S stretch
	R—N_3 (azides)	2120–2160	*s*	N_3 stretch
	>C=N— (Schiff bases, imines and oximes)	1640–1690	*v*	C=N stretch
	R—$\overset{+}{N}\equiv N$ (diazonium salts)	2240–2280	*s*	$\overset{+}{N}\equiv N$ stretch
	R—N=N—R (azo compounds)	1575–1630	*v*	N=N stretch
Sulphur compounds	—SH	2550–2600	*w*	S—H stretch; less affected by hydrogen bonding
	>C=S	1050–1200	*s*	C=S stretch
	—C(=S)—N(—)—H	~3400	*m*	N—H stretch; lowered to 3150 cm^{-1} in solid state
		1100–1300	*s*	Amide I, C=S stretch
		1460–1550	*s*	Amide II, N—H bending
	>S=O	1040–1060	*s*	S=O sitretch
	>SO_2	1300–1350	*s*	Asym. SO_2 stretch
		1120–1160	*s*	Sym. SO_2 stretch
	—SO_2—N<	1330–1370	*s*	Asym. SO_2 stretch
		1160–1180	*s*	Sym. SO_2 stretch
	—SO_2—O—	1330–1420	*s*	Asym. SO_2 stretch
		1145–1200	*s*	Sym. SO_2 stretch
Phosphorus compounds	P—H	2350–2440	*s*	P—H stretch
	P—O—R	1030–1240	*s*	P—O—C stretch
	P=O	1250–1300	*s*	P=O stretch; in phosphate esters
	P(O)(OH)	2560–2700	*v*	O—H stretch; hydrogen bonded
		1180–1240	*s*	P=O stretch

*s = strong, m = medium, w = weak, v = vaiiable.

3.9 Fingerprint Region

It is not possible for any two different compounds (except enantiomers) to have exactly the same IR spectrum. Therefore, the IR spectrum of a compound is called its *fingerprint*. The region below 1500 cm^{-1} is called fingerprint region because every compound has unique absorption pattern in this region, just as every person has unique fingerprints. The fingerprint region contains many absorption bands caused by bending vibrations as well as absorption bands caused by C—C, C—O like in alcohols, ethers, esters, etc. and C—N (e.g. in amines, amino acids, amides, etc.) stretching vibrations. Since the number of bending vibrations in a molecule is much greater than its stretching vibrations, the fingerprint region is rich in absorption bands and shoulders. Thus, the superimposability of IR bands of the spectra of any two different compounds becomes impossible in this region. However, similar compounds may show very similar spectra above 1500 cm^{-1}.

The use of IR spectroscopy to confirm the identity of a compound with the help of an authentic sample is more reliable than taking mixed m.p. or comparing other physical properties. If two pure samples give different IR spectra under same conditions, they represent different compounds. If the samples give the superimposable IR spectra, they represent the same compound. For example, the IR spectra of two stereoisomeric steroids, androsterone and epiandrosterone, show three strong bands above 1500 cm^{-1}, viz. around 3600, 2950 and 1740 cm^{-1} due to O—H, C—H and C=O stretching vibrations, respectively. However, the absorption patterns in the fingerprint region of their IR spectra are quite different showing them to be different compounds. It should be noted that these two compounds differ only in the stereochemistry at C–3 where the OH group is axial in androsterone and equatorial in epiandrosterone.

3.10 Applications of Infrared Spectroscopy

Among all the properties of an organic compound, no single property gives as much information about the compound's structure as its infrared spectrum. Thus, IR spectroscopy is the most widely used method for structure determination of organic compounds. The basic reason why IR spectra are of such value to organic chemists is that molecular vibrations depend on interatomic distances, bond angles and bond strengths, rather than on bulk properties of the compound. Thus, these vibrational frequencies provide a molecular fingerprint which enables the identification of the compound either in the pure state or in mixtures. IR spectroscopy is especially used for detection of functional groups in organic compounds and for establishing the identity of organic compounds. Some important applications of IR spectroscopy to organic chemistry are summarized as follows.

(i) Detection of Functional groups

All functional groups absorb in a definite frequency region. Thus, the presence or absence of a band in a definite frequency region tells the presence or absence of a particular functional group in the compound. For example, the presence of a $\nu_{C=O}$ band in the region 1720–1740 cm^{-1} along with another band (usually two bands) in the region 2700–2900 cm^{-1} shows the presence of an aldehydic carbonyl group in the compound. The normal absorption frequency of a particular group may change within its definite range. This change indicates the nature of the factor, such as inductive effect, conjugation, hydrogen bonding, steric effect and field effect, influencing the group frequency, thus giving an idea of the structure of the compound. A discussion on characteristic absorptions in various classes of organic compounds is given in Section 3.8, and characteristic group frequencies are summarized in Table 3.1. These are very useful for detecting various functional groups with the help of IR spectra.

(ii) Confirmation of the Identity of Compounds

The identity (and thus the purity) of a compound is often established by comparing its IR spectrum with that of an authentic sample. If the IR spectra, recorded under same conditions, are superimposable, they represent the same compound. The large number of bands, especially in the fingerprint region, are most useful for identification. This technique is more reliable than taking mixed m.p. or comparing other physical properties, particularly when only small quantities of a substance are available.

(iii) Estimation of the Purity of Samples

IR spectra of impure sample are usually blurred and have many bands which cannot be interpreted, whereas a pure compound gives a clear IR spectrum. For example, a sample of an alcohol containing a ketone as an impurity gives poor IR spectrum which shows additional absorption bands due to the carbonyl group. Similarly, the presence of a ketone as impurity in a sample of hydrocarbon can easily be detected because the hydrocarbons do not show any absorption band around 1715 cm^{-1} which is the characteristic band of a carbonyl group. The percentages of individual components in a mixture can be estimated by intensity measurements of specific absorption bands. Thus, the purity of a sample may be estimated by inspection of its IR spectrum and comparison with a reference spectrum.

(iv) Study of Hydrogen Bonding

As discussed in Section 3.7(iii), IR spectroscopy is useful in detecting hydrogen bonding, in estimating the strength of hydrogen bonds and in distinguishing intermolecular and intramolecular hydrogen bondings.

(v) Calculation of Force Constants

The frequencies of IR absorptions are commonly used to calculate the force constants of various bonds. For this purpose, Hooke's law as represented by Eq. (3.1) in Section 3.6 is applied.

(vi) Determination of Orientations in Aromatic Compounds

Absorptions in the region 675–900 cm^{-1} due to out-of-plane bending vibrations indicate the relative positions of substituents on the benzene ring. The position of absorption bands in this region depends on the number of adjacent hydrogen atoms on the ring. The *o*-disubstituted benzenes show a strong band in the region

735–770 cm^{-1} but no band in the region 690–710 cm^{-1}. The *m*-disubstituted benzenes show two strong bands in the regions 680–725 and 750–810 cm^{-1}, whereas *p*-disubstituted benzenes exhibit only one band in the region 800–860 cm^{-1}.

(vii) Study of the Progress of Reactions

In most of the cases, the progress of an organic reaction can be followed by IR spectroscopy. This is done by examining the IR spectra of portions of the reaction mixture withdrawn at certain time intervals. For example, in a reaction involving the oxidation of a secondary alcohol into a ketone, it is expected that the ν_{O-H} band near 3570 cm^{-1} will disappear and a new $\nu_{C=O}$ band will appear near 1715 cm^{-1} on completion of the reaction.

Besides the above applications, IR spectroscopy is often used to distinguish between *cis* and *trans* isomers (e.g. see Sections 3.8(iii) and 3.8(ix)), and between three, four, five and larger membered ring ketones (Section 3.8(ix)), lactones (Section 3.8(xi)) and lactams (Section 3.8(xiv)).

3.11 Interpretation of Infrared Spectra

There are no set rules for interpreting IR spectra. Organic chemists generally interpret IR spectra by inspecting and comparing the position, intensity and shape of bands with reference data available in tables of characteristic group frequencies. The presence or absence of an absorption band indicates the presence or absence of a particular functional group in a compound. For example, appearance of an absorption band near 3330 cm^{-1} is indicative of an intermolecularly hydrogen bonded O—H group (Figs. 3.7 and 3.8). Similarly, appearance of a band around 1700 cm^{-1} indicates the presence of a C=O group (Figs. 3.9 and 3.10). After tentative assignment of an absorption band to a particular group, it should be confirmed wherever possible by examination of other band(s) expected for that group. For example, the assignment of a carbonyl band to an aldehyde should be confirmed by the appearance of a band or a pair of bands in the region 2700–2900 cm^{-1} due to aldehydic ν_{C-H} (Fig. 3.9). Similarly, the assignment of a carbonyl band to an ester should be confirmed by the presence of a strong band due to $\nu_{C=O}$ in the region 1000–1300 cm^{-1} (Fig. 3.12).

The absence of characteristic absorption bands in the assigned regions for various functional groups shows the absence of such groups in the molecule. IR spectra of organic compounds are generally complex and it is not necessary from the interpretation point of view to assign each absorption band to a particular group. Usually, characteristic absorption bands of functional groups are used for their detection. It is rarely possible to deduce complete structure of a compound from its IR spectrum alone. In structure determination, IR spectroscopy is supplemented by chemical evidence and UV, NMR and mass spectral data.

The intensity and shape (or width) of bands are also important in the interpretation of IR spectra. For example, it would be wrong to pick out a weak band near 1700 cm^{-1} and assign it to a $\nu_{C=O}$ in a spectrum of a pure compound because C=O and similar highly polar groups absorb strongly due to stretching vibrations. Broad bands are generally indicative of hydrogen bonding.

For easy interpretation of an IR spectrum, it may be divided into the following regions.

(i) 3200–3650 cm^{-1}

The appearance of medium to strong absorption bands in this region shows the presence of hydroxyl or amino groups. These bands arise from $\nu_{O—H}$ or $\nu_{N—H}$ vibrations. The position, intensity and width of the bands indicate whether the group is free or intermolecularly hydrogen bonded or intramolecularly hydrogen bonded. A medium band due to ≡C—H stretching also appears near 3300 cm^{-1}.

(ii) 3000–3200 cm^{-1}

Absorption bands due to =C—H stretching and aromatic C—H stretching appear in this region. These bands are of medium intensity.

(iii) 2700–3000 cm^{-1}

In this region, usually a complex band or bands appear near 2850 cm^{-1} due to stretching vibrations of C—H bonds of saturated groups, i.e. —CH_3, $>CH_2$ or $>CH$— (Fig. 3.4). The appearance of weak but sharp bands near 2700–2900 cm^{-1} due to $\nu_{C—H}$ indicates the presence of aldehyde, methoxyl or N-methyl groups. A broad $\nu_{O—H}$ band present in the 2700–3000 cm^{-1} region is characteristic of hydrogen-bonded —COOH groups.

(iv) 2000–2700 cm^{-1}

Groups of the type $X \equiv Y$, $X{=}Y{=}X$, etc. absorb in this region and exhibit bands of variable intensities. For example, bands due to C≡C stretching appear in the region 2100–2260 cm^{-1} and that due to C≡N stretching appear in the region 2200–2260 cm^{-1}. Similarly, isocyanates absorb in the region 2240–2275 cm^{-1} due to N=C=O stretching. Besides these, $\nu_{O—H}$, $\nu_{N—H}$ and $\nu_{S—H}$ bands of carboxylic acid dimers, amine salts and thiols (or thiophenols), respectively also appear in this region (2000–2700 cm^{-1}).

(v) 1600–1900 cm^{-1}

Strong absorption bands in the upper part of this region are due to C=O stretchng. Aldehydes, ketones, carboxylic acids, esters, amides, acid anhydrides, acyl halides, etc. absorb strongly in this region due to C=O stretching. The position of $\nu_{C=O}$ band gives good indication of the nature of the attached groups and the molecular environment as summarized in the end of Section 3.8(xiv). Much weaker bands in the lower part of this region are often assignable to C=C stretching of alkenes. $\nu_{C=N}$ and δNH_2 bands also occur in the lower part of this region.

The region 1800–2000 cm^{-1} is often used for confirmation of the substitution pattern of a benzene nucleus but actually it is not useful because of weakness of these overtone bands ($2\gamma_{C—H}$ etc.).

(vi) 1000–1600 cm^{-1}

The most characteristic medium bands due to C=C stretching in aromatic compounds appear around 1600, 1580, 1500 and 1450 cm^{-1} (Fig. 3.6). The presence of aliphatic CH_2 and CH_3 groups is confirmed by the appearance of medium bands near 1450 cm^{-1} due to δCH_2 and δCH_3, and near 1375 cm^{-1} due to δCH_3 only. If the compound contains geminal methyl groups, then the 1375 cm^{-1} band appears as a doublet. This region is useful for identification of nitro compounds and also for confirming the presence of ethers and esters, and primary, secondary and tertiary alcohols. For example, the appearance of a strong band due to $\nu_{C=O}$ between 1000 and 1300 cm^{-1} confirms the presence of an ester group provided $\nu_{C=O}$ band is present in the expected region, e.g. 1735–1750 cm^{-1} for saturated acyclic esters (Fig. 3.12).

(vii) 667–1000 cm^{-1}

This region is especially useful for the determination of orientations in aromatic compounds as discussed in Section 3.10(vi).

3.12 Some Solved Problems

Problem 1. How many bands due to fundamental vibrations do you expect to observe in the IR spectrum of water?

Solution. Water molecule is a nonlinear molecule, hence its vibrational degree of freedom, i.e. fundamental vibrations will be equal to $3n$–6 and can be calculated as follows:

Number of atoms (n) = 3
Total degrees of freedom ($3n$) = $3 \times 3 = 9$
Rotational degrees of freedom = 3
Translational degrees of freedom = 3
Hence, vibrational degrees of freedom ($3n - 6$) = $9 - 3 - 3 = 3$

Thus, there will be three fundamental modes of vibration. These are symmetrical stretching, asymmetrical stretching and bending vibrations. All these modes of vibrations are IR active and are non-degenerate. Thus, three absorption bands corresponding to these three fundamental modes of vibration are expected in the IR spectrum of water (the observed absorption bands occur at 3652, 3756 and 1595 cm^{-1}).

Problem 2. Using IR spectroscopy, how will you distinguish the following isomeric compounds:

(i) $CH_3CH_2C{\equiv}CH$ and $CH_3C{\equiv}CCH_3$
(ii) CH_3CH_2OH and CH_3OCH_3
(iii) $(CH_3)_3N$ and $CH_3CH_2NHCH_3$

Solution. (i) $CH_3CH_2C{\equiv}CH$ will show strong absorption bands at ~3300, 2100 and 625 cm^{-1} due to $\equiv$C—H stretching, C$\equiv$C stretching and $\equiv$C—H bending vibrations, respectively. All these bands will be absent in the IR spectrum of $CH_3C{\equiv}CCH_3$. It should be noted that $\nu_{C\equiv C}$ band will also be asbsent in the IR spectrum of $CH_3C{\equiv}CCH_3$ because the symmetrical substitution makes the C$\equiv$C stretching frequency IR inactive.

(ii) CH_3CH_2OH will show absorption bands in the region 3200–3600 cm^{-1} due to stretching vibration of intermolecularly hydrogen-bonded O—H group and at ~1050 cm^{-1} due to $\nu_{C=O}$. These bands will be absent in the IR spectrum of CH_3OCH_3. It will show an absorption band around 1100 cm^{-1} due to C—O—C stretching.

(iii) $CH_3CH_2NHCH_3$ will show a medium band in the region 3310–3550 cm^{-1} due to N—H stretching. This band will be absent in IR spectrum of $(CH_3)_3N$.

Problem 3. How will you distinguish *o*-hydroxybenzaldehyde (salicylaldehyde) and *m*-hydroxybenzaldehyde on the basis of IR spectroscopy?

Solution. In salicylaldehyde, due to intramolecular hydrogen bonding, $\nu_{O—H}$ and $\nu_{C=O}$ bands are shifted to lower wavenumbers. Since it is intramolecular, change in concentration does not cause any shift in $\nu_{O—H}$ and $\nu_{C=O}$ absorption bands. In case of *m*-hydroxybenzaldehyde, $\nu_{O—H}$ and $\nu_{C=O}$ bands occur at a still lower wavenumber due to intermolecular hydrogen bonding. In this case, $\nu_{O—H}$ and $\nu_{C=O}$ bands shift to higher wavenumbers on dilution with a nonpolar solvent.

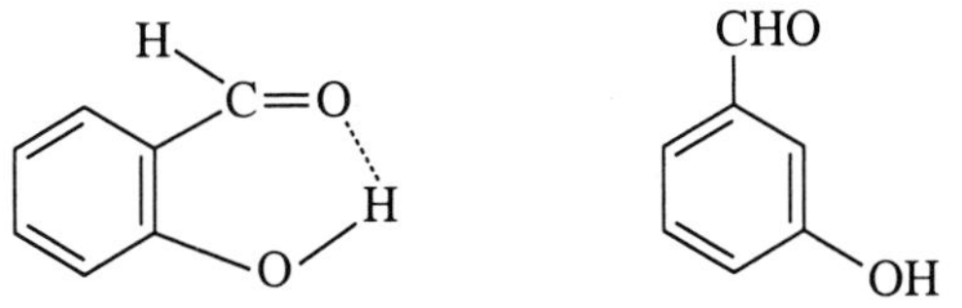

o-hydroxybenzaldehyde *m*-hydroxybenzaldehyde

Problem 4. A hydrocarbon containing 10% hydrogen shows the following bands in its IR spectrum:

(i) 3295 cm^{-1} (ii) 2130 cm^{-1} (iii) 625 cm^{-1}

Deduce the structure of the hydrocarbon.

Solution. Among various hydrocarbons, only alkynes show a band around 3300 cm^{-1} due to ≡C—H stretching. Thus, the appearance of a band at 3295 cm^{-1} clearly shows the presence of ≡C—H group in the hydrocarbon. The presence of a band at 2130 cm^{-1} due to C≡C stretching and another band at 625 cm^{-1} due to ≡C—H bending confirm that the hydrocarbon is an alkyne. 10% hydrogen content shows that it is propyne ($CH_3C\equiv CH$).

Problem 5. Give suitable explanation for the following:

(a) Stretching frequencies of C—C, C=C and C≡C fall in the regions which are in increasing order 800–1200 cm^{-1}, 1650–1670 cm^{-1} and 2100–2260 cm^{-1}, respectively.

(b) The ν_{O-H} band appears near 3570 cm^{-1}, whereas the ν_{O-D} band appears near 2630 cm^{-1}.

(c) 2-*t*-butylphenol has two ν_{O-H} bands—one at 3608 cm^{-1} and the other at 3643 cm^{-1}, whereas 2,6-di-*t*-butylphenol has only one band at 3642 cm^{-1}.

Solution. (a) According to Hooke's law (Section 3.6, Eq. (3.1)), the stretching frequency of a bond increases with an increase in the force constant, i.e., bond strength. Thus, the stretching frequencies of carbon-carbon single, double and triple bonds increase in the order of their increasing bond strengths, i.e. the increasing order of stretching frequencies is

$$\text{C—C} < \text{C=C} < \text{C}\equiv\text{C}$$

(b) According to Hooke's law (Section 3.6, Eq. (3.1)), the stretching frequency of a bond increases as the reduced mass of the bonded atoms decreases. Since hydrogen has lesser atomic mass than deuterium, the O—H stretching frequency (near 3570 cm^{-1}) is higher than the O—D stretching frequency (near 2630 cm^{-1}).

(c) In 2-*t*-butylphenol, two bands are present at 3608 and 3643 cm^{-1} showing that some molecules are intermolecularly hydrogen bonded, whereas in others OH is not hydrogen bonded due to steric hindrance by the bulky-*t*-butyl group.

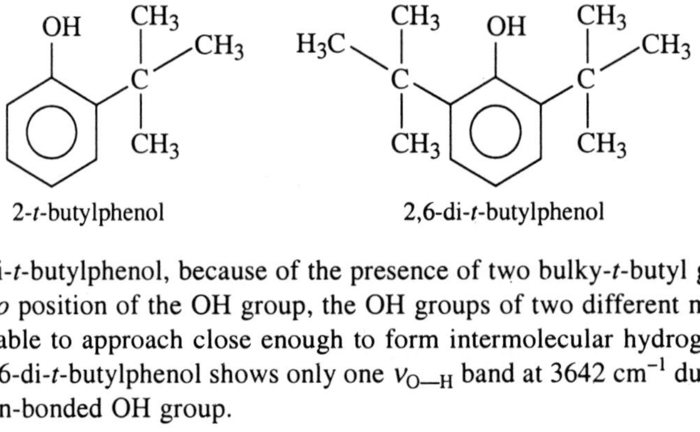

2-*t*-butylphenol

2,6-di-*t*-butylphenol

In 2,6-di-*t*-butylphenol, because of the presence of two bulky-*t*-butyl groups in the *ortho* position of the OH group, the OH groups of two different molecules are not able to approach close enough to form intermolecular hydrogen bond. Thus, 2,6-di-*t*-butylphenol shows only one ν_{O-H} band at 3642 cm^{-1} due to non-hydrogen-bonded OH group.

Problem 6. Giving reasons arrange the following compounds in order of decreasing frequency of carbonyl absorption in their IR spectra:

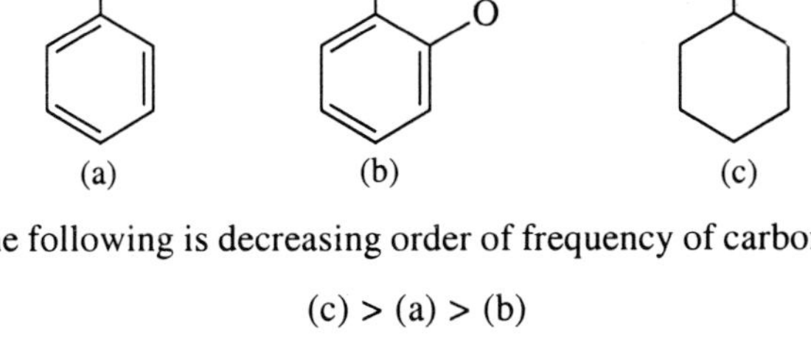

(a) (b) (c)

Solution. The following is decreasing order of frequency of carbonyl absorption

$$(c) > (a) > (b)$$

Cyclohexane carboxaldehyde (c) is a saturated aldehyde, hence will absorb around 1730 cm^{-1}. Due to conjugation of the C=O group with the double bonds of the benzene ring, the $\nu_{C=O}$ absorption of benzaldehyde (a) will be shifted to lower frequency (~1700 cm^{-1}). In salicylaldehyde (b), the lowering is due to conjugation as well as intramolecular hydrogen bonding (chelation), thus the $\nu_{C=O}$ absorption frequency is further lowered (~1665 cm^{-1}).

Problem 7. How will you distinguish the two members of each of the following pairs of compound by their IR spectra:

(a) $CH_3CH_2C{\equiv}CH$ and $CH_3CH_2C{\equiv}N$
(b) RNH_2 and $RCONH_2$
(c) CH_3COCH_3 and $CH_3CH{=}CHCH_2OH$
(d) $CH_2{=}CH{-}O{-}CH_3$ and CH_3CH_2CHO
(e) CH_3CH_2CHO and CH_3COCH_3

Solution. (a) $CH_3CH_2C{\equiv}CH$ will show strong absorption band at ~3300 and 625 cm^{-1} due to $\equiv$C—H stretching and bending vibrations, respectively. These bands will be absent in the IR spectrum of $CH_3CH_2C{\equiv}N$.

(b) $RCONH_2$ will show a strong $\nu_{C=O}$ band around 1690 cm^{-1} (in dilute solution) which will be absent in the IR spectrum of RNH_2.

(c) CH_3COCH_3 will show a strong $\nu_{C=O}$ band at ~1715 cm^{-1}. This band will be absent in $CH_3CH{=}CHCH_2OH$. Compound $CH_3CH{=}CHCH_2OH$ will show a strong ν_{O-H} band in the region 3200–3650 cm^{-1} and another medium $\nu_{C=C}$ band in the region 1620–1680 cm^{-1}.

(d) CH_3CH_2CHO will show a strong $\nu_{C=O}$ band near 1730 cm^{-1} and another weak band (generally appearing as a doublet due to Fermi resonance) in the region 2700–2900 cm^{-1} due to aldehydic ν_{C-H}. $CH_2{=}CH{-}O{-}CH_3$ will show a strong band in the range 1020–1075 cm^{-1} due to C—O—C stretching and another medium band near 1650 cm^{-1} due to C=C stretching.

(e) Acetone will show a strong $\nu_{C=O}$ band at ~1715 cm^{-1}, whereas this band appears at ~1730 cm^{-1} in propionaldehyde. In addition, CH_3CH_2CHO will show a band in the range 2700–2900 cm^{-1} due to aldehydic C—H stretching. This band generally appears as a doublet due to Fermi resonance.

Problem 8. How will you distinguish maleic acid and fumaric acid by their IR spectra.

Solution

H—C—COOH ‖ H—C—COOH	H—C—COOH ‖ HOOC—C—H
Maleic acid	Fumaric acid
$\nu_{C=O}$ 1720 cm^{-1} (*s*)	$\nu_{C=O}$ 1680 cm^{-1} (*s*)

In maleic acid, the bulky COOH groups are on the same side of the double bond, hence due to repulsive interactions the C=O group is forced out of the plane of the C=C bond. Thus, the conjugation is diminished resulting in the appearance of $\nu_{C=O}$ band at a higher frequency as compared to that in fumaric acid where the C=O group is in conjugation with the C=C bond.

Problem 9. A compound having molecular formula C_3H_6O gave the following spectral data:

(i) UV: λ_{max} 292 nm, ε_{max} 21
(ii) IR: 2720 cm^{-1} (*w*) and 1738 cm^{-1}(*s*)

Deduce the structure of the compound.

Solution. The compound shows λ_{max} 292 nm with ε_{max} 21 indicating that it is either an aldehyde or ketone and this band has arisen from $n \rightarrow \pi^*$ transition of its C=O group. The appearance of a strong IR absorption band at 1738 cm^{-1} is indicative of an aldehydic carbonyl group which is confirmed by the presence of another band at 2720 cm^{-1} due to aldehydic C—H stretching. Thus, the compound is CH_3CH_2CHO.

Problem 10. Fig. 3.16 shows the IR spectrum of a compound having molecular formula C_8H_8O. Interpret the spectrum and deduce the structure of the compound.

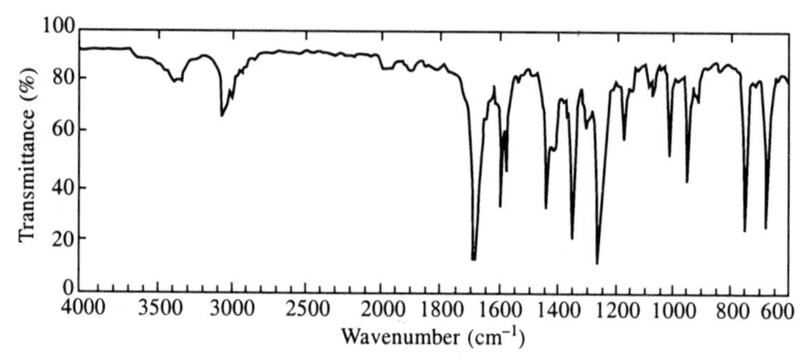

Fig. 3.16

Solution. The compound shows an absorption around 3010 cm^{-1} indicating the presence of a =C—H group. The presence of a phenyl nucleus is shown by the absorption bands at 1600, 1580 and 1450 cm^{-1}. The appearance of out-of-plane bending absorptions at 762 and 692 cm^{-1} indicates a monosubstituted benzene ring.

The presence of the 1580 cm^{-1} band indicates the presence of a substituent which can conjugate with the phenyl nucleus. The appearance of a strong band at 1690 cm^{-1} shows that the compoud is an aromatic ketone. It cannot be an acid because it contains only one oxygen atom; its molecular formula is C_8H_8O. Thus, the compound is acetophenone.

PROBLEMS

1. (a) In terms of transitions, how does IR spectroscopy differ from UV spectroscopy?
   (b) Arrange the following transitions in order of their increasing energy requirement:

Vibrational, electronic and rotational

2. Write notes on:

(a) Stretching and bending vibrations
(b) Fingerprint region
(c) Fermi resonance

3. Comment on the following:

(a) Hydrogen bonding raises the wavelength of IR absorption.
(b) Ethanol and methanol are good solvents for recording UV spectra but not for IR spectra.
(c) $\nu_{C=O}$ frequency for ethyl acrylate is lower than that for ethyl propionate. (Hint. Due to conjugation.)

4. Calculate the number of fundamental vibrations in the following molecules:

(i) Methane (ii) Ethanol (iii) Acetylene (iv) Ethylene (v) Oxygen

5. Assign the following IR absorption bands to a particular carboxylic acid derivative in each case:

(i) 1715–1750 cm^{-1} (ii) 1750–1815 cm^{-1} (iii) 1630–1690 cm^{-1}
(iv) 1740–1790 and 1800–1850 (two bands)

6. Explain the following:

(a) Concentrated solutions of alcohols in CCl_4 have an IR absorption band at about 3300 cm^{-1} but this band is shifted to a higher frequency on dilution.
(b) IR spectra have a large number of absorption bands as compared to UV spectra. (Hint. Due to large number of stretching and bending vibrations in the molecule.)
(c) *o*-Nitrophenol shows ν_{O-H} band at 3200 cm^{-1} in KBr pellet as well as in $CHCl_3$ solution, whereas the *para* isomer shows this band at different frequencies in the two media (in pellet at 3330 cm^{-1} and in $CHCl_3$ solution at 3520 cm^{-1}.) (Hint. The O—H group of *o*-nitrophenol is intramolecularly hydrogen bonded. Hence its frequency is not affected by the medium or change of concentration, whereas the *para* isomer is intermolecularly hydrogen-bonded in pellet and the hydrogen bonding diminishes in $CHCl_3$ solution due to relative dilution resulting in the shift of ν_{O-H} band to higher frequency, 3520 cm^{-1}.)

7. How will you distinguish the two members of each of the following pairs of compounds using IR spectroscopy:

(a) CH_3COOH and $HCOOCH_3$
(b) β-propiolactone and γ-butyrolactone
(c) $ClCH_2CH_2CH_2COOH$ and $CH_3OCH_2CH_2COCl$
(d) CH_3CH_2CHO and $CH_2{=}CH{-}CH_2OH$

8. How can IR and UV spectroscopy be used to determine when the following reaction is completed?

$$H_3C{-}\underset{\underset{CH_3}{|}}{C}{=}CH{-}\overset{\overset{O}{\|}}{C}{-}CH_3 \xrightarrow{\left[H_3C{-}\overset{\overset{CH_3}{|}}{CH}{-}O\right]_3 Al} H_3C{-}\underset{\underset{CH_3}{|}}{C}{=}CH{-}\overset{\overset{OH}{|}}{CH}{-}CH_3$$

(Hint. The UV spectrum of the starting material will show λ_{max} at 237 and 310 nm due to conjugated C=C and C=O groups respectively. These bands will disappear on completion of the reaction. Similarly, the IR absorption band in the region 1665–1685 cm^{-1} due to conjugated C=O group (present in the starting material) will disappear on completion of the reaction. In addition, a new IR absorption band will appear around 3330 cm^{-1} due to hydrogen-bonded O—H group.)

9. Giving reasons arrange the following compounds in order of increasing wavenumber of carbonyl absorption in their IR spectra:

(a) Acetophenone, *p*-nitroacetophenone and *p*-aminoacetophenone. (Hint. A group with +*M* effect will decrease and that with –*M* effect will increase the bond order of C=O bond.)
(b) Cyclobutanone, cyclohexanone and cyclopentanone.

10. Discuss the factors which affect the IR absorption frequency of a functional group.

11. Explain why esters of *o*-chlorobenzoic acid show two C=O stretching absorptions?

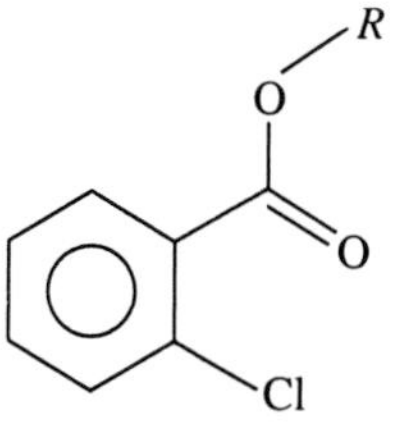

(a) 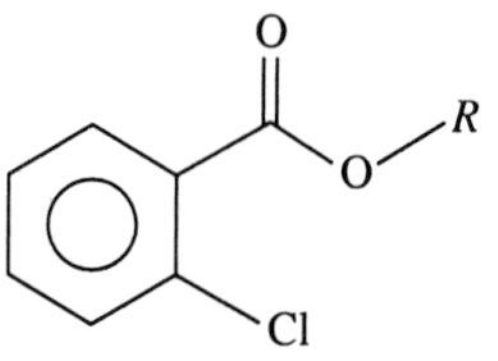 (b)

(Hint. In the rotational isomer (a) of *o*-chlorobenzoic ester, Cl is near the C=O group and its negative field repels the non-bonding electrons of the carbonyl oxygen (field effect). Thus, the force constant of C=O bond is increased resulting in shifting of the $\nu_{C=O}$ band to a higher frequency. Such shift does not occur in the isomer (b). Normally, both the isomers are present, thus two $\nu_{C=O}$ absorptions are observed.)

12. Deduce the structures of isomeric compounds (A) and (B) having molecular formula C_3H_6O and the following IR spectral data:

(A) 1710 cm^{-1} (B) ~3300 cm^{-1} and 1640 cm^{-1}

13. How will you distinguish the following pairs of compounds on the basis of IR spectroscopy?

(a) CH_3CH_2COCl and $ClCH_2COCH_3$
(b) β-propiolactam and γ-butyrolactam
(c) Benzene and cyclohexane
(d) Salicylic acid and *m*-hydroxybenzoic acid

14. Which of the following compounds do you expect to have higher ν_{O-H} frequency in dil. $CHCl_3$ solution and why?

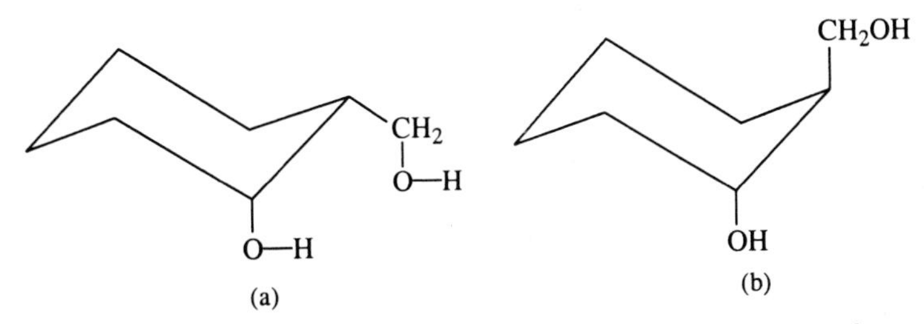

(Hint. In (a) intramolecular hydrogen bonding will decrease ν_{O-H} frequency.)

15. Giving reasons arrange the following compounds in order of increasing wavelength of carbonyl absorption in their IR spectra:

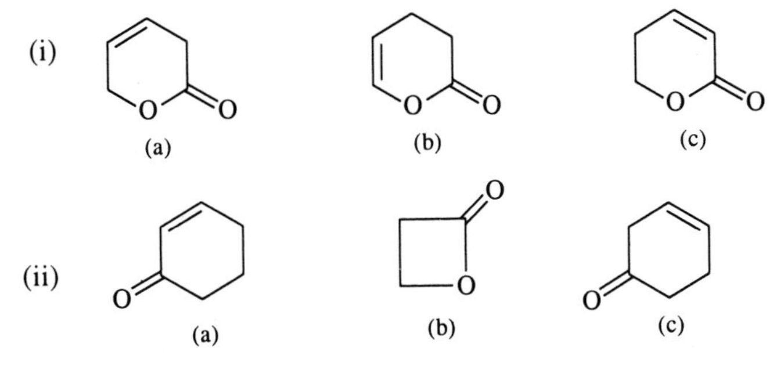

16. An organic compound (A) having molecular formula C_3H_7NO shows IR absorption bands at 3413 (*m*), 3236 (*m*), 2899–3030 (*m*), 1667 (*s*), 1634 (*s*) and 1460 cm^{-1} (*s*). Give the probable structure of (A).

17. Giving examples discuss how inductive and mesomeric effects influence the carbonyl absorption frequency?

18. Explain why a compound in the vapour state has higher stretching frequency than that in the liquid or solid state? (Hint. Due to intermolecular interactions, especially when polar groups capable of hydrogen bonding are present, the force constants of bonds are decreased.)

19. Giving reasons arrange the following in increasing order of their ν_{O-H} frequency: ethanol, trichloroacetic acid, acetic acid and chloroacetic acid.

20. The IR spectrum of methyl salicylate shows absorption bands at 3300, 3050, 2990, 1700, 1590 and 1540 cm^{-1}. Which of these bands are due to which of the following groups?

$$C{=}O,\ CH_3,\ O{-}H \text{ and aromatic ring}$$

21. Fig. P3.1 shows the IR spectrum of a pure liquid (A) which is easily oxidized by alkaline $KMnO_4$ into a carboxylic acid. Deduce the structure of (A).

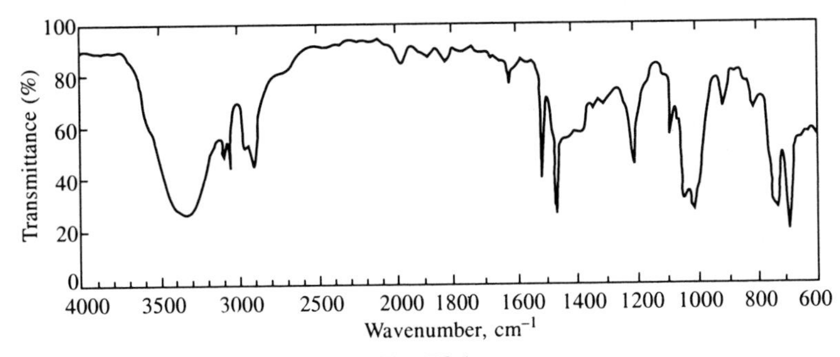

Fig. P3.1

22. Fig. P3.2 shows the IR spectrum of a neat liquid with molecular formula C_2H_6O. Assign its structure.

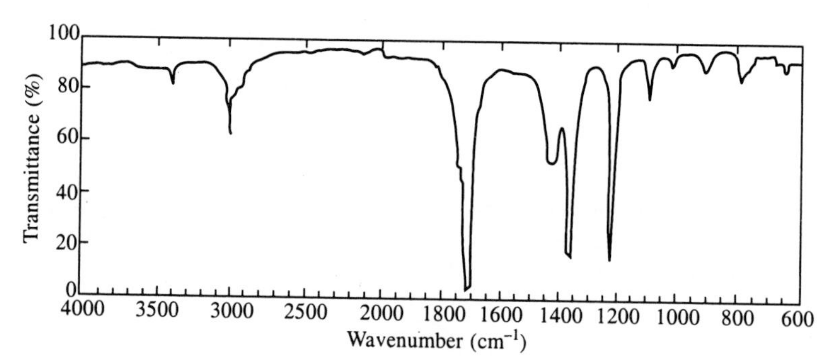

Fig. P3.2

23. Ethyl acetate shows $\nu_{C=O}$ absorption at 1735 cm^{-1}, whereas phenyl acetate at 1770 cm^{-1}. Explain.

24. Arrange the following compounds in order of increasing absorption frequency of their carbonyl groups. Give reason for your answer.

C_6H_5CHO, CH_3CHO, CH_3COCl and CH_3COCH_3

25. Using IR spectroscopy, how will you distinguish:

(a) Intermolecular and intramolecular hydrogen bonding
(b) *cis*-cinnamic acid and *trans*-cinnamic acid
(c) Axial and equatorial O—H group. (Hint. Due to 1,3-diaxial interaction, the stretching vibration of the axial O—H group is hindered and thus its absorption frequency is raised.)

26. Indicate which of the following vibrations will be IR active or inactive?

	Molecule	Mode of vibration
(i)	$CH_2{=}CH_2$	C=C stretching
(ii)	$CH_2{=}CH_2$	C—H stretching
(iii)	N_2	N≡N stretching
(iv)	SO_2	Symmetrical stretching
(v)	CO_2	Symmetrical stretching
(vi)	CH_3–CH_3	C—C stretching

27. 2-hydroxy-3-nitroacetophenone shows two $\nu_{C=O}$ bands at 1692 and 1658 cm^{-1}. Explain.
(Hint. In some molecules, there is intramolecular hydrogen bonding between OH and C=O, while in other between OH and NO_2 group.)

28. IR spectrum of a neat liquid with molecular formula C_4H_8O is given in Fig. P3.3. Interpret the spectrum and assign the structure to the compound.

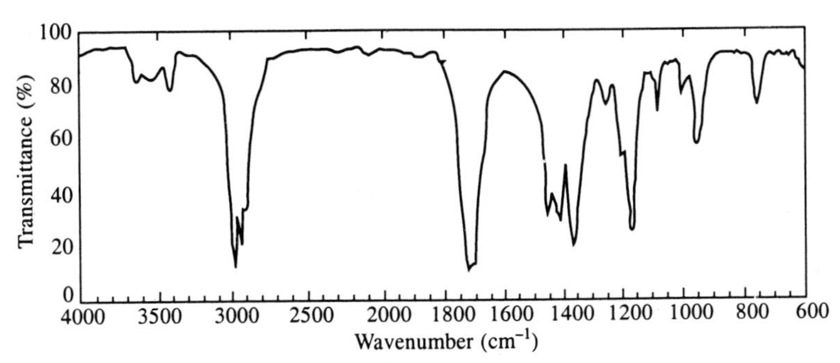

Fig. P3.3

29. Explain why $\nu_{C=O}$ frequency for *m*-chlorobenzoic acid is higher than that for *p*-chlorobenzoic acid? (Hint. Due to +*M* effect, Cl decreases the force constant of C=O bond from the *para* position but not from the *meta* position.)

30. A compound with molecular formula C_7H_8O gives the following IR spectral data. Deduce the structure of the compound.
IR bands at ~3300 (*s*), ~3040 (*m*), 2800–2950 (*w*), 1606 (*m*), 1582 (*m*), 1500 (*m*), 1450 (*w*), 1380 (*w*), 1185 (*s*) 780 (*s*) and 692 cm^{-1} (*s*).

31. IR spectrum of a neat liquid with molecular formula C_7H_6O is given in Fig. P3.4. Interpret the spectrum and assign the structure to the compound.

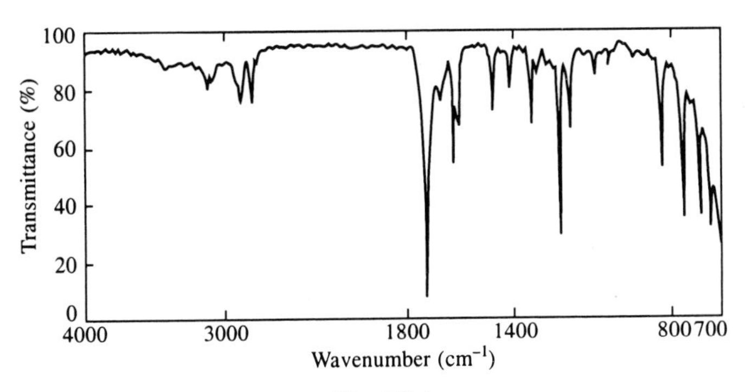

Fig. P3.4

References

1. A.D. Cross, Introduction to Practical Infrared Spectroscopy, 3rd Ed., Butterworth, London, 1969.
2. C.N.R. Rao, Chemical Applicaton of Infrared Spectroscopy, Academic Press, New York and London, 1963.
3. D.H. Williams and I. Fleming, Spectroscopic Methods in Organic Chemistry, McGraw-Hill, London, 1966.
4. F. Schienmann, Ed., An Introduction to Spectroscopic Methods for the Identification of Organic Compounds, Vol. 1, Pergamon Press, Oxford, 1970.
5. H.A. Szymanski, Correlation of Infrared and Raman Spectra of Organic Compounds, Hertillon Press, 1969.
6. J.M. van der Mass, Basic Infrared Spectroscopy, Heyden, London, 1969.

7. J.R. Dyer, Applications of Absorption Spectroscopy of Organic Compounds, Prentice-Hall, Englewood Cliffs, N.J., 1965.
8. K. Nakanishi and P.H. Solomon, Infrared Absorption Spectroscopy—Practical, 2nd Ed., San Francisco, Holden-Day, 1977.
9. L.J. Bellamy, The Infrared Spectra of Complex Organic Molecules, Methuen, London, Vol. 1, 3rd Ed., 1975; Vol. 2, 1980.
10. N.B. Colthup, L.H. Daly and S.E. Wiberley, Introduction to infrared and Raman Spectroscopy, 3rd Ed., Academic Press, New York and London, 1990.
11. R.M. Silverstein, G.C. Bassler and T.C. Morrill, Spectrometric Identification of Organic Compounds, 3rd Ed., Wiley, New York, 1974.
12. R.T. Conley, Infrared Spectroscopy, 2nd Ed., Allyn and Bacon, Boston, 1972.
13. S. Sternhell and J.R. Kalman, Organic Structures from Spectra, Wiley, Chichester, 1986.

4

Raman Spectroscopy

4.1 Introduction

Infrared and Raman spectroscopy are closely related as both originate from transitions in vibrational and rotational energy levels of the molecule on absorption of radiations. Since different methods of excitation are used, the spectroscopic selection rules* are different. The intensity of IR absorption depends on the change in dipole moment of the bond, whereas Raman intensity depends on the change in polarizability of the bond accompanying the excitation. Thus, an electrically symmetrical bond (i.e. having no dipole moment) does not absorb in IR region (i.e. the transition is forbidden) but it does absorb in Raman scattering (i.e. the transition is allowed). In other words, an electrically symmetrical bond is Raman active but IR inactive. However, an electrically unsymmetrical bond may be IR active and Raman inactive or both IR and Raman active.

IR and Raman spectroscopy are complementary. For example, studies on bond angles, bond lengths and other structural confirmations require Raman data in addition to IR studies.

4.2 Raman Effect and Origin of Raman Spectroscopy

When a beam of monochromatic radiation is passed through a transparent substance, a fraction of radiation is scattered at right angles to the direction of the beam by the molecules or aggregates of molecules present in the path of the beam. An examination of the scattered beam with the help of a spectroscope shows that the frequency of the scattered radiation is generally the same as that of the incident radiation. This type of scattering is known as Rayleigh scattering.

Sir C.V. Raman (1928) discovered that 'when a beam of strong radiation of a definite frequency is passed through a transparent substance (gas, liquid or solid), the radiation scattered at right angles has not only the original frequency but also some other frequencies which are generally lower and occasionally higher than that of the incident radiation'. This is known as *Raman scattering*

---

*Spectroscopic selection rules are derived from theoretical arguments and they permit to predict which transitions are allowed and which are forbidden. If there are a number of energy levels, why is spectrum relatively simple? The answer is related to selection rules according to which only some of the transitions are allowed.

or *Raman effect.** The spectral lines resulting from lower frequencies than that of the incident radiation are called *Stokes lines* and those from higher frequencies are called *anti-Stokes lines* (Fig. 4.1). The spectral lines whose frequencies have been modified in Raman effect are called *Raman lines*. Thus, Stokes and anti-Stokes lines are Raman lines. Raman spectra are manifestation of Raman effect which is accompanied by transitions in vibrational and rotational energy levels of the molecule. Just as in IR spectra, the positions of spectral lines (or bands) in Raman spectra are also reported in wavenumbers (cm^{-1}).

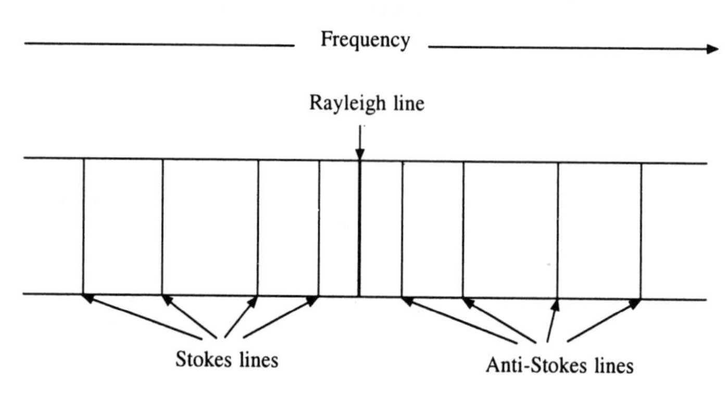

Fig. 4.1 Rayleigh, Stokes and anti-Stokes lines

The incident light is associated with energy $h\nu_i$, a part of which is used up for causing transitions from lower to higher vibrational and rotational energy levels, so the scattered radiation has a lower energy content $h\nu_s$ and thus a new line (Raman line) appears in the spectrum. Raman also discovered that the frequency difference $\Delta\nu$ between the incident frequency ν_i and any scattered frequency (say ν_s) is constant and characteristic of the substance exposed to radiation and is completely independent of the frequency of the incident radiation ν_i. $\Delta\nu$ is known as *Raman frequency shift* or *Raman shift* and is given by the relation

$$\Delta\nu = \nu_i - \nu_s$$

where ν_i and ν_s are the frequencies (in wavenumbers, cm^{-1}) of the incident radiation and of a particular radiation scattered by the given substance, respectively, $\Delta\nu$ is positive for Stokes line and negative for anti-Stokes lines. Although Raman shifts $\Delta\nu$ correspond to IR absorption or emission, IR and Raman spectra of a substance are not always identical.

(i) Characteristics of Raman Lines and Raman Shifts

(a) Raman lines (Stokes and anti-Stokes lines) are symmetrically displaced about the parent line (Rayleigh line) (Fig. 4.1). With increase in temperature, there is a decrease in separation of individual Raman lines from the parent line.

---

*Although such a scattering was theoretically predicted by Smekal in 1923, Raman was the first to observe it experimentally in 1928.

(b) The intensity of Stokes lines is always greater than that of the corresponding anti-Stokes lines.

(c) Raman shifts $\Delta\nu$ generally fall in the IR region (100-3000 cm^{-1}) and they are expressed in wavenumbers (cm^{-1}).

(d) Raman shifts represent the frequencies of absorption bands of the substance under investigation.

(e) A striking feature of Raman shifts is that they are identical (within limits of experimental error) with the absorption frequencies obtained from IR spectrum of the substance under investigation and they do not depend on the frequency of the incident (exciting) radiation (Table 4.1).

Table 4.1 Raman and infrared spectral data of benzene

Frequencies of incident (exciting) radiations ν_i (cm^{-1})	Frequencies of scattered radiations ν_s (cm^{-1})	Raman shifts $\Delta\nu$ (cm^{-1})	Frequencies obtained from infrared spectra (cm^{-1})
22,937	21,940	997	970
	21,755	1182	1155
	21,340	1597	1612
	19,880	3057	3076
22,994	21,997	997	970
	21,521	1473	1481
24,704	23,713	997	970
	23,520	1184	1154
	53,100	1604	1612
	21,642	3062	3076

4.3 Theories of Raman Effect and Raman Spectroscopy

(i) Classical Theory

The classical electromagnetic wave theory gives only an elementary and imperfect explanation of Raman effect. However, it leads to an understanding of a basic concept of Raman spectroscopy. A distortion takes place in a molecule when it is placed in an electric field. This is due to the attraction of positively charged nuclei towards negative pole of the field and of electrons towards positive pole. Such separation of charge centres causes induced electric dipole moment and the molecule becomes polarized. The magnitude of the induced dipole moment μ depends on the magnitude of the applied electric field E and on the ease with which the molecule can be distorted, i.e. the polarizability α of the molecule.

Thus,

$$\mu = \alpha E \tag{4.1}$$

Polarizability decreases with increasing electron density, increasing bond strength and decreasing bond length.

According to the electromagnetic theory of light, the variation of the electric field of strength E, associated with a radiation of frequency ν, with time t is given by

$$E = E_0 \sin 2\pi\nu t \tag{4.2}$$

where E_0 (a constant) is the equilibrium value of the field strength. Eqs. (4.1) and (4.2) show that

$$\mu = \alpha E_0 \sin 2\pi\nu t \tag{4.3}$$

The field E will induce a dipole moment which varies according to Eq. (4.3). Therefore, the radiation of frequency ν will induce, in the molecule with which it interacts, a dipole which oscillates with the same frequency ν, and thus emits, i.e. scatters radiation of its own oscillation frequency ν. This explains Rayleigh scattering for which the incident and scattered radiations have the same frequency.

In the above discussion, we have not considered the vibration and rotation of molecules which also affect their polarizability and consequently the induced dipole moment as well as scattering will also be influenced.

(a) Effect of Vibration

Let us suppose that the molecule is diatomic. Then, as the two nuclei vibrate along the line joining them, the polarizability of the molecule will vary. The variation in polarizability α with small displacement x from the equilibrium position is given by

$$\alpha = \alpha_0 + \beta \frac{x}{A} \tag{4.4}$$

where α_0 is the equilibrium polarizability, β the rate of variation of the polarizability with displacement and A the vibrational amplitude.

If the molecule undergoes simple harmonic oscillations, the variation of x with time is given by

$$x = A \sin 2\pi\nu_\upsilon t \tag{4.5}$$

where ν_υ is the vibrational frequency of the diatomic molecules for small displacements. Therefore, from Eqs. (4.4) and (4.5), we get

$$\alpha = \alpha_0 + \beta \sin 2\pi\nu_\upsilon t \tag{4.6}$$

From Eqs. (4.3) and (4.6), the variation of the induced dipole moment is given as

$$\mu = \alpha_0 E_0 \sin 2\pi\nu t + \beta E_0 \sin 2\pi\nu t \sin 2\pi\nu_\upsilon t$$

Using trigonometric relation, we get

$$\mu = \alpha_0 E_0 \sin 2\pi\nu t + \frac{1}{2} \beta E_0 [\cos 2\pi(\nu - \nu_\upsilon)t - \cos 2\pi(\nu + \nu_\upsilon)t] \tag{4.7}$$

Eq. (4.7) shows that the induced dipole moment oscillates not only with the frequency ν of the incident radiation but also with the frequencies $\nu - \nu_\upsilon$ and

$\nu + \nu_\upsilon$ which are lower and higher, respectively, than that of the incident radiation by an amount equal to the vibrational frequency ν_υ of the diatomic molecule. Of these frequencies of the oscillating dipole, ν gives rise to Rayleigh scattering and the other two, i.e. $\nu - \nu_\upsilon$ and $\nu + \nu_\upsilon$ give rise to Raman scattering: the scattered radiations having these frequencies result in the Stokes and anti-Stokes lines, respectively. Thus, according to this theory, Raman shift $\Delta\nu$, is equal to the vibrational frequency ν_υ of the diatomic molecule

$$\Delta\nu = \nu_\upsilon$$

It is clear from the above discussion that Raman effect is due to the variation of polarizability with time during the course of vibration (Eq. (4.7)).

(b) Effect of Rotation

During the course of rotation, the orientation of the molecule with respect to the electric field of radiation changes. Thus, if the molecule has different polarizabilities in different direction, i.e. if it is optically anisotropic, then its polarization will vary with time. Analogous to Eq. (4.6), the variation of polarizability in the course of a rotation may be represented by

$$\alpha = \alpha_0 + \beta' \sin 2\pi(2\nu_r)t \tag{4.8}$$

where ν_r is the frequency of rotation. Since a rotation through an angle of π brings the diatomic molecule in a position in which its polarizability is the same as initially, the rate at which the polarizability changes is twice as great as the rotation. Therefore, instead of ν_r, $2\nu_r$ has been used in Eq. (4.8). Substituting the value of α from Eq. (4.8) in Eq. (4.3), we get

$$\mu = \alpha_0 E_0 \sin 2\pi\nu t + \beta' E_0 \sin 2\pi\nu t \sin 4\pi\nu_r t$$

$$= \alpha_0 E_0 \sin 2\pi\nu t + \tfrac{1}{2} \beta' E_0 [\cos 2\pi(\nu - 2\nu_r)\tau - \cos 2\pi(\nu + 2\nu_r)t] \tag{4.9}$$

Eq. (4.9) shows that Raman effect is due to the variation of polarizability with time during the course of rotation and that the frequencies of Raman lines should be $\nu - 2\nu_r$ for the Stokes lines and $\nu + 2\nu_r$ for anti-Stokes lines. Thus, Raman shift $\Delta\nu$ should be equal to twice the frequency of rotation ν_r of the molecule

$$\Delta\nu = 2\nu_r$$

In conclusion, if the vibration or rotation does not change polarizability of a molecule, then $\beta = 0$ or $\beta' = 0$ and in such cases, dipole oscillates only at the frequency of the incident radiation. Thus, for a molecule to be Raman active, it must produce some change in a component of molecular polarizability during the course of molecular vibration or rotation.

(ii) Quantum Theory

Raman effect is easily understandable on the basis of quantum theory of radiation. This theory solved the problem of intensities of Raman lines and prediction of selection rules correctly, whereas the classical theory could not solve these problems.

According to quantum theory, Raman effect is considered to be the outcome of the collisions between the light photons and molecules. Thus, applying the principle of conservation of energy, we can write

$$E_p + \tfrac{1}{2} mv^2 + h\nu_i = E_q + \tfrac{1}{2} mv^2 + h\nu_s \tag{4.10}$$

where E_p and E_q are the intrinsic energies of the molecules before and after collision, respectively, m is the mass of the molecule, ν and ν' are the velocities of the molecule before and after collision, respectively, and ν_i and ν_s are the frequencies of the incident and scattered photon, respectively.

As the collision does not cause any appreciable change of temperature, it may be assumed that the kinetic energy of the molecule remains practically unchanged in the process. Hence Eq. (4.10) can be simplified as

$$E_p + h\nu_i = E_q + h\nu_s$$

Therefore
$$h(\nu_s - \nu_i) = E_p - E_q$$

or
$$\nu_s = \nu_i + \frac{E_p - E_q}{h} \tag{4.11}$$

The three possibilities which arise from Eq. (4.11) are:

(i) If $E_p = E_q$ then $\nu_s = \nu_i$ which represents unmodified lines, i.e. Rayleigh scattering.

(ii) If $E_p > E_q$ then $\nu_s > \nu_i$ which refers to anti-Stokes lines. In this case, the molecule transfers some of its energy to the incident photon and thus the scattered photon has higher energy than the incident photon.

(iii) If $E_p < E_q$ then $\nu_s < \nu_i$ which refers to Stokes lines. In this case, the molecule absorbs some energy from the incident photon and thus the scattered photon has lower energy than the incident photon.

The quantum theory of Raman effect solved the problem of intensities of Raman lines as follows:

The molecules of a material medium are distributed among a series of quantum states of energies E_1, E_2, E_3 etc. Assuming that the statistical distribution of the molecules in these different quantum states is governed by Boltzmann's law, the number of molecules N_p in the particular state E_p is given by

$$N_p = CNg_p e^{-E_p/kT} \tag{4.12}$$

where C is a constant, N the total number of molecules, g_p the statistical weight of the state and k the Boltzmann's constant.

From Eq. (4.12), we can draw the following conclusions regarding the relative intensities of Raman lines:

(a) The smaller the value of E_p, the larger the value of $e^{-E_p/kT}$ and thus the greater the number N_p. Stokes lines are caused by molecules of low energy value which are more numerous than those of higher energy. Thus, statistical distribution results in Stokes transitions being more frequent than the anti-

Stokes, resulting in more intense Stokes lines than the anti-Stokes lines as experimentally observed.

(b) With increase of temperature, the kinetic energy of molecules increases and more molecules will be raised to higher energy states by inelastic collisions. Referring N_p in Eq. (4.12) to molecules in higher energy states, we see that their relative number will increase with increase in temperature because as T increases, the value of $e^{-E_p/kT}$ becomes greater. With increase in the number of high energy molecules, the number of low energy molecules will correspondingly diminish. Thus, the intensity of anti-Stokes lines increases as the temperature is raised because they are produced by molecules of high energy value.

Similar to the conclusions drawn from the classical theory of Raman effect, it has also been shown by the quantum theory of Raman effect that for exhibiting Raman effect, a molecule must produce some change in a component of molecular polarizability during the course of molecular vibration or rotation. The discussion on the quantum mechanics leading to the above conclusion is beyond the scope of this book, as it involves difficult calculus of matrices.

The polarizability of a molecule can be considered to be made up of three components in directions at right angles to each other. These components give the dimensions of so-called polarization ellipsoid. If there is a change in any of the three dimensions of the polarization ellipsoid, i.e. if there is a change in polarizability in any one direction during the course of a molecular vibration, then that vibration will interact with radiation to exhibit Raman effect. If the ellipsoid is spherical, then rotation will not affect the polarizability. Hence, for exhibiting rotational Raman effect the ellipsoid should not be spherical, i.e. its three dimensions should not be equal.

For a diatomic molecule (e.g. H_2, O_2, N_2, HCl, HBr, CO etc.) whether homonuclear or not, the polarizability ellipsoid will not be spherical and it will also change its dimensions in the course of a molecular vibration. Hence, all such molecules will exhibit both rotational and vibrational Raman spectra. If there is no change in any of the dimensions of the polarization ellipsoid, then no vibrational Raman effect will be observed. This is the case with certain vibrations of polyatomic molecules, e.g. frequencies of bending and asymmetrical stretching vibrations of CO_2 are not observed in the Raman spectrum because these vibrations are not accompanied by any change in polarizability. Similarly, spherically symmetrical molecules, such as CH_4, CCl_4 etc. exhibit no rotational Raman spectrum.

4.4 Zero-Point Energy

According to classical theory of Raman effect, when the vibrational quantum number is zero, the molecule has no vibrational energy and hence does not oscillate. Molecules in their lowest ($v = 0$) levels should thus not be able to display Raman scattering. It has been found experimentally that such molecules do actually exhibit Raman effect. The explanation for this fact is provided by quantum theory of Raman effect. According to quantum mechanics, molecules possess a finite zero-point energy E_0 of vibration even in their lowest energy levels, i.e. the molecules can never cease to vibrate. Thus, Raman scattering is possible for molecules even when vibrational quantum number is zero.

The zero-point energy E_0 is represented by

$$E_0 = \frac{1}{2} h\nu \tag{4.13}$$

where ν is the frequency of oscillator in sec^{-1}. Eq. (4.13) is sometimes written as

$$E_0 = \frac{1}{2} hc\nu$$

where ν is now the frequency in wavenumbers, i.e. cm^{-1} and c the velocity of light.

It is seen from Eq. (4.13) that even in the vicinity of the absolute zero of temperature, the lowest possible value of the vibrational energy of a molecular oscillator cannot be less than the zero-point energy of $\frac{1}{2} h\nu$ per oscillator whose frequency is ν sec^{-1}. The existence of this zero-point energy is an aspect of uncertainty principle. It might be imagined that at the absolute zero, internal motions of molecules cease entirely, so that positions of the constituent atoms could be identified exactly. But this is not the case because the molecule still has vibrational energy equal to $\frac{1}{2}h\nu$ for each oscillator and atoms are in a state of vibration even at the absolute zero and thus their precise position cannot be defined.

4.5 Vibrational Raman Spectra

When the polarizability of a molecule changes in the course of a vibration, the vibrational frequency of this vibration will be present in the Raman spectrum. The quantum mechanical theory has shown that the selection rule for vibrational Raman spectrum is the same as that for vibrational spectra (IR spectra)

$$\Delta v = \pm 1$$

Thus, a Raman vibrational transition will occur only from one to the next upper (Stokes) or to the next lower (anti-Stokes) level. At ordinary temperatures, most of the molecules are in their lowest vibrational state, i.e. $v = 0$. Hence, majority of transitions will be of type $v = 0$ to $v = 1$. A small portion of the molecules will initially occupy the $v = 1$ level and these can undergo $v = 1$ to $v = 2$ (Stokes) or $v = 1$ to $v = 0$ (anti-Stokes) transition. In either case, the intensities of the resulting lines must be low because of small number of molecules involved. Thus, the vibrational Raman spectrum at ordinary temperatures consists of strong Stokes and weak anti-Stokes lines symmetrically displaced about the parent (Rayleigh) line (Fig. 4.1). At higher temperatures, the number of molecules that are initially in higher vibrational levels increases, hence the intensity of anti-Stokes lines also increases.

Just as in other form of vibrational spectra (IR spectra), an isotope effect is also observed in vibrational Raman spectra, e.g. in case of the three isotopic molecules H_2, HD and D_2, the observed Raman frequency shifts are 4156, 3631 and 2992 cm^{-1}, respectively.

4.6 Pure Rotational Raman Spectra

The selection rule for rotational Raman transitions is different from that for purely rotational changes as in far IR. The selection rule for Raman effect for diatomic molecules in ground state is

$$\Delta J = 0, \pm 2$$

where J is rotational quantum number and can have integral values including zero.

When $\Delta J = 0$, the frequency of Raman line will be the same as that of the incident radiation. Thus, it corresponds to Rayleigh scattering, i.e. there is no Raman shift.

When $\Delta J = \pm 2$, it corresponds to Stokes lines ($\Delta J = +2$) and anti-Stokes lines ($\Delta J = -2$).

Equation for the energy of a rigid rotator is

$$E_r = \frac{h^2}{8\pi^2 I} J(J+1) \tag{4.14}$$

where E_r is the energy of rigid rotator and I the moment of inertia.

Using Eq. (4.14) for $\Delta J = \pm 2$, the values of rotational Raman shifts of Stokes lines will be represented by

$$\Delta \nu = \frac{h}{8\pi^2 Ic} [(J+2)(J+3) - J(J+1) = 2B(2J+3) \tag{4.15}$$

where B (rotational constant) $= \dfrac{h}{8\pi^2 Ic}$ cm^{-1}.

Exactly similar equation, but of opposite sign, can be derived for anti-Stokes rotational lines as

$$\Delta \nu = -2B(2J+3)$$

Thus, rotational Raman shift can be written in the form

$$\Delta \nu = \pm 2B(2J+3) \tag{4.16}$$

where $J = 0, 1, 2 \ldots.$

The values of $\Delta\nu$ for $J = 0, 1, 2, 3, \ldots$ are $6B$, $10B$, $14B$, $18B \ldots$, respectively (Eq. (4.16)). Thus, the numerical value of frequency separation of successive lines in the rotational Raman spectrum is $4B$ compared to $2B$ for IR and UV spectra. It should be noted that the displacement of the first line ($J = 0$) on each side of the exciting (Rayleigh) line is larger than the other frequency separations of $4B$; if $J = 0$ in Eq. (4.16), then Raman shift $\Delta\nu$ becomes $6B$. Thus, a rotational Raman spectrum will consist of two series of lines (Stokes and anti-Stokes), one set on each side of the exciting line $\bar{\nu}_{ex}$. The first line in each case will be displaced by a frequency of $6B$ and the subsequent lines being separated by $4B$. Experimental results are in acccordance with these anticipations. Fig. 4.2 shows rotational energy levels and allowed transitions of a diatomic molecule and Fig. 4.3 shows its rotational Raman spectrum.

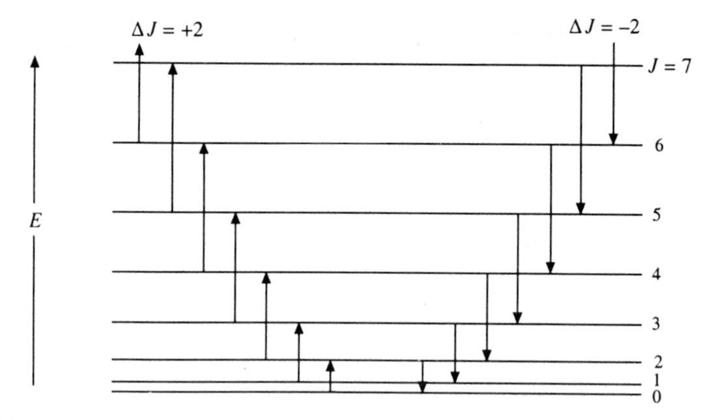

Fig. 4.2 Rotational energy levels of a diatomic molecule and allowed transitions

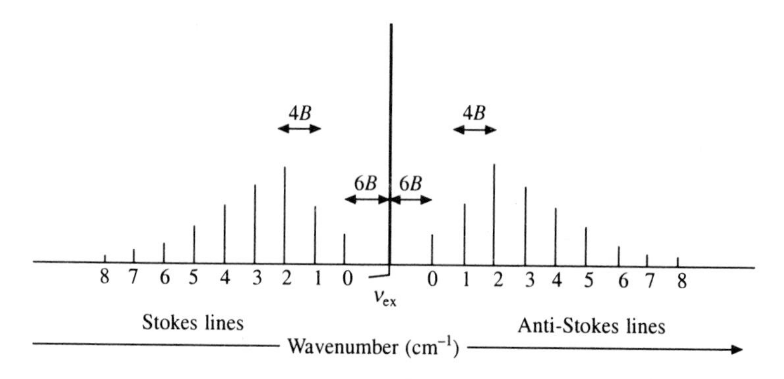

Fig. 4.3 Rotational Raman spectrum of a diatomic molecule arising from the transitions shown in Fig. 4.2. Spectral lines are numbered according to their lower *J* values

If a molecule has a centre of symmetry (e.g. H_2, O_2 and CO_2), then the effects of nuclear spin is observed in both the Raman and electronic spectra of these molecules. Thus, for O_2, CO_2 etc. every alternate rotational level is absent, e.g. in the case of O_2, every level with even J values is missing and so every transition labelled $J = 0, 2, 4, \ldots$ in Fig. 4.2 will be missing from the spectrum, i.e. spectral lines 0, 2, 4, . . . in Fig. 4.3 will be absent.

4.7 Types of Molecules and Rotational Raman Spectra

The rotation of a three dimensional body is quite complex and for convenience, it may be resolved into rotational components about three mutually perpendicular directions through the centre of gravity—the principal axes of rotation. Thus, the body has three principal moments of inertia, one about each axis, usually designated I_A, I_B and I_C.

Molecules may be classified as follows according to the relative values of their three principal moments of inertia.

(i) Linear Molecules

In case of linear molecules, $I_B = I_C$ and $I_A \cong 0$. All such molecules like H_2, O_2, CO_2, HCl etc. exhibit rotational Raman spectra because their end-over-end rotations produce change in polarizability. However, rotation about the bond axis produces no change in polarizability.

(ii) Symmetric Top Molecules

In symmetric top molecules, $I_B = I_C$; $I_A \neq 0$. All such molecules like CH_3F and $CHCl_3$, exhibit rotational Raman spectra because rotation about the top axis (e.g. C—F bond axis in case of CH_3F) produces no change in the polarizability but end-over-end rotations produce such a change.

(iii) Spherical Top Molecules

Molecules like CH_4, CCl_4 etc. are spherical top. These have all the three moments of inertia identical

$$I_A = I_B = I_C$$

Owing to their symmetry, rotation alone can produce no change in the polarizability and hence no rotational Raman spectrum is observable.

(iv) Asymmetric Top Molecules

Majority of substances belong to this class of molecules. These have all the three moments of inertia different

$$I_A \neq I_B \neq I_C$$

Simple examples are water and vinyl chloride. All such molecules exhibit rotational Raman spectra. However, their Raman spectra are complicated because all rotations of these molecules are Raman active, i.e. they produce change in the polarizability.

4.8 Vibrational-Rotational Raman Spectra

Theoretically it is possible for vibrational and rotational transitions to occur simultaneously in a Raman transition. In such a case, the selection rules are identical with those for separate rotational and vibrational transitions, i.e.

$$\Delta J = 0, \pm 2 \text{ and } \Delta v = \pm 1$$

Since ΔJ may be zero, a Raman line representing Q branch will be observed. (In a nomenclature, spectral lines are designated O, P, Q, R and S branches corresponding to ΔJ values of $-2, -1, 0, +1$ and $+2$, respectively.) The frequency of this line is referred to as $\Delta \nu_0$ (it has same value for all J) and is identical with that for the pure vibrational transition. If $\Delta \nu_0$ is the frequency shift for the purely vibrational transition (Q branch), the Raman shifts for the accompanying rotational transitions will be represented by

$$\Delta J = +2 : \Delta \nu = \Delta \nu_0 + 2B(2J + 3)$$

where $J = 0, 1, 2 \ldots$, and

$$\Delta J = -2 : \Delta \nu = \Delta \nu_0 - 2B(2J + 3)$$

where $J = 0, 1, 2, \ldots$.

Thus, the central vibrational Raman lines (Q branch) will be accompanied by two wings or branches. The branch on the low frequency side for which ΔJ is –2 is known as O branch, and that on the high frequency side for which ΔJ is +2 is called S branch.

4.9 Polarization of Raman Lines

The lines in some Raman spectra are found to be plane-polarized to different extents even if the exciting radiation is completely depolarized. The state of polarization of a Raman line is measured by a quantity known as *depolarization factor* or *degree of depolarization* ρ which is the ratio of the horizontal and vertical components when the incident light is vertically polarized. For determining the polarization of Raman lines, essentially the same experimental arrangement (Fig. 4.4) is used which is employed for obtaining a Raman spectrum. In addition, a suitably oriented double image prism is used in front of the slits of the spectrograph, which separates the vertical and horizontal components in the scattered light so that two images are simultaneously photographed.

Theoretically, if the degree of depolarization ρ is less than or equal to 0.86, then the vibration concerned is symmetric and the Raman line is described as polarized, while if $\rho > 0.86$, the line is depolarized and the vibration is non-symmetric. Usually, the order of increasing symmetry of molecules is as follows:

linear molecules < symmetric top molecules < spherical top molecules

Note that the higher the molecular symmetry, smaller will be the degree of depolarization of the Raman line for a particular type of vibration.

The degree of polarization of spectral lines can be readily estimated by noting how the intensity of each line varies when a piece of polaroid or other analyzer is put into the scattered radiation, first with its polarizing axis parallel to xy plane (where z is the direction of the incident beam) and secondly, perpendicular to this plane. The degree of depolarization ρ lies between 0 and 1. It has been found that:

(a) the depolarization factor ρ varies from 0 to 0.86 for the vibrational Raman lines, whereas it has a constant value 0.86 for the rotational lines.
(b) sharp and strong lines are ordinarily characterized by low depolarization factor, whereas weak and diffuse ones by high depolarization factor ρ.
(c) since the polarization of a Raman line is mainly decided by the symmetry of the oscillation, corresponding lines in molecules having similar structures have nearly the same degree of depolarization.

Polarization of Raman lines provides a method by which some observed Raman lines can be assigned to their appropriate molecular vibration. In general, a symmetric vibration results in a polarized or partially polarized Raman line, whereas a non-symmetric vibration gives a depolarized line.

4.10 Rule of Mutual Exclusion

This rule states that 'if a molecule has a center of symmetry, then Raman active vibrations are IR inactive and IR active vibrations are Raman inactive. If there is no center of symmetry, then some (but not necessarily all) vibrations may be both Raman and IR active.'

Thus, the Raman and IR spectra having no common bands show that the molecule has a centre of symmetry but here caution is necessary because a vibration may be Raman active but too weak to be observed. If some of the bands are present in both the Raman and IR spectra, it is certain that the molecule has no center of symmetry.

4.11 Instrumentation

Raman spectroscopy is essentially emission spectroscopy. The experimental arrangement for recording Raman spectra is quite simple in principle. An intense monochromatic radiation is passed through a sample and the light scattered at right angles to the incident (or exciting) beam is analyzed by a spectrophotometer. Schematic diagram of a Raman spectrometer is given in Fig. 4.4.

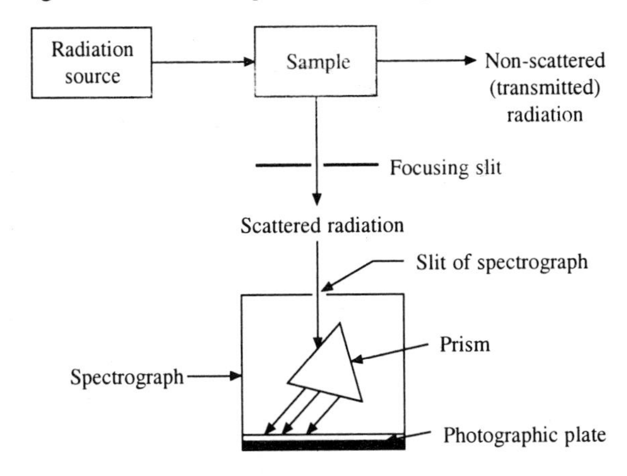

Fig. 4.4 Schematic diagram of a Raman spectrophotometer

The main components of a Raman spectrometer are:

(i) Radiation source
(ii) Filters
(iii) Sample (Raman) tube
(iv) Spectrograph

(i) Radiation Source

Since Raman lines are weak, it is essential to use a radiation of high intensity. The intensity of Raman lines is proportional to the fourth power of the frequency of incident (or exciting) radiation. Thus, the excitation frequency which is high enough, but not so much as to cause photodecomposition of the sample, is used. The mercury arc lamp is the most useful source of radiation. The most commonly used radiation in Raman spectroscopy is the line corresponding to 4358 Å which is obtained from the mercury arc lamp by use of suitable filters.

Nowadays laser is almost an ideal source of radiation for Raman spectroscopy and it is largely displacing the traditional mercury arc lamp. Helium-neon laser source has become very common; laser beam of 6328 Å wavelength is generally employed. An argon-ion laser with lines at 4880 and 5145 Å are used, especially when higher intensity of Raman lines is required.

Before the application of laser in Raman spectroscopy, it suffered from the following disadvantages:

(a) Samples had to be colourless, clear and non-fluorescent liquids.
(b) The low intensity of Raman lines required relatively concentrated solutions.
(c) Much larger volume of sample solutions were needed than that for IR spectroscopy.

These disadvantages were major factors in limiting the use of Raman spectroscopy. The scope of Raman spectroscopy is greatly widened with the application of laser as the exciting source because of its following advantages:

(a) It is a single, intense frequency source, hence no filtering is necessary.
(b) The line width of laser line is smaller than the mercury exciting line, hence it gives better resolution.
(c) Because of highly coherent character of laser radiation, it is easier to focus.
(d) A large number of exciting frequencies of laser are available, thus it is possible to study coloured substances without causing any electronic transitions. This is particularly useful in the study of solutions of inorganic salts which are generally coloured.

(ii) Filters

In case of non-monochromatic incident radiation, there will be overlapping of Raman shifts making the interpretation of the spectrum difficult. Thus, monochromatic incident radiations are necessary. Filters are used to obtain monochromatic radiation. They are generally made of nickel oxide or quartz. Sometimes, a coloured solution, e.g. aqueous solution of potassium ferricyanide or solution of iodine in carbon tetrachloride, is used as a monochromator.

(iii) Sample (Raman) Tube

Various types of sample tubes are in use for Raman spectroscopy (Fig. 4.5). The shape and size of the tube depends on the intensity of incident radiation, nature of the sample and its available amount. For gases, relatively bigger tubes are required.

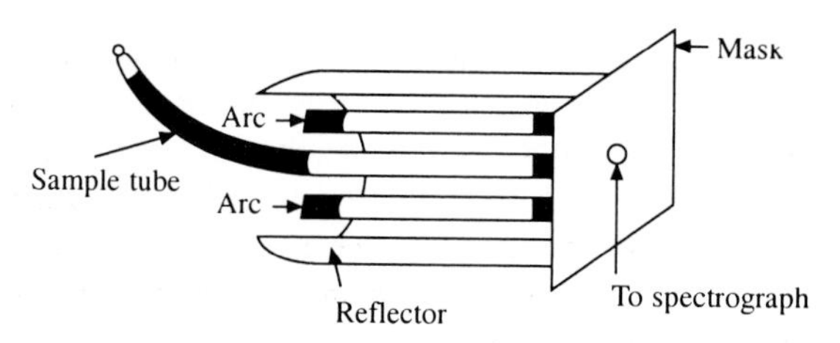

Fig. 4.5 Sample (Raman) tube

(iv) Spectrograph

The spectrograph used in Raman spectroscopy should have large gathering power, hence, special prisms of high resolving power and short focus camera are used. A lens directs the scattered radiation upon the slit of the spectrograph and Raman lines can be obtained on a photographic plate. Raman spectrographs with automatic recorders are also available. They use photographic emulsions or photomultiplier tubes. However, due to low intensity of Raman lines, photographic method is preferred because it is more sensitive.

The intensity of Raman lines mainly depends on the polarizability of the molecules, intensity of the exciting radiation and concentration of Raman active species. Usually, Raman intensities are directly proportional to the concentration of Raman active species.

4.12 Sample Handling

Raman spectra have been recorded in the liquid, gas or solid phase. However, liquids are by far easy to handle. Generally, 10-100 ml of liquid sample is required, but cells of 1 ml or less can also be used. Solutions of at least 0.1 M strength are used. Rasetti has developed a technique for recording Raman spectra of gases under high pressure.

Any solvent which is suitable for recording UV spectra can also be used in Raman spectroscopy. Water is a good solvent for recording Raman spectra because its own Raman spectrum is very weak and unlike IR spectrophotometers, Raman spectrometers are not attacked by moisture.

4.13 Applications of Raman Spectroscopy

Besides its theoretical interest, Raman spectroscopy is of immense practical importance because of its many useful application in physics and chemistry. Raman effect can be easily controlled and made to appear as a part of visible spectrum because the positions of Raman lines depend only on the frequency of the incident radiation. Complementary character of Raman spectra with reference to IR spectra is very useful for structure determination of molecules. Studies on bond angles, bond strengths and other structural confirmations require Raman data in addition to IR studies. Some of the important applications of Raman spectroscopy are summarized as follows.

(i) Structure of Diatomic Molecules

The analysis of Raman spectrum of a diatomic molecule gives an idea about the nature of chemical bond existing between the atoms. We know that the greater the bond strength, higher is its vibrational frequency. The vibrational frequency of a molecule containing light atoms is higher than that containing heavier atoms (Eq. (3.1)). For example, the values of vibrational frequencies of H_2, N_2, O_2, HCl, HBr and HI obtained from their Raman spectra are 4156, 2992, 2331, 2880, 2558 and 2233 cm^{-1}, respectively.

Covalent diatomic molecules show intense Raman lines, whereas no Raman lines appear in case of electrovalent diatomic molecules. The intensity of Raman lines essentially depends on the extent to which the polarizability is affected by the oscillations. In covalent molecules, the bonding electrons are common to the two nuclei, thus the polarizability is considerably changed by nuclear oscillations resulting in intense Raman lines. On the other hand, in electrovalent molecules, bonding electrons are transferred from one nucleus to the other during the formation of the molecules. Thus, the polarizability of the molecule is little affected by nuclear oscillations, and hence no Raman lines will appear.

(ii) Structure of Triatomic Molecules

Let us discuss some molecules of the type AB_2 which are among the most thoroughly studied molecules. The questions to be answered are whether each molecule is linear or not and, if linear, whether it is symmetrical (*B*—*A*—*B*) or asymmetrical (*B*—*B*—*A*). We are taking the following examples for explaining the above points.

(a) Carbon dioxide (CO_2)

It shows two very strong absorption bands at 2349 and 668 cm^{-1} in its IR spectrum, whereas only one strong band at 1389 cm^{-1} in its Raman spectrum. Since none of these bands is present in both the spectra, the molecule of CO_2 must have a center of symmetry (O—C—O) according to the rule of mutual exclusion (Section 4.10). For triatomic molecules, this observation shows that the molecule is linear and symmetric. Thus, the molecular structure of CO_2 is completely determined which is of the type O—C—O. The symmetric structure of CO_2 is confirmed by its rotational Raman spectrum which consists only of alternate (odd) levels like that of O_2. The abovementioned three bands, two in the IR spectrum and one in the Raman spectrum, represent the three fundamental frequencies of CO_2. Carbon disulphide (CS_2) is another example of this type.

(b) Nitrous oxide (N_2O)

This molecule is isoelectronic with CO_2 and thus we might expect it to have a linear symmetric structure. But IR and Raman spectral analyses clearly show that N_2O, although linear, is not symmetrical. The three fundamental frequencies of N_2O are 2224, 1285 and 589 cm^{-1}. All the three appear in the IR spectrum and two of them, viz. 2224 and 1285 cm^{-1}, appear in the Raman spectrum also. The frequency 589 cm^{-1} could not be recorded due to weak intensity. Thus, acccording to the rule of mutual exclusion (Section 4.10), the molecule has no center of symmetry, i.e. it does not have the structure N—O—N, because two bands are common to both its IR and Raman spectra. Hence, N_2O has the unsymmetrical structure N—N—O. This conclusion is confirmed by the rotational Raman spectrum which consists of both (odd and even) sets of lines without any alternation in intensity. Other molecules of this type are HCN, ClCN, BrCN etc.

(c) Water (H_2O)

According to theory, all triatomic molecules having bent symmetrical structure should give rise to three Raman lines all of which should be present in the IR spectrum also. The IR spectrum of water exhibits very strong bands at 3756 and 1595 cm^{-1}. In the Raman spectrum, two bands have been recorded at 3605 and

1665 cm^{-1} which roughly correspond to the IR bands. This confirms that water has a symmetrical and bent structure

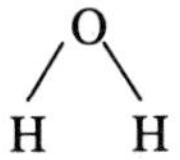

Water gives some other extra bands due to polymerized (through hydrogen bonding) molecules $(H_2O)_2$ and $(H_2O)_3$. Examples of other molecules having similar bent and symmetrical structures are D_2O, H_2S and SO_2.

(iii) Structure of Benzene

Raman spectrum of benzene shows two strong Raman lines at 995 and 1050 cm^{-1} due to C—C and C=C bonds, respectively. This supports the Kekulé structure of benzene.

(iv) Determination of Orientation

Raman spectroscopy has been used to determine the relative positions of substituents on benzene ring. For example, al! *meta*-disubstituted benzene derivatives exhibit an intense and strongly polarized Raman line at 995 cm^{-1} which is absent from the Raman spectra of *ortho* and *para* substituted derivatives. The Raman spectra of *ortho*-substituted compounds are richer in lines than their *para* isomers and the *para* isomer shows a line at about 625 cm^{-1} which is absent in the case of *ortho* isomer.

(v) Detection of Symmetry

According to the rule of mutual exclusion, if none of the bands are common to both the IR and Raman spectra, then the molecule must have a center of symmetry. For example, the IR spectrum of CO_2 shows two bands at 2349 and 668 cm^{-1}, whereas only one strong band at 1389 cm^{-1} in its Raman spectrum. Since none of these bands is common to both the spectra, the CO_2 molecule must have a center of symmetry, which it actually has. Similarly, if some of the bands are common to both the IR and Raman spectra, then the molecule has no center of symmetry, e.g. N_2O molecule; for discussion see Section 4.13(b)(ii).

(vi) Structure of Mercurous Salts

The Raman spectrum of aqueous solution of mercurous nitrate shows an additional line which is not present in the Raman spectra of other metal nitrates. This line may be due to the vibration of Hg—Hg covalent bond present in mercurous nitrate.

(vii) Study of Chloro Complexes of Mercury

The Raman spectrum of an aqueous solution of mercuric chloride and ammonium chloride in 1 : 2 molar ratio shows a strong line at 269 cm^{-1}. Comparison of this line with the strong line at 273 cm^{-1} for solid ammonium tetrachloromercurate (II) indicates the formation of $HgCl_4^{2-}$ in solution.

(viii) Nature of Bonding in Complexes

The IR spectra of symmetrical tetrahedral complexes ML_4 (e.g. $ZnCl_4^{2-}$, $CdCl_4^{2-}$

etc.) and octahedral complexes (e.g., SiF_6^{2-}, SF_6 etc.) exhibit no bands but the Raman spectra of such complexes exhibit strong lines. Thus, from the *M—L* bond stretching-force constant, useful information about the strength of the metal-ligand bond can be obtained. The oxy anions (e.g., PO_4^{3-}, SO_4^{2-} etc.) have much larger force constants. This is considered as an evidence for the presence of $d_\pi - p_\pi$ bonding between the central atom and oxygen atom in addition to σ bonding.

Besides the above applications of Raman spectroscopy in structure determination, it is also useful in many other fields. For example, in the study of electrolytic dissociation, hydrolysis and transitions of states; for the study of amorphous and crystalline regions in polymers; in quantitative analyses; in the study of biological systems etc.

It should be noted that Raman spectrometer is not attacked by moisture and the presence of water in a sample does not interfere. Thus, contrary to IR spectroscopy, studies by Raman spectroscopy can also be conducted in aqueous solutions.

4.14 Difference Between Raman and Fluorescence Spectra

The phenomenon of emission of radiation of a lower frequency than that of the incident radiation absorbed is known as *fluorescence*. The main differences between Raman and fluorescence spectra are:

Raman spectra	Fluorescence spectra
1. Spectral lines have frequencies lesser and greater than the frequency of incident radiation.	1. Frequencies of the spectral lines are always lesser than that of the incident radiation.
2. Frequencies of Raman lines depend on incident frequencies but not on the nature of the scatterer.	2. Frequencies of fluorescence lines depend on the nature of the scatterer.
3. Raman lines are weak in intensity, so concentrated solutions are preferred as samples to give enough intensity.	3. The intensity of fluorescence lines are relatively high and samples of concentrations as low as 1 part in 10^9 are used.
4. Raman lines are strongly polarized.	4. Fluorescence lines are not polarized.

4.15 Difference Between Raman and IR Spectra

Although Raman shifts fall in the IR region of the electromagnetic spectrum, Raman spectra are quite different from infrared spectra.

Raman spectra	Infrared spectra
1. These originate from scattering of radiation by vibrating and rotating molecules.	1. These originate from absorption of radiation by vibrating and rotating molecules.
2. Some change in molecular polarizability during the molecular vibration or rotation is essential for a molecule to exhibit Raman spectrum.	2. Some change in dipole moment during the molecular vibration or rotation is essential for a molecule to exhibit IR spectrum.

3. Raman lines are weak in intensity, hence concentrated solutions are preferred as samples to give enough intensity.	3. Generally, dilute solutions are preferred.
4. Water can be used as solvent because its own Raman spectrum is very weak and Raman spectrometer is not attacked by water.	4. Water cannnot be used as solvent because it very strongly absorbs in the IR region and the optical components of IR spectrometer are attacked by water.
5. Optical systems of Raman spectrometer are made of glass or quartz.	5. Optical systems of IR spectrometer are made of NaCl, NaBr, KCl, KBr etc.
6. Sometimes photochemical reactions take place in the frequency regions of Raman lines and so create difficulties.	6. Photochemical reactions do not take place.
7. Method is accurate but not very sensitve.	7. Method is accurate as well as very sensitive.
8. These are recorded by using a beam of monochromatic radiation.	8. These are recorded by using a beam of radiation having a large number of frequencies in the IR region.
9. Homonuclear diatomic molecules are Raman active.	9. Homonuclear diatomic molecules are IR inactive.
10. Vibrational frequencies of large molecules can be measured.	10. Vibrational frequencies of large molecules cannot be measured.
11. Pure substances are required for studies by Raman spectra.	11. Studies by the IR spectra do not require a high degree of purity.

4.16 Some Solved Problems

Problem 1. A sample gives a Stokes line at 4458 Å when the exciting radiation of wavelength 4358 Å is used. Deduce the wavelength of the anti-Stokes line in Å.

Solution. Raman shift $\Delta\nu$ is given by

$$\Delta\nu = \nu_i - \nu_s$$

where ν_i is the wavenumber of incident (exciting) radiation and ν_s the wavenumber of scattered radiation.

For Stokes line $\nu_s < \nu_i$. In the present case

$$\nu_i = \frac{1}{4358 \text{ Å}} = \frac{1}{4358 \times 10^{-8} \text{ cm}} = 22{,}946 \text{ cm}^{-1}$$

and

$$\nu_s = \frac{1}{4458 \text{ Å}} = \frac{1}{4458 \times 10^{-8} \text{ cm}} = 22{,}432 \text{ cm}^{-1}$$

Thus, shift is

$$\Delta \nu = \nu_i - \nu_s = 22{,}946 - 22{,}432 = 514 \text{ cm}^{-1}$$

As the Stokes and anti-Stokes lines have the same wavenumber displacement with respect to the exciting line ν_i, and for anti-Stokes lines $\nu_s > \nu_i$. Thus

$$\nu_{s(\text{anti-Stokes})} = \Delta \nu + \nu_i = 514 + 22{,}946 = 23{,}460 \text{ cm}^{-1}$$

The corresponding wavelength is

$$\lambda = \frac{1}{23460 \text{ cm}^{-1}} = 4263 \text{ Å}$$

Problem 2. The wavelength of exciting radiation in an experiment is 4358 Å and that of Stokes line is 4567 Å. Deduce the positions of Stokes and anti-Stokes lines for the same substance when the exciting radiation of wavelength 4047 Å is used.

Solution. Raman shift is the characteristic of the substance and it does not depend on the frequency of the exciting radiation. Thus, we shall calculate the Raman shift $\Delta \nu$ from the given data and then use it for the second exciting radiation of the wavelength 4047 Å.

$$\Delta \nu = \nu_i - \nu_s = \frac{1}{4358 \text{ Å}} - \frac{1}{4567 \text{ Å}}$$

$$= \frac{1}{4358 \times 10^{-8} \text{ cm}} - \frac{1}{4567 \times 10^{-8} \text{ cm}}$$

$$= 22{,}946 \text{ cm}^{-1} - 21{,}896 \text{ cm}^{-1} = 1050 \text{ cm}^{-1}$$

The wavenumber ν_i' of the second exciting radiation

$$\nu_i' = \frac{1}{4047 \text{ Å}} = 24{,}710 \text{ cm}^{-1}$$

Thus, the wavenumber of

$$\text{Stokes line} = \nu_i' - \Delta \nu = 24{,}710 - 1050 = 23{,}660 \text{ cm}^{-1}$$

$$\text{anti-Stokes line} = \nu_i' + \Delta \nu = 24{,}710 + 1050 = 25{,}760 \text{ cm}^{-1}$$

The corresponding wavelengths are

$$\lambda_{\text{Stokes}} = \frac{1}{23{,}660 \text{ cm}^{-1}} = 4226.5 \text{ Å}$$

$$\lambda_{\text{anti-Stokes}} = \frac{1}{25{,}760 \text{ cm}^{-1}} = 3882 \text{ Å}$$

(Note that the positions of Stokes and anti-Stokes lines can be expressed in terms of wavelength, wavenumber or frequency.)

Problem 3. When 435.8 nm line of mercury arc lamp was used as the source of radiation, a Raman line was observed at 444.7 nm. What is the Raman shift?

Solution. Raman shift $\Delta\nu$ is given by

$$\Delta\nu = \nu_i - \nu_s$$

where ν_i is the wavenumber of incident radiation and ν_s the wavenumber of scattered radiation.

$$\Delta\nu = \nu_i - \nu_s = \frac{1}{435.8 \text{ nm}} - \frac{1}{444.7 \text{ nm}}$$

$$= \frac{1}{435.8 \times 10^{-7} \text{ cm}} - \frac{1}{444.7 \times 10^{-7} \text{ cm}}$$

$$= 22{,}946 \text{ cm}^{-1} - 22{,}487 \text{ cm}^{-1} = 459 \text{ cm}^{-1}$$

Problem 4. By what factor will the scattered intensity of a given band be increased on changing the excitation frequency from a He—Ne laser (632.8 nm) to an argon laser (488.0 nm)?

Solution. We know that the intensity of Raman lines is proportional to the fourth power of the frequency of exciting radiation.

The frequency of He—Ne laser

$$\nu = \frac{c}{\lambda \text{ (in cm)}} = \frac{2.998 \times 10^{10}}{632.8 \times 10^{-7}} = 47{,}377 \times 10^{10} \text{ cps}$$

The frequency of argon laser $= \dfrac{2.998 \times 10^{10}}{488.0 \times 10^{-7}} = 61{,}434 \times 10^{10}$ cps

Thus, the factor by which the intensity will increase

$$\left(\frac{61{,}434 \times 10^{10} \text{ cps}}{47{,}377 \times 10^{10} \text{ cps}}\right)^4 = 2.9$$

Problem 5. Outline the fundamental modes of vibration of CO_2 and predict which of these modes will be IR active and which will be Raman active.

Solution. Since CO_2 is a linear molecule, it will have $3n - 5$ or $3 \times 3 - 5 = 4$ fundamental modes of vibration outlined as follows:

	← → O=C=O	↑ O=C=O ↓ ↓	− + − O=C=O	→ ← → O=C=O
Mode of vibration	Symmetrical stretching	In-plane bending	Out-of-plane bending	Asymmetrical stretching
Symbol	ν_1	ν_2	ν_2	ν_3
Infrared	Inactive	Active	Active	Active
Raman	Active	Inactive	Inactive	Inactive

The symmetrical stretching vibration ν_1 produces no change in the dipole moment of the molecule, hence it is IR inactive. On the other hand, it produces change in polarizability, hence it is Raman active. The bending vibrations ν_2 occur at the same frequency and are equivalent (degenerate) and produce change in dipole moment, hence are IR active. Similarly, the asymmetrical vibrations do not produce change in polarizability, hence are Raman inactive. Thus, the CO_2 molecule shows three fundamental bands, two in the IR spectrum and one in the Raman spectrum.

Problem 6. Fig. 4.6 shows IR and Raman spectra of tetrachloroethylene. What conclusion can be drawn about the structure of the molecule?

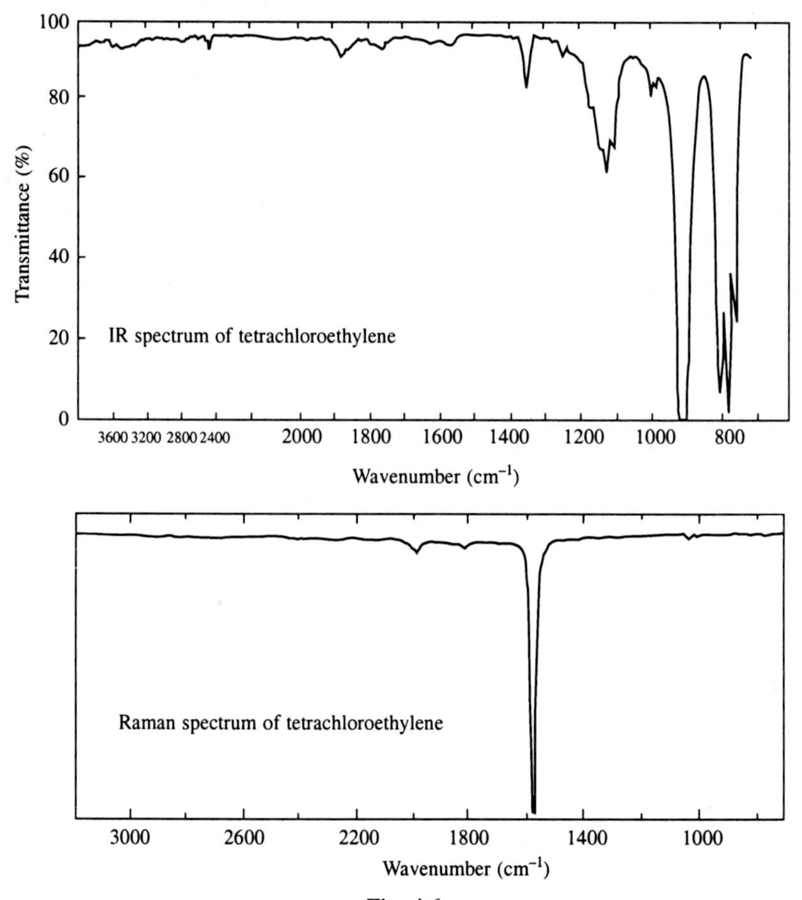

Fig. 4.6

Solution. It is clear from the IR and Raman spectra of tetrachloroethylene that there is no band common to both of these spectra. Thus, according to rule of mutual exclusion, it can be concluded that tetrachloroethylene has a center of symmetry and planar structure:

```
Cl       Cl
  \     /
   C=C
  /     \
Cl       Cl
```

Problem 7. On irradiation with 4358 Å mercury line, acetylene shows a Raman line at 4768 Å which is attributed to its symmetrical stretching vibration. Calculate the fundamental frequency for this vibration.

Solution

$$\Delta \nu = \nu_i - \nu_s = \frac{1}{4358 \text{ A}} - \frac{1}{4768 \text{ A}}$$

$$= \frac{1}{4358 \times 10^{-8} \text{ cm}} - \frac{1}{4768 \times 10^{-8} \text{ cm}}$$

$$= 22{,}946 \text{ cm}^{-1} - 20{,}973 \text{ cm}^{-1} = 1973 \text{ cm}^{-1} \text{ (wavenumber)}$$

Vibrational frequency ν = Wavenumber $\times c$

$$= 1973 \text{ cm}^{-1} \times 2.998 \times 10^{10} \text{ cm/sec} = 5.92 \times 10^{13} \text{ cps}$$

Problem 8. Both N_2O and NO_2 exhibit three different fundamental vibrational frequencies, and for the two molecules some modes are observed in both the IR and Raman spectra. The bands in case of N_2O show only simple PR structure (no *Q* branches), whereas in case of NO_2 the bands shows complex rotational structure. What can be deduced about the structure of each molecule?

Solution. Since some bands are common to both the IR and Raman spectra in case of both the compounds, both of these have no center of symmetry (according to the rule of mutual exclusion). The bands in case of N_2O show only simple *PR* branches without *Q* branches. This proves that the molecule has a linear structure. Thus, N_2O has a linear structure without a center of symmetry, i.e. N—N—O. On the other hand, NO_2 has a non-linear structure because its bands show complex rotational structure. Thus, NO_2 has a non-linear structure without a center of symmetry, as

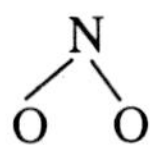

PROBLEMS

1. What is Raman effect? Discuss the characteristics of Raman lines.
2. How is the polarizability of molecules affected by their vibration and rotation?
3. Explain Raman effect on the basis of quantum theory. How is this theory superior to the classical theory of Raman effect?
4. Suggest a spectroscopic method to prove whether an unknown molecule has a center of symmetry.
5. The wavelength of the incident radiation in an experiment is 5460 Å and the Stokes line appears at 5520 Å. Calculate the wavelength of the anti-Stokes line.
6. By what factor will the scattered intensity of a given band be reduced on changing from an argon laser (488.0 nm) to a neodymium-doped laser (1065 nm)?
7. The Hg line at 22,938 cm^{-1} is frequently used in Raman spectroscopy. Does this radiation possess sufficient energy to dissociate H_2 molecule? The dissociation energy of H_2 is 428.4 kJ mole^{-1}.

(Hint. Calculate the energy associated with the radiation and see whether it is greater than the dissociation energy of H_2.)

8. Explain the following:

(a) Homonuclear diatomic molecules are IR inactive but Raman active.
(b) In Raman spectra, Stokes lines are far more intense than the anti-Stokes lines which are sometimes too weak to be observed.
(c) Water can be used as solvent in Raman spectrometry but not in IR spectrometry.
(d) Covalent diatomic molecules give intense Raman lines, whereas no Raman lines appear in case of electrovalent diatomic molecules.

9. For carbon disulphide, all vibrations that are Raman active are IR inactive and vice-versa, whereas for nitrous oxide the vibrations are simultaneously Raman and IR active. What can be deduced about structure of CS_2 and N_2O?

10. A compound having molecular formula $C_2H_2Cl_2$ shows IR and Raman bands but none of these bands is common to both the spectra. Assign the structure to the compound.

11. What are the differences between Raman and infrared spectra?

12. What is zero-point energy? How does it help in explaining the Raman scattering for molecules even when their vibrational quantum number is zero (v_0)?

13. Giving reasons, indicate which of the following molecules will display rotational Raman spectra:

(i) H_2 (ii) CH_4 (iii) $CHCl_3$ (iv) CCl_4 (v) H_2O

14. Write notes on the following:

(a) Polarization of Raman lines
(b) Mutual exclusion principle
(c) Raman shift

15. The Raman spectrum of CCl_4 obtained by irradiation with the 435.8 nm Hg line exhibits lines at 439.9, 441.8, 444.6 and 450.7 nm. Calculate the Raman frequencies in cm^{-1}.

(Hint. It should be noted that all of these are Stokes lines, because they have higher wavelengths (lesser frequency) than the exciting line.)

16. With the help of IR and Raman spectroscopy, how will you show that the following molecules have the structures as indicated:

(a) SO_2, symmetrical and bent
(b) CO_2, linear with a center of symmetry
(c) N_2O, linear without a center of symmetry

17. Giving reasons, point out which of the following molecules will exhibit no rotational Raman spectra and which will exhibit both the rotational and vibrational Raman spectra:

(a) O_2 (b) CO (c) HBr (d) CCl_4 (e) H_2

18. Arrange the following in order of their decreasing Raman frequency shifts:

D_2, H_2 and HD

19. Write notes on:

(a) Vibrational-rotational Raman spectra
(b) Use of laser in Raman spectroscopy
(c) Differences between Raman and fluorescence spectra

20. Giving reasons, indicate which of the following types of molecules will exhibit rotational Raman spectra:

(a) Linear molecules (b) Symmetric tops
(c) Spherical tops (d) Asymmetric tops

21. Which of the following vibrations will be Raman active and Raman inactive:

	Molecule	Mode of vibration
(i)	CS_2	Symmetrical stretching
(ii)	CO_2	Asymmetrical stretching
(iii)	CO_2	Bending
(iv)	$CH_2{=}CH_2$	C=C stretching
(v)	SO_2	Bending

22. A molecule X_2Y_2Z, in which X, Y and Z are different atoms, exhibits IR and Raman bands. None of these bands are common to both spectra. Deduce the structure of the molecule.
(Hint. The molecule must have a center of symmetry.)

23. When excited by mercury line at 435.8 nm, the spectral trace of a compound contains Raman lines at 850, 1584, 1605 and 3047 cm^{-1}. At what wavelengths will these Raman lines appear when the compound is irradiated with a He—Ne laser line at 632.8 nm.
(Hint. Note that here 850, 1584, 1605 and 3047 cm^{-1} are $\Delta\nu$.)

24. Fig. P4.1 shows the IR and Raman spectra of a compound having molecular formula C_4H_6. Deduce the structure of the compound.

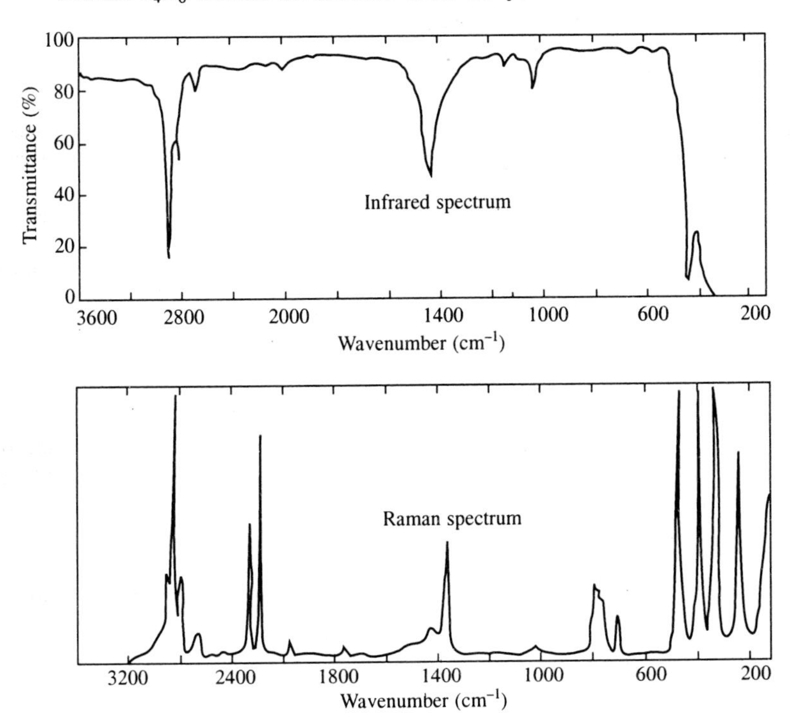

Fig. P4.1

(Hint. The Raman spectrum exhibits two bands in the frequency region of C≡C, i.e. near 2100-2260 cm^{-1}. The absence of these bands from the IR spectrum shows that the C≡C is symmetrically substituted.)

References

1. A. Anderson, The Raman Effect, Marcel Dekker Inc., Vol. 1, Principles, 1971 and Vol. 2, Applications, 1973.
2. G. Herzberg, Infrared and Raman Spectra, Van Nostrand, New York, 1945.
3. H.A. Szymanski, Ed., Raman Spectroscopy: Theory and Practice, Plenum, New York, Vol. 1, 1961 and Vol. 2, 1970.
4. M.C. Tobin, Laser Raman Spectroscopy, Wiley-Interscience, 1971.
5. N. Colthup, L. Daly and S. Wiberley, Introduction to Infrared and Raman Spectoscopy, Academic Press, New York and London, 1964.
6. R.N. Jones, J.B. DiGiorgio, J.J. Illiott and G.A. Nonnenmacher, J. Org. Chem., 1965, 30, 1822, Raman Spectoscopy.
7. S. Bhagvantam, Scattering of Light and the Raman Effect, Chemical Publishing Co., New York, 1942.
8. S. Glasstone, Theoretical Chemistry, Affiliated East-West Press Pvt. Ltd., 1973.

5

Proton Nuclear Magnetic Resonance (PMR or ^{1}H NMR) Spectroscopy

5.1 Introduction

Similar to the UV and IR spectroscopy, nuclear magnetic resonance (NMR) spectroscopy is also an absorption spectroscopy in which samples absorb electromagnetic radiation in the radio-frequency region (3 MHz to 30,000 MHz) at frequencies governed by the characteristics of the sample. As the name itself implies, NMR spectroscopy involves nuclear magnetic resonances which depend on the magnetic property of atomic nuclei. Thus, NMR spectroscopy deals with nuclear magnetic transitions between magnetic energy levels of the nuclei in molecules. NMR signals were first observed in 1945 independently by Prucell at Harvard and Bloch at Stanford. The first application of NMR to the study of structure was made in 1951 and ethanol was the first compound thus studied. In 1952, Prucell and Bloch won the Nobel Prize in Physics for their discovery.

There are approximately 100 isotopes for which NMR spectroscopy is possible, but the most commonly used by organic chemists are proton nuclear magnetic resonance (PMR or ^{1}H NMR) spectroscopy and carbon-13 nuclear magnetic resonance (^{13}C NMR) spectroscopy. This chapter deals with PMR spectroscopy.

5.2 Theory

Nuclei of some isotopes possess a mechanical spin, i.e. they have angular momentum. The total angular momentum of a spinning nucleus depends on its spin, or spin number I (also called as nuclear spin quantum number). Each proton and neutron has its own spin and I is a resultant of these spins, i.e. vector combination of proton and neutron spins. Unfortunately, the laws governing this combination are not yet known, hence the spin of a particular nucleus cannot be predicted in general. However, the observed spins can be rationalized and some empirical rules have been formulated which are given in Table 5.1. The spin number I may have values 0, $\frac{1}{2}$, 1, $\frac{3}{2}$, $\frac{5}{2}$ etc. depending on the mass number and atomic number of the atom as shown in Table 5.1. Nuclei composed of even number of protons and neutrons have no net spin ($I = 0$) because their spins are paired off.

The isotopes with either odd mass number or odd atomic number possess a mechanical spin and only such isotopes can exhibit a nuclear magnetic resonance spectrum.

Table 5.1 Spin number of some isotopes

Mass number	Atomic number	Spin number I		Examples
Odd	Odd or even	$\frac{1}{2}$	Half-integer	^{1}H, ^{13}C, ^{15}N, ^{19}F, ^{31}P
		$\frac{3}{2}$		^{11}B, ^{35}Cl, ^{37}Cl, ^{79}Br, ^{81}Br
		$\frac{5}{2}$		^{127}I, ^{17}O
Even	Even	0 (no spin)		^{12}C, ^{16}O, ^{32}S, ^{34}S
Even	Odd	1	Integer	^{14}N, ^{2}H (or D)
		3		^{10}B

It is well known that circulation of a charge generates an electric current which is associated with a magnetic field. Since all atomic nuclei have a positive charge, spinning nuclei generate a magnetic dipole (i.e. north and south poles) along the axis of rotation, i.e. the nuclear axis. Thus, spinning nuclei behave like a tiny bar magnet with a magnetic moment μ.

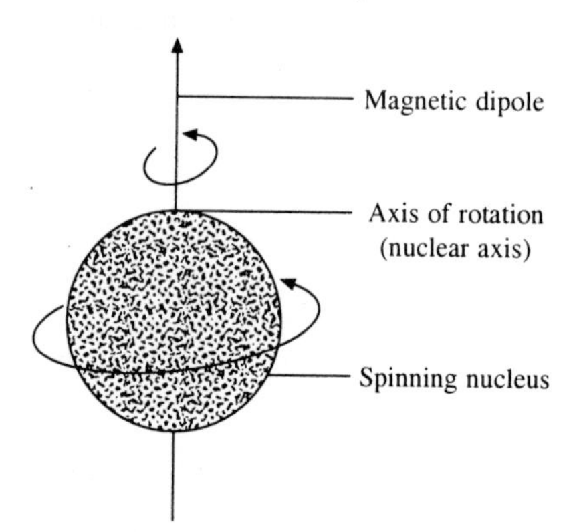

Fig. 5.1 Spinning nucleus and generated magnetic dipole

In the absence of an external magnetic field, the magnetic nuclei (nuclear magnetic dipoles) are randomly orientated, i.e. the nuclear spin is of no consequence (Fig. 5.2).

When a magnetic nucleus with spin number I is placed in a uniform magnetic field H_0, its magnetic dipole or magnetic moment may assume any one of $2I + 1$ orientations with respect to the direction of the applied magnetic field H_0 and the system is said to be quantized. The most important nuclei for organic chemists are ^{1}H and ^{13}C. For both of these $I = \frac{1}{2}$. Hence, number of orientations for their magnetic dipoles will be $2 \times \frac{1}{2} + 1 = 2$ (because number of orientations = $2I + 1$).

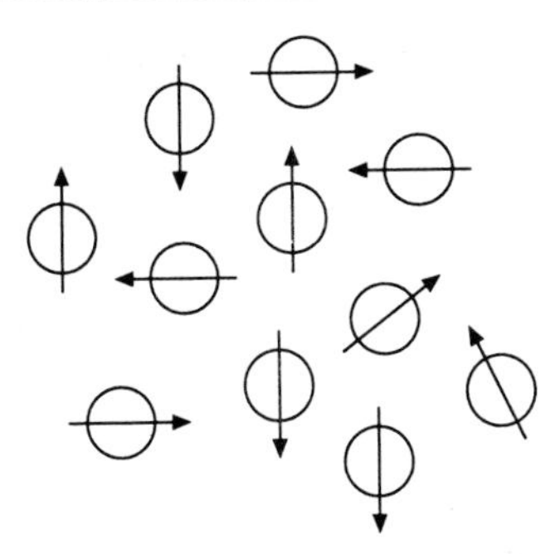

Fig. 5.2 Orientation of nuclear magnetic dipoles in the absence of an external magnetic field

Thus, the magnetic dipoles of nuclei with $I = \frac{1}{2}$, e.g. ^{1}H and ^{13}C will align parallel or antiparallel to the applied magnetic field, i.e. with or against the applied magnetic field, respectively (Fig. 5.3).

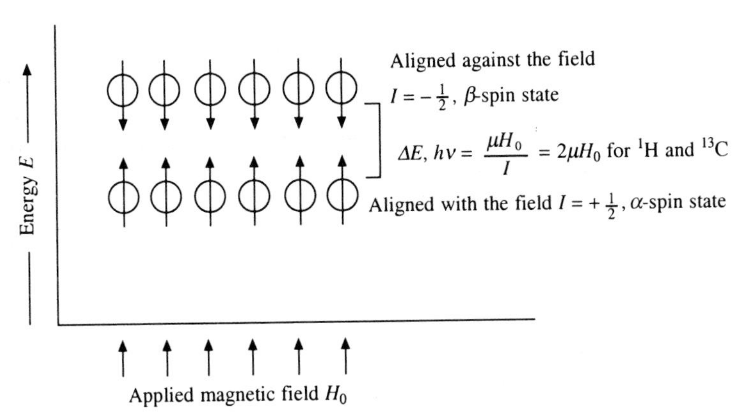

Fig. 5.3 Orientation of nuclear magnetic dipoles in an external magnetic field H_0

The alignment with the applied magnetic field is of lower energy and corresponds to the α-spin state $(+\frac{1}{2})$ of the nucleus, and the alignment against the applied magnetic field is of higher energy and corresponds to the β-spin state $(-\frac{1}{2})$ of the nucleus.

The energy difference ΔE (i.e. the energy required for a transition) has been shown to be a function of the applied magnetic field H_0. The following fundamental NMR equation correlates the electromagnetic frequency ν for the transition in a given field H_0

$$\nu = \frac{\gamma H_0}{2\pi} \tag{5.1}$$

or

$$\nu \propto H_0$$

where γ is magnetogyric ratio (or gyromagnetic ratio) which is a fundamental nuclear constant.

The magnetic moment μ of a spinning nucleus behaving like a tiny bar magnetic is directly proportional to its spin number I. It has been shown that

$$\gamma = \frac{2\pi\mu}{hI} \tag{5.2}$$

where h is Planck's constant and γ, the magnetogyric ratio, is the proportionality constant between μ and I.

From Eqs. (5.1) and (5.2)

$$\nu = \frac{\gamma H_0}{2\pi} = \frac{2\pi\mu}{hI} \cdot \frac{H_0}{2\pi} = \frac{\mu H_0}{hI}$$

or

$$h\nu = \Delta E = \frac{\mu H_0}{I} \tag{5.3}$$

For ^{1}H and ^{13}C, $I = \frac{1}{2}$. Hence

$$h\nu = \Delta E = 2\mu H_0$$

or

$$\nu = \frac{2\mu H_0}{h} \tag{5.4}$$

Equation (5.3) shows that the energy required for a transition ΔE is directly proportional to the strength of the applied magnetic field (because μ/I is constant for a given nucleus). This is shown graphically in Fig. 5.4. The stronger the field, greater will be the tendency of the nuclear magnetic dipoles to remain aligned with it and higher will be the energy required for a transition.

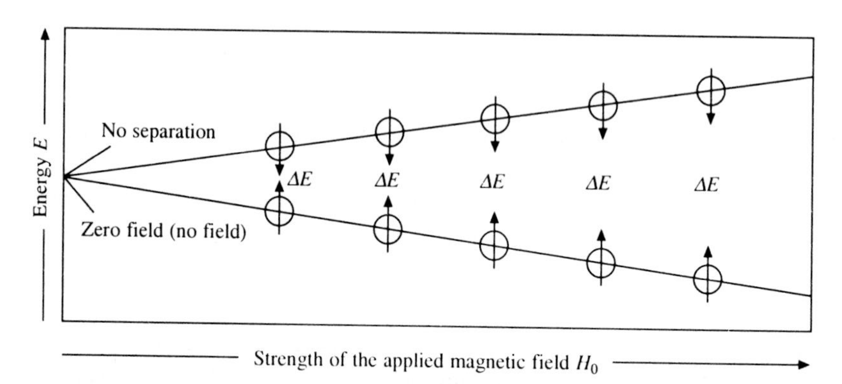

Fig. 5.4 Relationship between the transition energy ΔE and the applied magnetic field H_0

(i) Process of Absorption of Energy

If the axis of the nuclear magnet is not oriented exactly parallel or antiparallel to the applied magnetic field, then there will be a certain force by the applied magnetic field to so orient it. Since the nucleus is spinning, the effect is that its rotational axis draws out a circle perpendicular to the applied field, i.e. the nucleus starts precessing. This precession is similar to the gyroscopic motion of a common top which precesses if spun with an initial axis of rotation different from earth's gravitational field.

The nuclei aligned in such a way that their magnetic axes make an angle with the axis of the applied magnetic field H_0 are responsible for the process of absorption or emission of energy, i.e. for the NMR phenomenon. Fig. 5.5 shows a nucleus precessing in a magnetic field H_0.

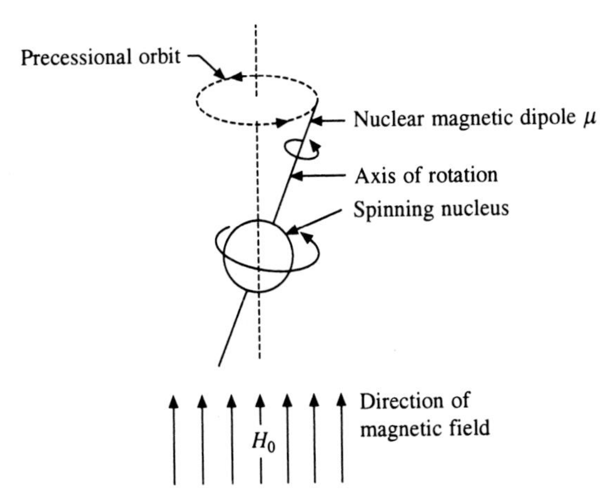

Fig. 5.5 A nucleus precessing in a magnetic field H_0

The precessional angular velocity $\omega_0 = \gamma H_0$.
From the fundamental NMR Eq. (5.1)

$$\gamma H_0 = 2\pi\nu$$

Therefore,

$$\omega_0 = 2\pi\nu$$

The value of this frequency ν inserted is called *precessional frequency.* The precessional angular velocity is quantized. Thus, the difference between angular velocities in the ground state and excited state corresponds to a precise frequency (i.e. energy) equal to the precessional frequency. Thus, the precessional frequency of spinning nucleus is exactly equal to the frequency of electromagnetic radiation necessary to induce a transition from one nuclear spin state to another. The transition corresponds to a change in the angle that the nuclear magnet makes with the applied magnetic field. This change can be brought about through the application of electromagnetic radiation whose magnetic vector component H_1 is rotating in a plane perpendicular to the applied magnetic field H_0 (Fig. 5.6).

When the frequency of the rotating magnetic field and the precessional frequency of the nucleus become equal, they are said to be in resonance, and absorption or emission of energy by the nucleus can occur.

The transition from one spin state to the other is called *flipping* of the precessing nucleus. The energy involved in this transition is about 10^{-6} kcal/mole. The energy required for resonance depends on the strength of the applied external magnetic field and on the isotope brought into resonance (Eq. (5.3)). A frequency of 60 MHz is needed at a magnetic field H_0 of 14,092 gauss for ^{1}H nuclei

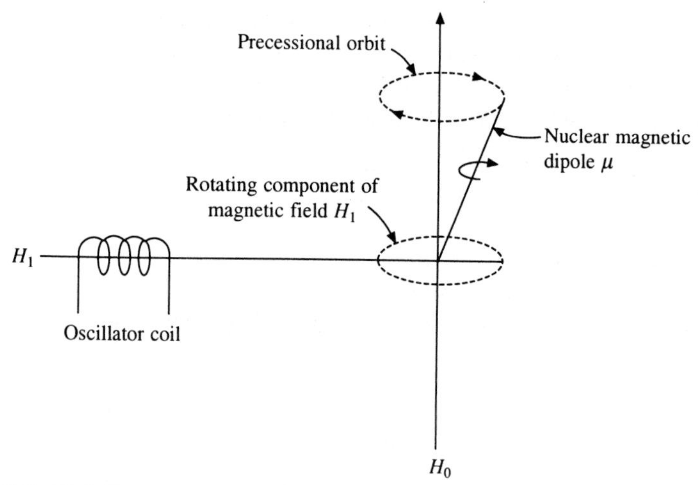

Fig. 5.6 Rotating component of magnetic field H_1 generated by an oscillator

(protons) to bring them into resonance (or any other desired combination in the same ratio; this comes from Eq. (5.4). At the same field strength, an electromagnetic radiation of frequency 15 MHz brings ^{13}C nuclei into resonance. These frequencies are in the radio-frequency region of the electromagnetic spectrum. A field strength of 14,092 gauss can be expressed as its equivalent 60 MHz. For flipping the nucleus to its higher energy level, most commonly the radio frequency (oscillator frequency) is kept constant and H_0 is varied, although this can also be done by varying the radio frequency and keeping H_0 constant.

According to the theory of electromagnetic radiation, the probability of absorption or emission of energy is equal, i.e. the probability of an upward and a downward transition is equal. Also, a spontaneous transition from a high energy state to a lower energy state is negligible in the radio-frequency region because ΔE is very low (about 10^{-6} kcal/mole). Thus, for all practical purposes NMR is a ground-state phenomenon. Hence, if two possible spin states in a collection of nuclei were exactly equally populated, the probability of an upward transition (absorption) would be exactly equal to a downward transition (emission) and there would be no NMR effect. Under ordinary conditions in a magnetic field, however, there is slight excess of nuclei in the lower spin state (low energy orientation), i.e. the nuclei take up Boltzmann distribution (under ordinary conditions the Boltzmann factor is about 0.001%). It is this very small excess of nuclei in the lower energy state which gives rise to net absorption of energy in the radio-frequency region, i.e. the NMR phenomenon.

As the collection of nuclei continually absorbs radio-frequency radiation, the excess of nuclei originally in a lower energy state may diminish and so the intensity of the absorption signal may diminish or vanish entirely. When the population of nuclei between the two spin states become equal, there will be no NMR effect, such a phenomenon is known as *saturation*.

The radiationless transitions by which a nucleus in an upper spin state returns to a lower spin state are called *relaxation processes*. These maintain an excess of nuclei in a lower energy state which is the necessary condition for the observation of NMR phenomenon. The two types of relaxation processes are:

(i) Spin-spin (or transverse) relaxation
(ii) Spin-lattice (or longitudinal) relaxation

(a) Spin-spin (or Transverse) Relaxation

It involves mutual exchange of spins by two precessing nuclei in close proximity to each other. Each precessing nucleus is associated with a magnetic vector component rotating in a plane perpendicular to the main field. When two nuclei are in close proximity, this small rotating magnetic field is the same as is required to induce a transition in the neighbouring nucleus, i.e. the transfer of energy from one high energy nucleus to another. There is no net loss of energy and this relaxation process shortens the lifetime of an individual nucleus in the higher spin state but does not contribute to the maintenance of the required excess of nuclei in a lower spin state.

The spectral line width is inversely proportional to the lifetime of the excited state (i.e. higher energy state). The shorter the lifetime of the excited state greater is line width. An efficient relaxation process involves shorter time T_1 and results in broadening of the absorption peak. The spin-spin relaxation contributes to line broadening. Solids and very viscous liquids usually provide properly oriented nuclei in lower spin state which may exchange spins in higher spin states, hence spin-spin relaxation times are very short. Thus, the spin-spin relaxation causes line broadening of such a magnitude that NMR spectra of solids become of little interest to organic chemists.

(b) Spin-lattice (or Longitudinal) Relaxation

It involves the transfer of energy from the nucleus in its higher energy state to its environment, i.e. to the molecular lattice (framework of molecules). The molecular lattice contains precessing nuclei all of which are undergoing translational, rotational and vibrational motions. Hence, a variety of small magnetic fields is present in the lattice. A particular small magnetic field, properly oriented in the lattice, can induce transition in a particular precessing nucleus from an upper state to a lower state. The energy in this process is transferred to the components of the lattice as additional translational, rotational and vibrational energy. Thus, the temperature of the system rises slightly, that is why samples are heated up during recording of NMR spectrum. The total energy of the system remains unchanged. This relaxation process maintains an excess of nuclei in a lower state which is the necessary condition for the observation of NMR phenomenon.

The spin-lattice relaxation process also contributes to the width of a spectral line. In solids and viscous liquids, molecular motions are greatly restricted, so properly oriented nuclei which may effect spin-lattice relaxation are present relatively infrequently. Thus, most solids and viscous liquids exhibit very long spin lattice relaxation times. Relaxation times for most non-viscous liquids and solids in solution are of the order of one second; this gives rise to a natural line width of about 1 cps. Other factors, like the presence of paramagnetic molecules (e.g. O_2) or ions in the sample also cause the line broadening. Resonance signals for protons attached to an element that has an electric quadrupole moment* will frequently be broadened.

5.3 Instrumentation

The schematic diagram of a NMR spectrometer containing the following components is given in Fig. 5.7:

(i) *A strong magnet with homogeneous field*: The strength of its field can be varied continuously and precisely over a relatively narrow range with the help of the sweep generator.
(ii) A radio-frequency oscillator.
(iii) A radio-frequency receiver and detector.
(iv) A recorder, calibrator and integrator.
(v) *A sample holder*: It spins the sample to increase the homogeneity of the magnetic field on the sample, and keeps the sample in the proper position with respect to the main magnetic field, the radio frequency oscillator and receiver coils.

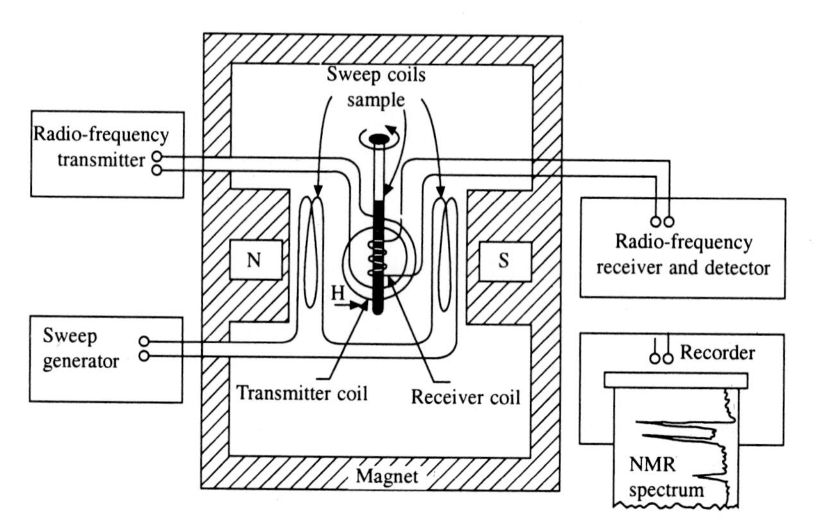

Fig. 5.7 Schematic diagram of a NMR spectrometer

The sample under investigation is taken in a glass tube and placed in the sample holder. Most commonly, NMR spectrometers irradiate the sample with a beam of constant radio frequency obtained from the radio-frequency oscillator, while the magnetic field strength is varied with the help of the sweep generator.

*Nuclei with a spin number ≥ 1 have a nonspherical charge distribution. This asymmetry is described by an electrical quadrupole moment.

ΔE varies as the H_0 varies (Eq. 5.3). At the field strength, when ΔE becomes equal to the energy of the incident radio frequency, absorption of energy takes place and transition from a lower spin state to a higher spin state occurs. This causes a tiny electric current to flow in the coil of the radio frequency receiver which is amplified and recorded as a signal on the chart paper by the recorder. A NMR spectrum is recorded as a plot of a series of peaks (signals) corresponding to different applied field strengths against their intensities. Each peak represents a set (a kind) of protons (in case of a PMR spectrum). The areas under the peaks (the intensities of the peaks) are directly proportional to the number of protons they represent. An electronic integrator traces a series of steps whose heights are proportional to the peak areas, i.e. the number of protons represented by that particular peak. A typical PMR spectrum is given in Fig. 5.8.

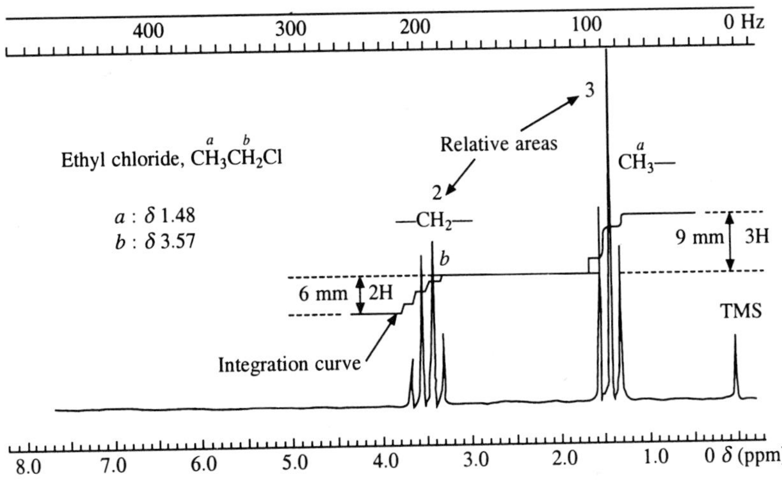

Fig. 5.8 PMR spectrum of ethyl chloride in $CDCl_3$ at 60 MHz

PMR spectra are usually run at 60 MHz (corresponding to the field of 14,092 gauss). Now high resolution instruments* which operate at 100 MHz (corresponding to the field of 23,486 gauss) or even higher (as high as 500 MHz) are available. By measuring frequency shifts from a reference marker (usually tetramethylsilane, TMS), an accuracy of ±1 Hz can be achieved.

5.4 Sample Handling

Ordinarily, about 0.4 ml of a neat liquid or 10-50 mg of a liquid or solid dissolved in 0.4 ml of a deuterated solvent is used. The sample is contained in a glass tube with 5 mm outside diameter and about 15 cm length. The ideal solvent should contain no protons, be inexpensive, low boiling, nonpolar and inert. Carbon tetrachloride is an ideal solvent if the compound under study is sufficiently soluble in it. Almost all of the common solvents are available in

*Instruments having ability to discriminate among the individual absorptions.

the deuterated form, e.g. deuterated chloroform CD_3Cl (chloroform-*d*), hexadeuteroacetone CD_3COCD_3 (acetone-d_6), hexadeuterodimethyl sulphoxide CD_3SOCD_3 (DMSO-d_6), hexadeuterobenzene C_6D_6 (benzene-d_6), D_2O etc. Note that deuterium (^2H or D) has a nuclear magnetic dipole and thus, it should exhibit a NMR signal in the spectrum. Since it does so only under different applied field strength and oscillator frequency, its NMR signal does not appear in the PMR spectra.

5.5 Shielding, Deshielding and Chemical Shift

Under the influence of the applied magnetic field, electrons surrounding a nucleus start to circulate perpendicular to the applied magnetic field H_0, and so they generate a secondary magnetic field called *induced magnetic field* (σH_0) which opposes the applied magnetic field in the region of the nucleus, e.g. proton (Fig. 5.9). Thus, the nucleus experiences a weaker magnetic field H_{eff} than the applied magnetic field H_0, and it is said to be *shielded*. This type of shielding is termed diamagnetic shielding and its effect is termed as *shielding effect*, i.e.

$$H_{eff} = H_0 - \sigma H_0$$

where σ is screening or shielding constant.

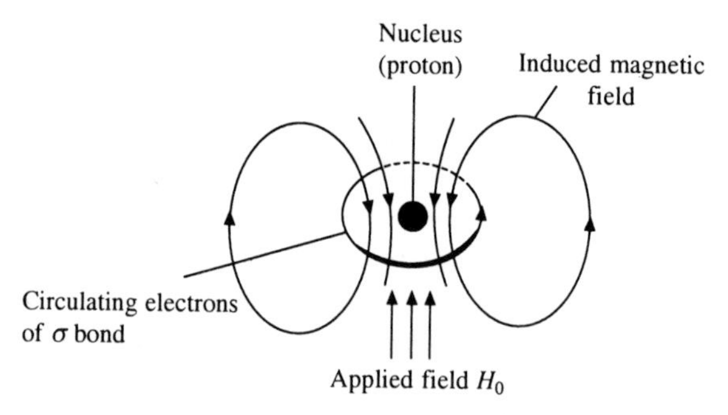

Fig. 5.9 Diamagnetic shielding of nucleus by circulating electrons

The magnitude of the induced field is directly proportional to the magnitude of the applied field H_0. The higher the electron density around the proton, the higher is the diamagnetic shielding. Circulation of electrons about nearby nuclei generates an induced magnetic field that can either oppose or reinforce the applied field at the proton, depending on its location in the induced magnetic field (see Section 5.7(ii)). If the induced field opposes the applied field in the region of proton, then the proton is *shielded* as mentioned above. If the induced field reinforces the applied field, then the field experienced by the proton is greater than the applied field. Such a proton is said to be deshielded and this effect is termed as *deshielding effect*. Compared to a naked proton, a shielded proton requires a higher applied field strength, whereas a deshielded proton requires a lower field strength for transition. Thus, shielding shifts the absorption position upfield, whereas deshielding shifts the absorption position downfield and these effects are termed as shielding and deshielding effects, respectively.

Such shifts in the NMR absorption positions are called *chemical shifts* because they arise from the circulation of electrons in chemical bonds. The chemical shift is expressed as the difference between the absorption position of a particular proton and the absorption position of a reference proton. Due to varying electronic environment of the proton or group of protons, their absorption signals appear at different field values. Thus, signals in PMR spectra give information about the different kinds of protons and their environments in molecules.

Why are the NMR absorption positions expressed relative to a reference compound?

The exact frequency values or field values cannot be calibrated with an accuracy of about ± 1 Hz out of about 60×10^6 Hz because the instrument required must be able to discriminate frequencies of the order of one part in 10^8. Thus, the absolute positions of absorptions cannot be obtained directly from the instrument as in UV and IR spectroscopy. However, relative proton frequencies can be determined with an accuracy of ± 1 Hz. For this reason, positions of absorption signals are always expressed relative to a reference compound (most commonly tetramethylsilane, TMS), i.e. as chemical shifts. For practical reasons, the signal from a naked proton is not used as the reference point.

Why is TMS a good reference compound in NMR spectroscopy?

Tetramethylsilane $(CH_3)_4Si$(TMS) is the most commonly used reference compound because of the following advantages:

(i) It is chemically inert, hence does not react with compounds under study.
(ii) It is volatile (b.p. 27°C), hence precious samples may be easily recovered after recording the spectra.
(iii) It gives a single, sharp and intense absorption peak because all its twelve protons are equivalent. Thus, very small quantity (1-2 drops) are needed.
(iv) Its protons absorb at higher field than that of almost all organic compounds, hence overlapping of signals does not occur. The protons of TMS are more shielded due to $+I$ effect of Si which increases electron density around them, hence they absorb at upfield.
(v) It is not involved in intermolecular association with the sample or solvent, hence the absorption position of its protons remain unchanged.
(vi) It is soluble in most of the organic liquids.

TMS is usually used as an internal reference. Since it is insoluble in water and D_2O, it cannot be used as internal reference with these solvents. However, it can be used as an external reference, i.e. it is sealed in a capillary and immersed in such solutions.

Sometimes the DSS (sodium 2,2-dimethyl-2-silapentane-5-sulphonate, $(CH_3)_3SiCH_2CH_2CH_2SO_3Na$), is used as an internal reference in aqueous solutions. The disadvantage of this reference compound is that it is nonvolatile and has absorptions other than CH_3Si. Any other water-soluble compound may be used as a standard in aqueous solutions, e.g. acetone, dioxane, *t*-butyl alcohol etc.

5.6 Measurement of Chemical Shift: NMR Scale

We can express the chemical shifts in terms of Hz by setting the TMS peak at 0 Hz at the right-hand edge. The magnetic field decreases towards left. When chemical shifts ν are given in Hz, the applied frequency must be specified (e.g. 60, 90, 100, 200 etc. MHz) because the chemical shift in Hz is directly proportional to the strength of the applied field H_0 and, therefore, to the applied frequency. The value of chemical shift ν in Hz is $\nu_s - \nu_r$, where ν_s and ν_r are the absorption frequencies of the sample and the reference in Hz, respectively.

Instruments with different field strengths (e.g. 60, 90, 100, 200 etc. MHz) are available, hence it is desirable that chemical shifts be expressed in some form independent of the field strength. The chemical shifts are commonly expressed in δ unit which is a proportionality and thus dimensionless. It is independent of the field strength. Chemical shift values in Hz, i.e. ν are converted into δ units as follows:

$$\delta\,(\text{or ppm}) = \frac{\text{Chemical shift in Hz}}{\text{Oscillator frequency in Hz}} \times 10^6$$

Oscillator frequency is characteristic of the instrument, e.g. a 60 MHz instrument has an oscillator frequency 60×10^6 Hz. The factor 10^6 is included in the above equation simply for convenience, i.e. to avoid fractional values. Since δ, which is dimensionless, is expressed in parts per million, expression ppm is often used.

Thus, a peak at 60 Hz (ν 60) from TMS at an applied frequency 60 MHz would be at δ 1.00 or 1.00 ppm

$$\delta\,(\text{or ppm}) = \frac{60}{60 \times 10^6} \times 10^6 = 1.00$$

The same peak at an applied frequency of 100 MHz would be at 100 Hz (ν 100) but would still be at δ 1.00 or 1.00 ppm

$$\delta\,(\text{or ppm}) = \frac{100}{100 \times 10^6} \times 10^6 = 1.00$$

The δ unit has been criticized because δ values increase in the downfield direction; the reply is that these are really negative numbers. In the other commonly used unit, a value of 10.00 is assigned to TMS peak. This unit expresses chemical shifts in τ values as

$$\tau = 10.00 - \delta$$

It should be noted that δ is treated as a positive number. Shifts at higher field than TMS are rare. If such shifts are present, their δ values are shown with a negative sign and τ values increase numerically, e.g. $\delta - 1.00$ will be equal to τ 11.00. Fig. 5.10 shows NMR scale at 60 MHz and 100 MHz.

Example 1. Protons of a compound exhibit a NMR signal at δ 2.5. What will be the value of chemical shift of these protons in Hz if the spectrum is recorded on a 60 MHz spectrometer?

Solution

$$\delta = \frac{\text{Chemical shift in Hz}}{\text{Oscillator frequency in Hz}} \times 10^6$$

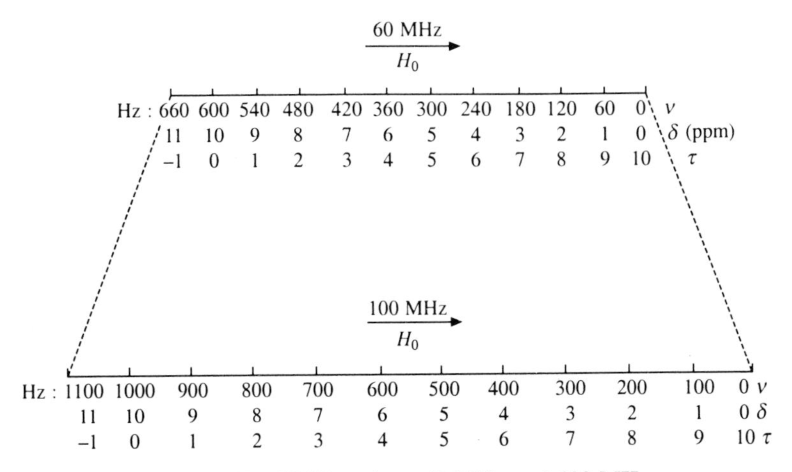

Fig. 5.10 NMR scale at 60 MHz and 100 MHz

Suppose the chemical shift in Hz is x. Therefore,

$$2.5 = \frac{x}{60 \times 10^6} \times 10^6$$

Hence $$x = 2.5 \times 60 = 150 \text{ Hz}$$

Example 2. If the observed chemical shift of a proton is 200 Hz from TMS and instrument frequency is 60 MHz, what is the chemical shift in terms of δ? Express it in τ value.

Solution

$$\delta = \frac{\text{Chemical shift in Hz}}{\text{Oscillator frequency in Hz}} \times 10^6 = \frac{200}{60 \times 10^6} \times 10^6 = 3.33$$

$$\tau = 10.00 - \delta = 10.00 - 3.33 = 6.67$$

The approximate chemical shift ranges of important chemical classes are given in Chart 5.1.

5.7 Factors Affecting Chemical Shift

Any factor which is responsible for shielding or deshielding of a proton will affect its chemical shift. The following factors affect the chemical shift:

(i) Electronegativity-inductive effect
(ii) Anisotropic effects
(iii) Hydrogen bonding
(iv) van der Waals deshielding

(i) Electronegativity-Inductive Effect

The degree of shielding depends on the electron density around the proton. The higher the electron density around a proton, the higher the shielding and higher

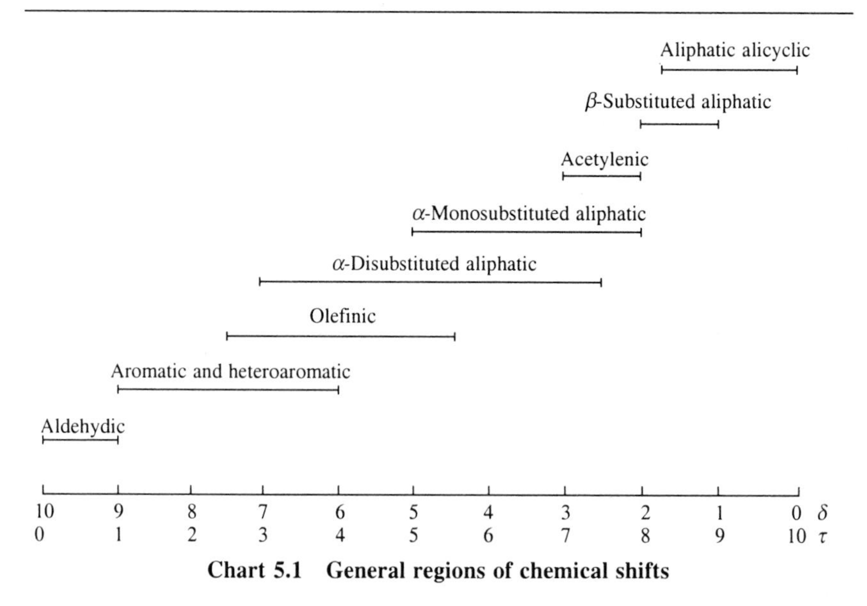

Chart 5.1 General regions of chemical shifts

is the field (lower the δ value) at which the proton absorbs. Thus, the electron density around a proton successfully correlates with its chemical shift.

A nearby electronegative atom withdraws electron density (due to $-I$ effect) from the neighborhood of the proton, so the NMR signal of such deshielded proton (the proton surrounded by less electron density) will appear downfield (higher δ value). Thus, the greater the electronegativiy of the atom, the greater is the deshielding of the proton. For example, the chemical shifts (in δ unit) of protons of methyl halides (CH_3F, CH_3Cl, CH_3Br, CH_3I : 4.26, 3.05, 2.68, 2.16, respectively) are in accordance with the electronegativity of the halogen attached to the methyl group, i.e. the greater the electronegativity of the halogen attached to the methyl group, the lower is the field (higher δ values) at which the PMR signal of its protons appears. Similarly, the chemical shifts of protons of the methyl group attached to carbon, nitrogen and oxygen (e.g. H_3C—, H_3C—N<, CH_3O—: $\delta \simeq 0.9$, $\simeq 2.2$, $\simeq 3.5$, respectively) are understandable in view of the electronegativity of C, N and O. The NMR signal of a proton appears at a lower field (higher δ value) as the number of electronegative atoms or groups attached to the carbon containing the proton increases. This is because of increasing deshielding of the concerned proton. For example, the chemical shifts of CH_4, CH_3Cl, CH_2Cl_2 and $CHCl_3$ protons are δ 0.33, 3.05, 5.28 and 7.24, respectively.

As the distance from the electronegative atom increases, its deshielding effect on the proton decreases, and thus the proton signal appears at a relatively higher field (lower δ value). For example, protons of the methyl groups in CH_3Cl absorb at δ 3.05, whereas the protons of the methyl group in CH_3CH_2Cl absorb at δ 1.48.

(ii) Anisotropic Effects

As discussed above, electronegativity correlates with chemical shifts. However, in some cases, e.g. in acetylenic, olefinic, aldehydic and aromatic protons, chemical shifts cannot be explained only on the basis of electronegativity. The carbon atom in acetylene is more electronegative than that in ethylene but the acetylenic protons are more shielded than the ethylenic protons, thus acetylenic protons absorb at $\delta 2.35$, whereas ethylenic absorbs at $\delta 4.60$. Such anomalies are explained on the basis of anisotropic (direction dependent) effects produced by circulation of π electrons under the influence of the applied magnetic field. These effects depend on the orientation of the molecule with respect to the applied field. Anisotropic effects are in addition to the induced magnetic field generated by the circulation of σ electrons. Generally, the induced magnetic field generated by circulating π electrons is stronger than that generated by σ electrons.

(a) Acetylenic Protons

Acetylene is a linear molecule. A small fraction of rapidly tumbling (moving disorderly) molecules are aligned parallel to the applied magnetic field. Hence, the electronic circulation within the cylindrical π electron could generate an induced magnetic field which opposes the applied field at the acetylenic proton (Fig. 5.11).

Thus, the acetylenic proton is additionally shielded and its signal moves higher field than expected from the electronegativity of the acetylenic carbons.

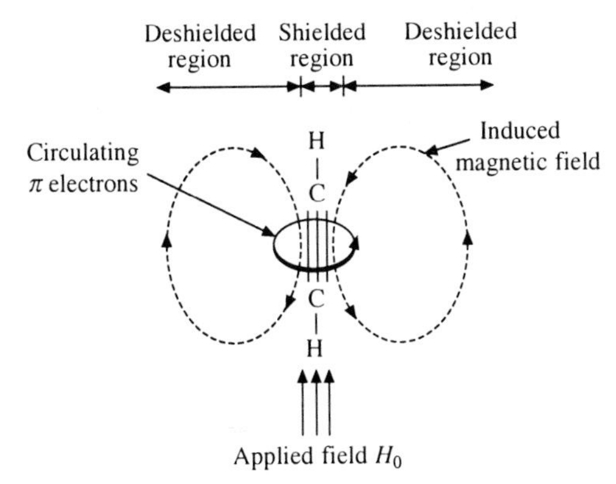

Fig. 5.11 Shielding of acetylenic protons, the molecule aligned parallel to the applied field H_0

When acetylene molecules are aligned perpendicular to the applied field. the acetylenic proton falls in the deshielded region and it is deshielded (Fig. 5.12). The magnitude of this deshielding is far less than that of the shielding (Fig. 5.11) because electrons are much more free to circulate in the direction shown in Fig. 5.11 than the direction shown in Fig. 5.12.

This is understandable in the light of the fact that π electrons of the triple bond are symmetrical about the bond axis, and the circulation as shown in Fig. 5.12 will disturb the symmetry. Although only a small fraction of tumbling molecules are aligned parallel to the applied magnetic field, the overall average chemical shift is affected by them, i.e. the acetylenic protons are much more

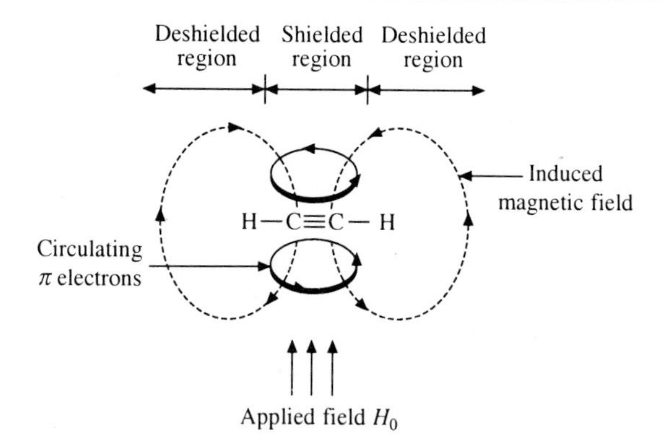

Fig. 5.12 Deshielding of acetylenic protons, the molecule aligned perpendicular to the applied field H_0

shielded than expected from the electronegativity of the acetylenic carbons and they absorb at higher field.

(b) Olefinic Protons

When an alkene molecule is oriented perpendicular to the applied magnetic field H_0, the induced magnetic field generated by circulating π electrons has the same direction at the olefinic protons as the applied magnetic field (Fig. 5.13).

Thus, the induced magnetic field reinforces the applied field resulting in deshielding of the olefinic protons. Consequently, they absorb at lower field than expected from the electronegativity of olefinic carbons.

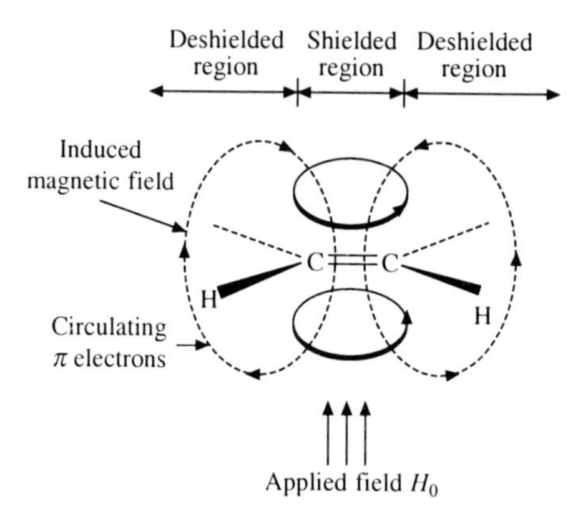

Fig. 5.13 Deshielding of olefinic protons

(c) Aldehydic Protons

When an aldehydic group is oriented perpendicular to the applied magnetic field H_0, the circulation of π electrons generates an induced magnetic field which reinforces the applied magnetic field at the aldehydic proton (Fig. 5.14) resulting in its deshielding similar to that of olefinic protons. Thus, aldehydic protons absorb at much lower field (δ ~9.5) due to the combined effects of the high electronegativity of oxygen and anisotropic effects produced by the π electrons of the carbonyl group.

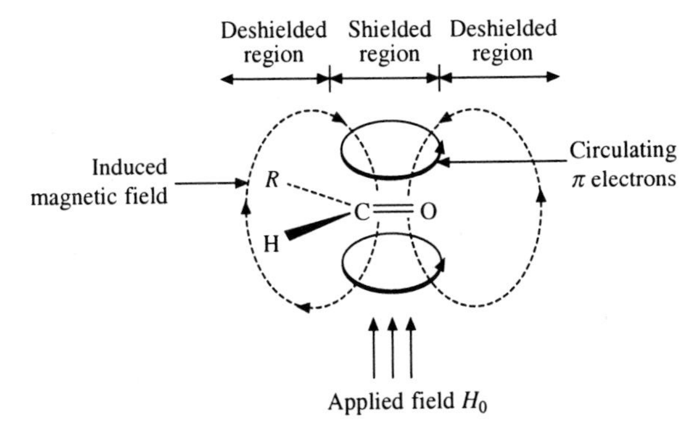

Fig. 5.14 Deshielding of aldehydic protons

(d) Aromatic Protons

Aromatic rings contain cyclic electron clouds of delocalized $4n + 2\pi$ electrons (Hückel rule). When a magnetic field is applied perpendicular to the plane of the aromatic ring, circulation of π electrons produces a ring current which induces a magnetic field perpendicular to the plane of the ring. This induced field is in the same direction as the applied field outside the ring but inside the ring it opposes the applied field (Fig. 5.15). Thus, aromatic protons, e.g. the protons of benzene, are highly deshielded, and hence appear at lower field. This is called the *ring-current effect* and is used as a very strong evidence for aromaticity.

Fig. 5.15 shows that a proton held above or below the plane of the aromatic ring should be shielded due to the ring current effect. This has actually been found to be the case for some of the methylene protons in 1,4-polymethylenebenzenes, e.g. [10]-paracyclophane.

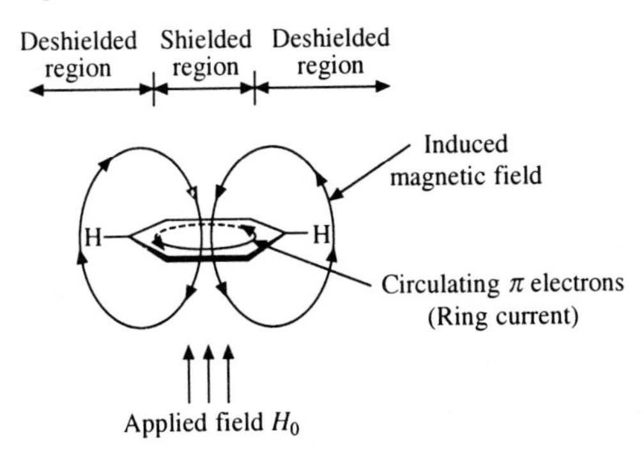

Fig. 5.15 Deshielding of aromatic protons (ring-current effects)

Some of the annulenes furnish interesting example of shielding and deshielding by ring currents. Protons outside the ring of [18]-annulene are strongly deshielded (δ 8.9), whereas those inside the ring are strongly shielded (δ –1.8).

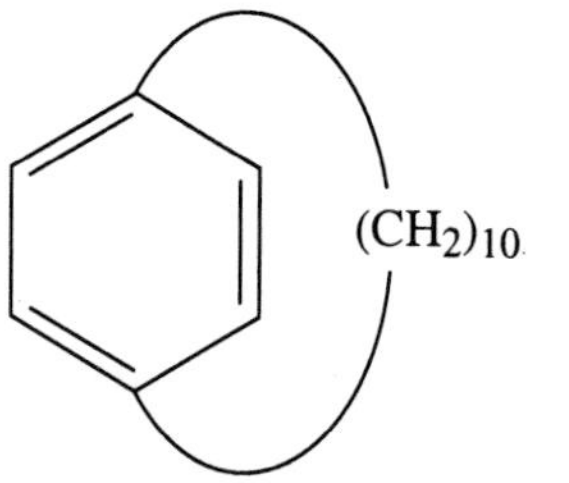

$(CH_2)_{10}$

[10] paracyclophane

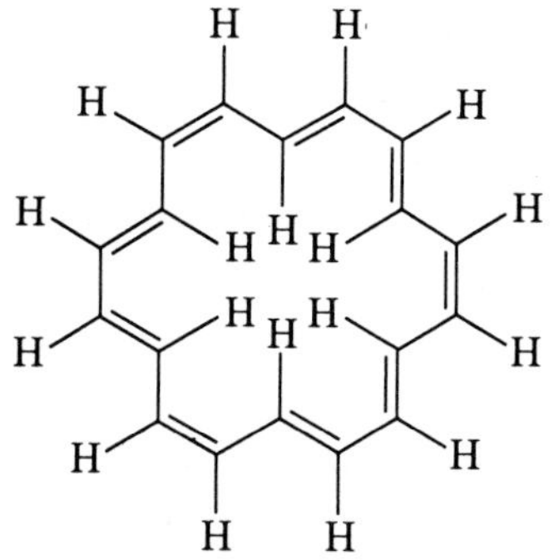

[18]-Annulene; 18π electrons

The shielding and deshielding resulting from aromatic ring currents are stronger than that resulting from the π electrons of olefinic bonds. Thus, olefinic protons absorb between δ 4.6 and 6.4, whereas aromatic protons absorb between δ 6.0 and 8.5.

From the above discussion, it is clear that the space around a double bond or an aromatic ring can be divided into shielding and deshielding regions (Fig. 5.16) and that protons present in these regions will absorb at a relatively high and low field, respectively. The boundary between shielding and deshielding regions resembles the surface of a double cone as shown in Fig. 5.16. For a carbonyl group, the situation is similar to alkenes (Fig. 5.16 (a)).

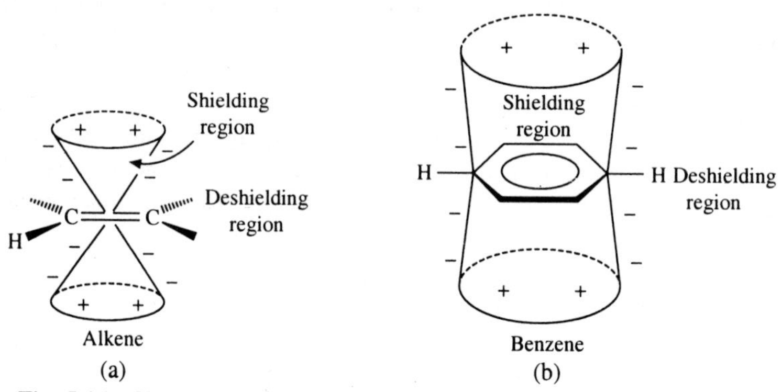

Fig. 5.16 Shielding (+) and deshielding (–) zones of π-bonded systems

The σ electrons of C—C bond also produce anisotropic effects but these are less powerful than that produced by circulating π electrons, and the axis of the C—C bond is the axis of the deshielding cone in the former (Fig. 5.17 (a)). This explains why the protons in the sequence RCH_3, R_2CH_2 and R_3CH appear progressively downfield. The tertiary proton (R_3CH) falls in the deshielding cones of three C—C bonds, secondary protons (R_2CH_2) in the deshielding cones of two C—C bonds and the primary protons (RCH_3) fall in the deshielding cone of only one C—C bond. Thus, the increasing order of their deshielding is $RCH_3 < R_2CH_2 < R_3CH$, i.e. the tertiary proton will absorb at the lowest and the primary at the highest field (Table 5.2). With the help of the deshielding cone (Fig. 5.17(b), it can also be explained why an equatorial proton of a conformationally rigid six-membered ring always appears downfield by 0.1-0.7 ppm than the axial proton on the same carbon.

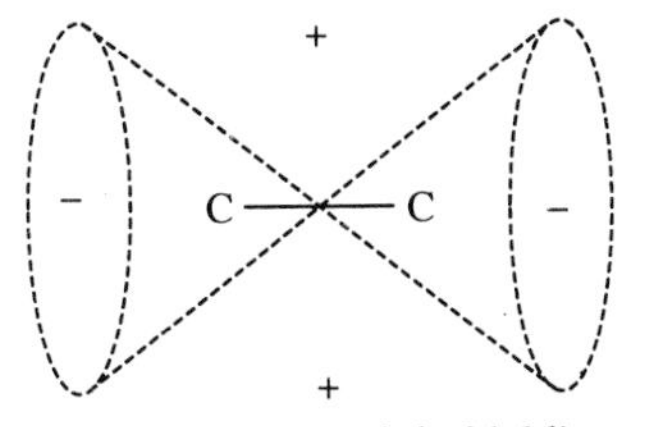

(a) Shielding (+) and deshielding (−) zones of C—C bond

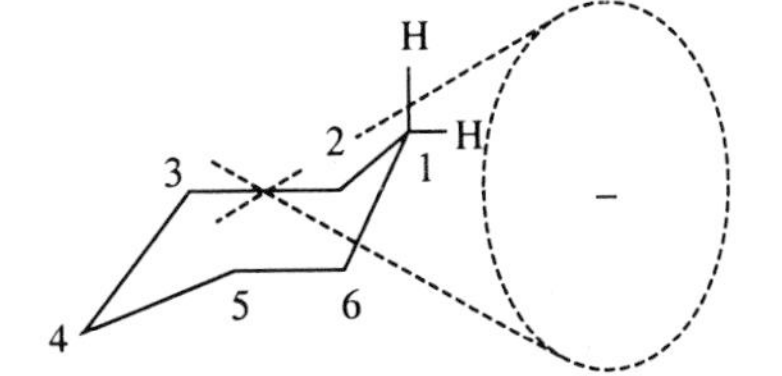

(b) Deshielding of equatorial proton of a rigid six-membered ring

Fig. 5.17 Shielding (+) and deshielding (−) zones of C—C bond

The axial and equatorial protons on C_1 are oriented similarly with respect of C_1—C_2 and C_1—C_6 bonds but the equatorial proton is within the deshielding cone of the C_2—C_3 bond (and C_5—C_6 bond).

It should be noted that anisotropic effects are field effects operating through space, whereas inductive effects operate through the chemical bonds.

(iii) Hydrogen Bonding

Hydrogen bonded protons absorb at a lower field than the non-hydrogen bonded protons. Due to high electronegativity of the atom to which the proton is hydrogen bonded, the electron density around it is decreased as compared to that around the non-hydrogen bonded proton. Thus, the hydrogen bonded protons are highly deshielded and absorb at a lower field than the non-hydrogen bonded protons. This downfield shift of the absorption depends on the strength of the hydrogen bonding. The stronger the hydrogen bonding, the lower will be the field at which the proton absorbs. Intermolecular and intramolecular hydrogen bondings can easily be distinguished by PMR spectroscopy because the latter show no shift in absorption position on changing the concentration of the sample, whereas the absorption position of the former is concentration-dependent. For example, the absorption position of the hydroxyl proton of ethanol is shifted to upfield on diluting the sample with a nonpolar solvent (e.g. carbon tetrachloride) due to breaking of intermolecular hydrogen bonds. Since intramolecular hydrogen bonds are not broken on dilution, intramolecularly hydrogen bonded protons show almost no change in their absorption position on dilution.

Hydrogen bonding explains why and how the chemical shift of the hydroxylic proton depends on concentration, temperature and solvent.

(iv) van der Waals Deshielding

In crowded molecules, some protons may occupy sterically hindered position resulting in van der Waals repulsion. In such a case, electron cloud of a bulky group (hindering group) will tend to repel the electron cloud surrounding the proton. Thus, the proton will be deshielded and will absorb at slightly lower field than expected in the absence of this effect. For example, the proton $\overset{a}{H}$ in a conformationally rigid cyclohexanone chair system (I) present in a steroid skeleton will resonate at lower field when $R = CH_3$ than when $R = H$.

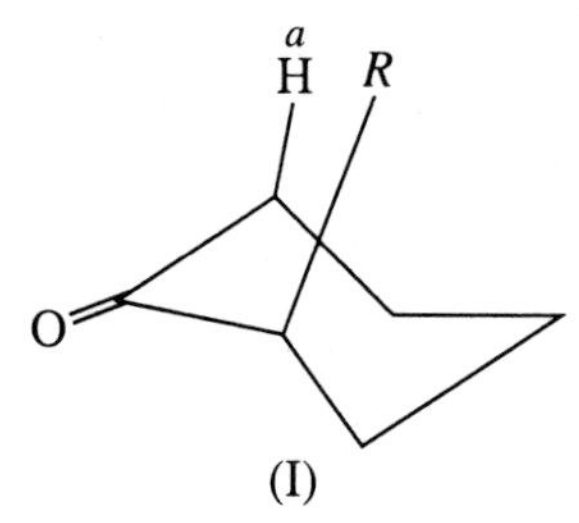

(I)

Chemical shifts of protons in various structural environments are given in Table 5.2. It should be noted that in otherwise equivalent environments, the order of δ values of methyl, methylene and methyne protons is as follows:

$$-\overset{|}{\underset{|}{C}}-H > -\overset{|}{C}H_2 > -CH_3$$

Table 5.2 Chemical shifts of protons in various structural environments

	Type of proton*	Chemical shift, δ (ppm)
Primary	$R\mathbf{CH_3}$	0.9
Secondary	$R_2\mathbf{CH_2}$	1.3
Tertiary	$R_3\mathbf{CH}$	1.5
Vinylic	$C{=}\mathbf{CH}$	4.6–5.9
Acetylenic	$C{\equiv}C{-}\mathbf{H}$	1.8–3.1
Allylic	$C{=}C{-}\mathbf{CH_3}$	1.7
Aromatic	$Ar{-}\mathbf{H}$	6–8.5
Benzylic	$Ar{-}C{-}\mathbf{H}$	2.2–3
Alcohols	$\mathbf{HC}{-}OH$	3.4–4
Ethers	$\mathbf{HC}{-}OR$	3.3–4
Esters	$RCOO{-}\mathbf{CH}$	3.7–4.1
Esters	$\mathbf{HC}{-}COOR$	2–2.2
Acids	$\mathbf{HC}{-}COOH$	2–2.6
Carbonyl compounds	$\mathbf{HC}{-}C{=}O$	2–2.7
Aldehydic	$RC\mathbf{H}O$	9–10
Alcoholic	$RO\mathbf{H}$	1–5.5
Phenolic	$ArO\mathbf{H}$	4–12
Enolic	$C{=}C{-}O\mathbf{H}$	15–17
Carboxylic	$RCOO\mathbf{H}$	10.5–12
Amino	$RN\mathbf{H_2}$, $ArN\mathbf{H_2}$	1–5
Thiols	$RS\mathbf{H}$	1.1–1.5
Thiophenols	$ArS\mathbf{H}$	3–4
Amine salts	$R_3\overset{+}{N}\mathbf{H}$	7.1–7.7
	$Ar\overset{+}{N}\mathbf{H_3}$	8.5–9.5
Amines	$\mathbf{HC}{-}NR_2$	2.1–3
Thioethers	$\mathbf{HC}{-}SR$	2.1–2.8
Fluorides	$\mathbf{HC}{-}F$	4–4.5
Chlorides	$\mathbf{HC}{-}Cl$	3–4
Bromides	$\mathbf{HC}{-}Br$	2.5–4
Iodides	$\mathbf{HC}{-}I$	2–4

*Indicated in bold.

5.8 Number of PMR Signals: Equivalent and Nonequivalent Protons

The number of signals in a PMR spectrum shows how many kinds of protons are present in a molecule. This is because protons with the same chemical environment absorb at the same field strength, whereas protons with different chemical environments absorb at different field strengths. The protons with the same chemical environment are said to be chemically equivalent. Chemically equivalent protons occupy chemically equivalent positions, i.e. they are in identical chemical environments. Chemically equivalent protons are chemical shift equivalent, i.e. they have the same chemical shift.

How can the chemical equivalence of protons be judged?

The simple method to judge the chemical equivalence of two or more protons, is to mentally replace each proton in turn by some other atom Z. If the replacement results in only one product or enantiomeric products, then the protons are chemically equivalent. We ignore conformers in judging the identity of products. For example, on replacement of a methyl proton by Z, ethyl chloride would give $CH_2Z—CH_2Cl$, whereas on replacement of a methylene proton it would give $CH_3—CH_2ZCl$. These are different products, hence we easily judge that the methyl and the methylene protons are not equivalent. When we replace any one of the methyl protons by Z, the same product $CH_2Z—CH_2Cl$ is obtained, hence all the three methyl protons are equivalent. Thus, we expect only one PMR signal for the three methyl protons and that is also the case.

Replacement of the either of the two methylene protons of ethyl chloride by Z gives enantiomeric products (a pair of enantiomers):

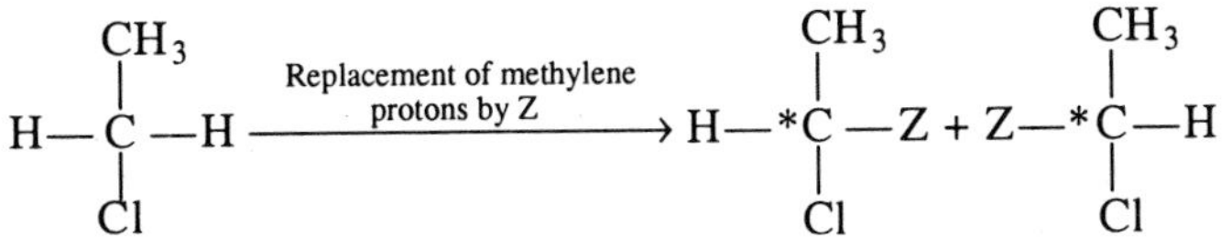

$$\text{H—}\underset{\text{Cl}}{\overset{\text{CH}_3}{\text{C}}}\text{—H} \xrightarrow[\text{protons by Z}]{\text{Replacement of methylene}} \text{H—*}\underset{\text{Cl}}{\overset{\text{CH}_3}{\text{C}}}\text{—Z} + \text{Z—*}\underset{\text{Cl}}{\overset{\text{CH}_3}{\text{C}}}\text{—H}$$

Enantiotopic protons
*C is chiral center

A pair of enantiomers

Such a pair of protons whose replacement gives a pair of enantiomers are called *enantiotopic protons*. These protons have the same chemical shift and exhibit only one PMR signal, i.e. these are equivalent protons.

On the other hand, a pair of protons whose replacement gives a pair of diastereomers are called *diastereotopic protons*. These protons do not have the same chemical shift and show different PMR signals, i.e. these are nonequivalent protons. For example, replacement of either of the two vinylic protons of 2-bromopropene by Z gives diastereomeric products (a pair of diastereomers, geometrical isomers):

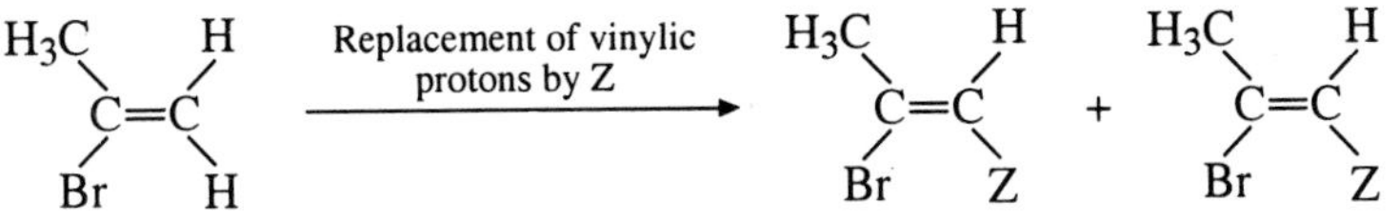

$$(H_3C)(Br)C{=}C(H)(H) \xrightarrow[\text{protons by Z}]{\text{Replacement of vinylic}} (H_3C)(Br)C{=}C(H)(Z) + (H_3C)(Br)C{=}C(Z)(H)$$

Diastereotopic protons
2-Bromopropene

A pair of diastereomers

Similarly, in 1,2-dichloropropane the two protons on C-1 are diastereotopic, hence are nonequivalent and show separate PMR signals.

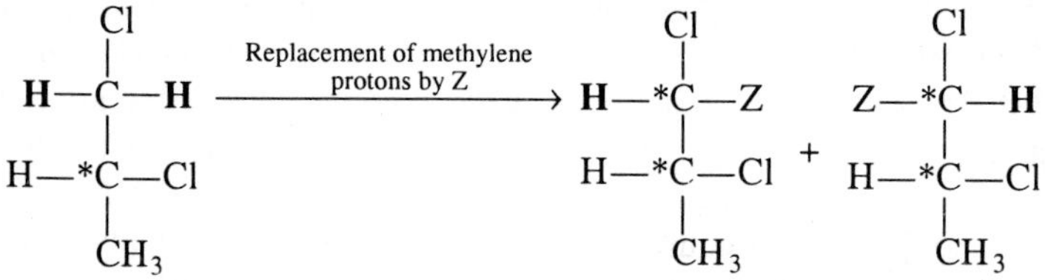

```
     Cl                                       Cl              Cl
     |         Replacement of methylene       |               |
  H—C—H              protons by Z          H—*C—Z          Z—*C—H
     |        ──────────────────────→         |       +       |
 H—*C—Cl                                  H—*C—Cl         H—*C—Cl
     |                                        |               |
    CH3                                      CH3             CH3
```

Diastereotopic protons
1,2-Dichloropropane

A pair of diastereomers

In view of the above discussion, now we are able to recognize various sets of equivalent protons (kinds of protons) and thus predict the number of PMR signals for molecules.

Example 1. Indicate the kinds of protons and number of PMR signals in the following compounds:

(a) $CH_3CH_2CH_2Cl$ (b) $CH_3CHClCH_3$ (c) CH_3COCH_3

Solution. (a) The compound has three kinds of protons labeled as *a*, *b* and *c*, hence it will exhibit 3 PMR signals.

$\overset{a}{CH_3}\overset{b}{CH_2}\overset{c}{CH_2}Cl$ Protons *c* are enantiotopic, hence equivalent
3 PMR signals
1-Chloropropane

(b) $\overset{a}{CH_3}\overset{b}{CH}Cl\overset{a}{CH_3}$
2 PMR signals
2-Chloropropane

(c) $\overset{a}{CH_3}CO\overset{a}{CH_3}$
1 PMR signal
Acetone

Example 2. Draw the structural formula of each of the following compounds and label all sets of equivalent protons. How many NMR signals do you expect from each of these compounds?

(a) *p*-xylene, (b) Vinyl chloride, (c) Cyclobutane,
(d) Diethyl ether, (e) Two isomers of $C_2H_4Cl_2$

Solution

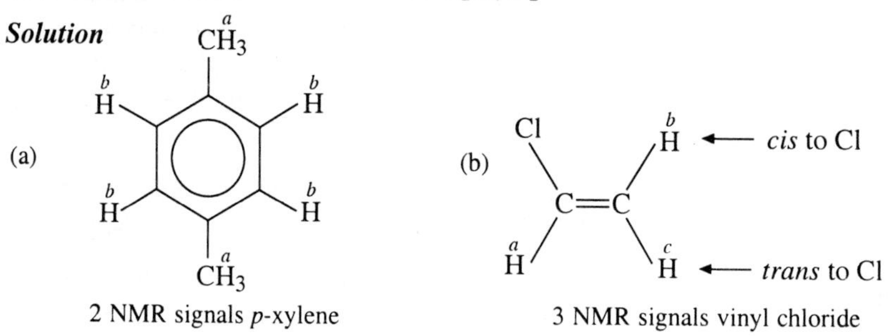

2 NMR signals *p*-xylene

3 NMR signals vinyl chloride

(c) $\overset{a}{C}H_2-\overset{a}{C}H_2$ / $\overset{a}{C}H_2-\overset{a}{C}H_2$ (ring)

1 NMR signal
Cyclobutane

(d) $\overset{a}{C}H_3\overset{b}{C}H_2O\overset{b}{C}H_2\overset{a}{C}H_3$

2 NMR signals
Diethyl either

(e) (i) $\overset{a}{H}_3C-\overset{a}{C}HCl_2$

2 NMR signals
1,1-Dichloroethane

(ii) $Cl\overset{a}{C}H_2-\overset{a}{C}H_2Cl$

1 NMR signal
1,2-Dichloroethane

Example 3. Three isomeric dimethylcyclopropanes *A*, *B* and *C* give 2, 3 and 4 NMR signals, respectively. Draw the stereochemical formulae for *A*, *B* and *C*.

Solution

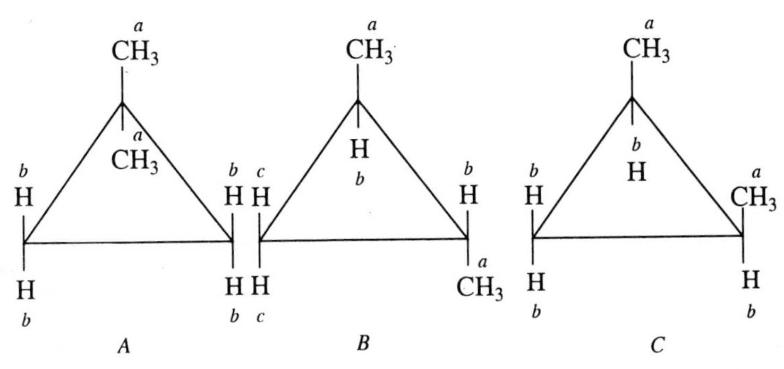

In the dimethylcyclopropane *C* the proton *c* is *cis* to CH_3 groups, whereas the proton *d* is *trans* to CH_3 groups. Hence these are nonequivalent.

Example 4. Draw the structural formula of each of the following compounds, label the kinds of protons and indicate the expected number of NMR signals.

(a) Methylcyclopropane
(b) Mesitylene
(c) Ethyl succinate
(d) 2,3-Dichloropropanoic acid

Solution

(a)

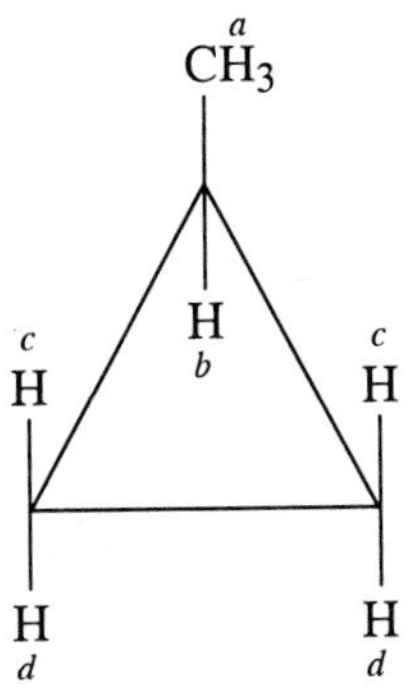

4 NMR signals
2-Methylcyclopropane

Protons *c* are *cis* to the methyl group and protons *d* are *trans* to it, hence they are diastereotopic (nonequivalent).

(b) Mesitylene: $\overset{a}{H_3C}$, $\overset{a}{CH_3}$, $\overset{a}{CH_3}$ on ring positions 1, 3, 5; $\overset{b}{H}$ on ring positions 2, 4, 6

2 NMR signals mesitylene

(c) $\overset{a}{CH_2}COO\overset{b}{CH_2}\overset{c}{CH_3}$

$\overset{a}{CH_2}COO\overset{b}{CH_2}\overset{c}{CH_3}$

3 NMR signals
Ethyl succinate

(d) $\overset{a}{COOH}$, $\overset{b}{H}-{}^{*}C-Cl$, $\overset{c}{H}-C-\overset{d}{H}$, Cl

4 NMR signals
2.3-Dichloropropanoic acid

C_2 is chiral. Protons $\overset{c}{H}$ and $\overset{d}{H}$ are diastereotopic, hence nonequivalent.

Strictly speaking, chemically equivalent protons must also be stereochemically equivalent. All the chemically equivalent protons are always chemical shift equivalent but the reverse is not always true. For example, enantiotopic protons are stereochemically nonequivalent (because they give enantiomers on replacement by some other atom or group), and thus they are also chemically nonequivalent (in strict sense), but they have the same chemical shift. This is because their environments are mirror images of each other and are not different enough for the PMR signals to be noticeably separated. Thus, enantiotopic protons behave as equivalent protons. It should be noted that PMR spectroscopy can neither distinguish between enantiotopic protons nor between enantiomers in achiral solvents. However, these can be distinguished by PMR spectroscopy in chiral solvents because chiral solvents interact differently with enantiotopic protons (or enantiomers) making them chemical shift nonequivalent.

In summary, identical protons (chemically equivalent protons in strict sense) are chemical shift equivalent in any environment, chiral or achiral. Enantiotopic protons are chemical shift equivalent only in achiral solvents. Diastereotopic protons are not chemical shift equivalent in any environment, i.e. chiral or achiral.

Magnetically equivalent protons have the same chemical shift and the same coupling constant J to every other nucleus in the spin system (see Section 5.14).

5.9 Peak Area and Proton Counting

In a PMR spectrum, various peaks (signals) represent different kinds of protons. The area under each peak (the intensity of the peak) is directly proportional to the number of protons causing that peak. Greater the number of protons which flip over at a particular frequency, greater is the energy absorbed and greater is the area under the absorption peak. For the determination of areas under peaks, modern NMR instruments are equipped with an electronic integrator which traces a series of steps at peaks whose heights are proportional to the areas of the respective peaks. By measuring the step heights, we arrive at a set of numbers which are in the same ratio as the numbers of different kinds of protons (Fig. 5.8). This set of numbers is converted into a set of smallest whole numbers just as it is done in calculating empirical formulae. The number of protons causing each signal is equal to the whole number for that signal or to some multiple of it. For example, the step heights in Fig. 5.8 are 6 mm and 9 mm; the corresponding set of smallest whole numbers will be 1 and 1.5, i.e. the ratio of protons of kinds b and a is:

$$6:9 = 1:1.5 = 2:3$$

If the molecular formula is known, for example, say C_2H_5Cl, then the number of b and a kinds of protons will be 2 and 3, respectively.

Alternatively, the number of each kind of protons can be counted as follows if the molecular formula C_2H_5Cl is known:

Because 6 + 9 = 15 mm is equal to 5H

Hence, $$1 \text{ mm} = \frac{5\text{H}}{15} = 0.333\text{H}$$

Thus, the number of protons of kind $a = 9 \times 0.333 = 2.997 = 3\text{H}$

$$b = 6 \times 0.333 = 1.998 = 2\text{H}$$

In either way we find a = 3H and b = 2H.

Example 5. A compound shows three signals a, b and c in its PMR spectrum, The heights of integration curve at these signals are 8.8, 2.9 and 3.8 units, respectively. If the molecular formula of the compound is $C_{11}H_{16}$ then count each kind of proton in it.

Solution. Because 8.8 + 2.9 + 3.8 = 15.5 units are equal to 16H

Hence, $$1 \text{ unit} = \frac{16\text{H}}{15.5} = 1.03\text{H}$$

Thus, the number of protons of kind

$$a = 1.03 \times 8.8 = 9.1$$

$$b = 1.03 \times 2.9 = 3.0$$

$$c = 1.03 \times 3.8 = 3.9$$

that is, the number of each kind of proton is

$$a = 9\text{H}, \quad b = 3\text{H}, \quad c = 4\text{H}$$

5.10 Spin-Spin Splitting: Spin-Spin Coupling

We have already studied that the number of signals in a PMR spectrum is equal to the number of kinds of protons present in the molecule. It has been found that only in some cases one kind of proton is represented by a single peak, e.g. *p*-xylene has two kinds of protons (aromatic and methyl) and shows two PMR signals consisting of single peaks, i.e. two singlets. On the other hand, in most of the cases, instead of a single peak (singlet) for one kind of protons, a group of peaks (a multiplet) is observed in the PMR spectrum. This is called the *splitting of NMR* signals or the *spin-spin splitting*. For example, CH_3CH_2Cl has two kinds of protons and shows two signals, one of which (due to CH_3) is split into three peaks (a triplet) and the other (due to CH_2) into four peaks (a quartet) as shown in Fig. 5.8. Now let us study the origin of a multiplet (a group of peaks) for a particular kind of proton, i.e. the splitting of an NMR signal.

The splitting of NMR signals is caused by spin-spin coupling which is indirect coupling of proton spins through the intervening bonding electrons. The field experienced by the proton is slightly increased if the neighboring proton (the proton on adjacent carbon or other atoms, i.e. the vicinal proton) is aligned with the applied field; or decreased if the vicinal proton is aligned against the applied field. The absorbing proton thus may experience each of the modified fields and its absorption is shifted up and downfields, and thus the signal is split into a group of peaks (a multiplet).

The nature of the instantaneous spin state is transmitted from one proton to another through the bonding electrons. In a given covalent bond, the net electronic spin magnetic moment is zero because the electron spins are paired. But a nuclear magnetic moment induces a small magnetic polarization of the nearest bonding electrons which in turn induces magnetic polarization of the bonding electrons, and so on through the next proton. Thus, the instantaneous spin orientation of one proton is transmitted to another. Coupling is generally not observable beyond three bonds unless there is ring strain as in small rings, or bridged systems, or bond delocalization as in aromatic and unsaturated systems. Intermolecular spin-spin coupling is not observed.

Possible spin orientations (alignments) of the methine (—CH—), methylene (—CH_2—) and methyl (—CH_3) protons are shown in Fig. 5.18. The two spin orientations of the methine proton shall affect the absorption position of the vicinal protons in two ways, and thus the signal of the latter is split into two peaks (a doublet) with the intensity ratio 1 : 1 because the probability of the two spin orientations is equal.

There are three different spin alignments possible (Fig. 5.18, (i)-(iii)) for the methylene protons which shall affect the absorption of the vicinal protons in three ways resulting in the splitting of the signal of the latter into three peaks (a triplet) with the intensity ratio 1 : 2 : 1. The middle peak of the triplet has twice the intensity of the side peaks because it arises due to two spin orientations (Fig. 5.18, (ii)) equivalent in energy.

Similarly, there are four different spin orientations possible (Fig. 5.8, (i)-(iv)) for the methyl protons which shall affect the absorption of the vicinal protons in four ways and split their signal into four peaks (a quartet) with the intensity ratio 1 : 3 : 3 : 1. Each of the middle two peaks of the quartet has thrice the intensity of each of the two outermost side peaks because it arises due to three orientations (Fig. 5.18, (ii) and (iii)) equivalent in energy. The relative intensities of component peaks in a multiplet are directly proportional to the number of nuclear spin orientations of equivalent energy causing different energy levels (Fig. 5.18).

Example 8. Draw the structure of each of the following compounds which meets the given requirements in its PMR spectrum:

(i) $C_3H_3Cl_5$; one doublet and one triplet
(ii) $C_4H_{10}O$; one singlet, one doublet and one septet
(iii) $C_4H_8O_2$; one singlet, one triplet and one quartet
(iv) C_3H_7Cl; one doublet and one septet
(v) $C_4H_8Cl_2O$, two triplets

Solution

(i) $C_3H_3Cl_5$

The compound shows one doublet and one triplet. This indicates the presence of —CH—CH_2— group in the molecule. Thus, following structure with molecular formula $C_3H_3Cl_5$ fulfils this condition:

$$Cl_3C\overset{a}{C}H_2\overset{b}{C}HCl_2$$

a (doublet) and *b* (triplet).

(ii) $C_4H_{10}O$

The presence of a doublet and a septet indicates $(CH_3)_2CH$— (isopropyl) group. There is one singlet in the spectrum which shows that a group of protons have no vicinal proton in the molecule. Thus, following structure with the above molecular formula meets the given requirements:

$$(\overset{a}{H_3C})_2\overset{b}{C}H-O-\overset{a}{C}H_3$$

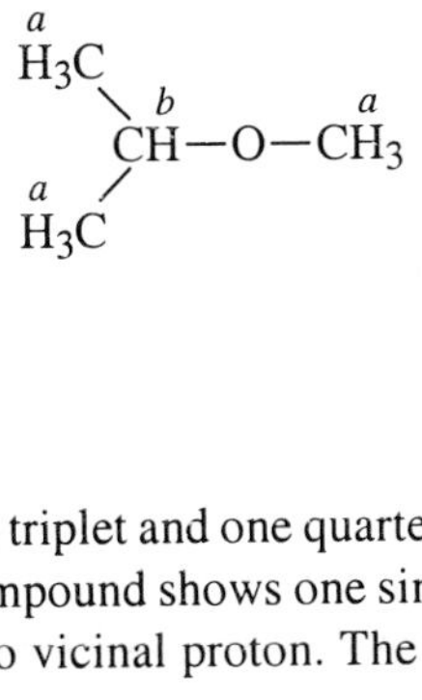

a (doublet) and *b* (septet).

(iii) $C_4H_8O_2$

The compound shows one triplet and one quartet which indicates the presence of —CH_2CH_3 group. The compound shows one singlet indicating that a group of three protons (—CH_3) has no vicinal proton. The following structures meet the above requirements:

$$\overset{a}{C}H_3COO\overset{b}{C}H_2\overset{c}{C}H_3 \quad \text{and} \quad \overset{c}{C}H_3\overset{b}{C}H_2COO\overset{a}{C}H_3$$

a (singlet), *b* (quartet), *c* (triplet).

(iv) C_3H_7Cl

The presence of a doublet and a septet indicate $(CH_3)_2CH$— group. Thus, the structure which meets the given requirements is

$$\overset{a}{H_3C}\diagdown\overset{b}{CH}\diagup\overset{a}{H_3C}$$

a (doublet) and *b* (septet).

(v) $C_4H_8Cl_2O$

This compound shows two triplets which indicates the presence of —CH_2CH_2— group where both the —CH_2— groups are non-equivalent. Thus, the structure which fits the given requirements and the molecular formula is

$$Cl\overset{a}{CH_2}\overset{b}{CH_2}—O—\overset{b}{CH_2}\overset{a}{CH_2}Cl$$

5.11 Coupling Constant (*J*)

The distance between the centres of two adjacent peaks in a multiplet is called *coupling constant* or *spin-spin coupling constant J* (Fig. 5.20). The values of coupling constants *J* are always quoted in Hz or cps and never in δ (ppm) or τ values. The value of *J* remains constant in different applied magnetic fields or radio frequencies used, whereas the values of chemical shifts (in Hz) are directly proportional to the applied magnetic fields or radio frequencies. This difference between spin-spin splitting and chemical shift affords a method for distinguishing between them. If the spectrum of a compound is recorded at different applied magnetic fields, then the separation of signals (in Hz) due to chemical shift change, whereas separation of two adjacent peaks (in Hz) in a multiplet remains always constant. Thus, if the separation between adjacent peaks does not change, then they are component peaks of a multiplet. On the other hand, if the separation between the peaks changes on changing the applied field, then they represent different signals. The values of coupling constants *J* between protons generally lie between 0 and 20 Hz.

The separations of peaks *J* in two coupled multiplets are exactly the same, i.e. spin-spin coupling is a reciprocal affair. For example, in the PMR spectrum of 1,1,2-trichloroethane (Fig. 5.20) two multiplets (one doublet and one triplet) are observed. The value of J_{ab} (6 Hz) in the doublet is exactly the same as in the triplet, where J_{ab} is the coupling constant for protons *a* ($\overset{a}{H}$) split by proton *b* ($\overset{b}{H}$) or for $\overset{b}{H}$ split by $\overset{a}{H}$. Drawing of a splitting diagram (Fig. 5.20) permits us to identify identical spacings between component peaks in the multiplets.

(i) Factors Affecting *J*

Coupling constant is a measure of the effectiveness of spin-spin coupling. The value of *J* depends on the number, type and geometrical orientation of bonds separating the coupled nuclei. We have already noted that *J* is independent of the applied magnetic field because splitting arises due to instantaneous spin states of the neighboring protons and not due to flipping of the spin states. A

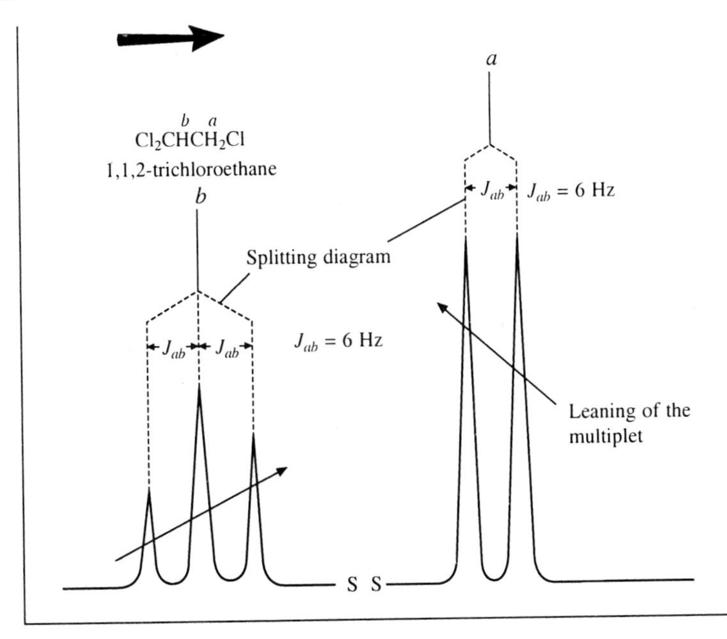

Fig. 5.20 Spin-spin splitting, coupling constant *J* leaning of the coupled multiplets towards one another and splitting diagram

coupling constant is designated as + or – to permit certain theoretical correlations, but the sign can be ignored except for calculations. The important factors which affect the magnitude of coupling constants in various types of couplings are discussed as follows.

(a) Geminal Coupling

Protons attached to the same carbon atom are called *geminal protons.* These are separated by two bonds, and when they are nonequivalent, they show spin-spin splitting. Geminal coupling constant J_{gem} is usually negative and increases algebraically on increasing the angle θ between the coupling protons ((II) and (III)). For example, in cyclohexane and cyclopentane rings, the angle θ is similar to the tetrahedral angle (θ = 109°) and J_{gem} is about –12 Hz which is comparable to that of acyclic saturated systems.

J_{ab} = 10-16 Hz

(II)

J_{ab} = 0-3 Hz

(III)

On decreasing the ring size to cyclopropane system, θ' is decreased with consequent increase in θ (II) which becomes > 109°. Thus, the J_{gem} of protons of methylene groups of a cyclopropane ring increases to about –3 Hz. In a terminal methylene group in which the carbon atoms are sp^2 hybridized and $\theta = 120°$ (III), J_{gem} further increases to zero or even becomes positive.

In a system RCH_2X, J_{gem} between CH_2 protons increases algebraically with increasing electronegativity of X. On the other hand, J_{gem} decreases algebraically with increasing electronegativity of a substituent attached to the carbon atom adjacent to the geminal protons in 1,1-dichlorocyclopropane. Such opposite effects of electronegativity illustrate that changes in J are not simply attributable to the direct inductive effect of the substituent.

(b) Vicinal Coupling

Protons attached to adjacent atoms are called *vicinal protons* (IV). These are separated by three bonds. Vicinal coupling constants J_{vic} which depend on the dihedral angle (angle of rotation) ϕ are largest when the angle ϕ is 0 or 180°, and

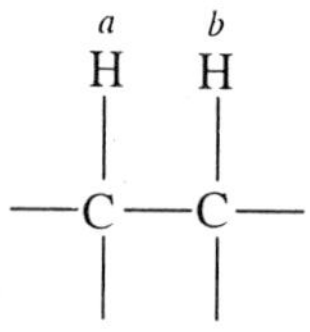

J_{ab} = ~6-8 Hz
(IV)

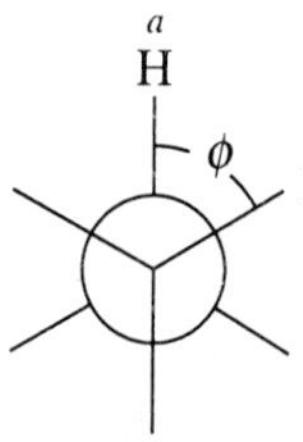

Gauche (ϕ = 60°)
(V)

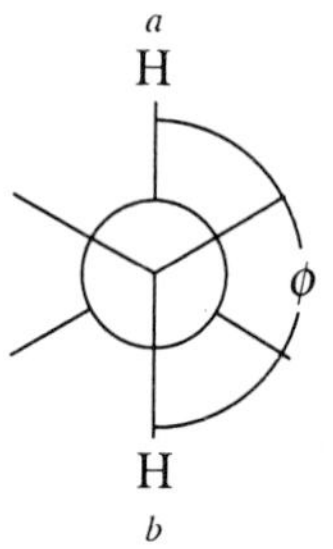

Anti (ϕ = 180°)
(VI)

has small negative value near 90°. For axial-axial protons in cyclohexanes, where dihedral angle is about 180° (VI), the J_{vic} is approximately 8 Hz, whereas for axial-equatorial and equatorial-equatorial protons, where dihedral angle is about 60° (V), the J_{vic} is about 2 Hz. The relationship between J_{vic} and dihedral angle ϕ is given approximately by theoretically derived Karplus equations as

$$J_{vic} = 10 \cos^2 \phi, \text{ for values of } \phi \text{ between 0 and } 90° \quad (5.5)$$

$$J_{vic} = 15 \cos^2 \phi, \text{ for values of } \phi \text{ between 90 and } 180° \quad (5.6)$$

Karplus relationships (Eqs. (5.5) and (5.6)) are very useful for determining the stereochemistry of organic compounds.

For isomeric olefins, J_{trans} is always greater than J_{cis}, and it is usually observed that $J_{cis} = \sim\frac{2}{3} J_{trans}$. Thus, it is possible to determine the configuration of geometrical isomers of a disubstituted olefin. For monosubstituted olefins (VII), $J_{trans} > J_{cis} > J_{gem}$. As an example, experimental data for styrene (VIII) are given as follows:

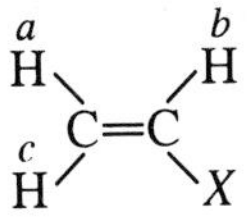

J_{ab} (*cis*) = 6-12 Hz
J_{bc} (*trans*) = 12-18 Hz
J_{ac} (gem) = 0-3 Hz
(VII)

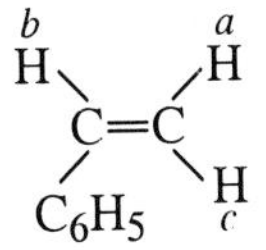

J_{ab} (*cis*) = 10.6 Hz
J_{bc} (*trans*) = 17.4 Hz
J_{ac}(gem) = −1.4 Hz
(VIII)

The J_{vic} decreases with increasing electronegativity of X in a freely rotating system —CH—CH—X.

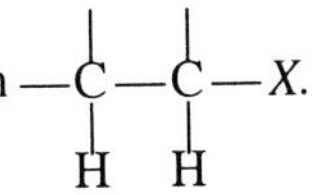

(c) Long-range Coupling

The magnitude of J decreases sharply with distance. It is about 1 Hz for coupling through four covalent bonds. In special cases, observable coupling through five covalent bonds has been reported. Coupling between protons separated by more than three bonds may occur in olefins, acetylenes, aromatics, heteroaromatics, and strained ring systems (small or bridged rings). Such proton-proton couplings beyond three bonds are called *long-range couplings*. Some appreciable long-range couplings are as follows:

(1) Allylic coupling. Allylic coupling constants are about 0 to 3 Hz (IX). In conjugated polyacetylenic chains, coupling may occur through as many as nine bonds.

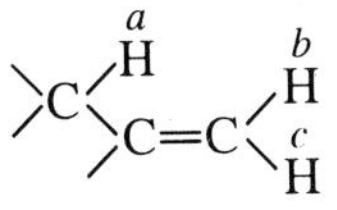

J_{ab} = 0-3 Hz
(IX)

(2) Homoallylic coupling. As might be expected, homoallylic coupling (H—C—C=C—C—H) constants are usually very small (about 0-2 Hz).

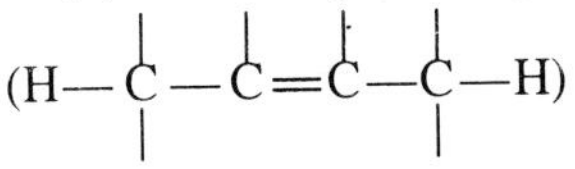

(3) Aromatic coupling. *Meta* coupling in benzene ring is 1-3 Hz, and *para* 0-1 Hz. *Ortho* coupling in benzene ring is 6-10 Hz. It should be noted that the *ortho* coupling is not a long-range coupling because here the coupled protons are separated by only three bonds. Coupling constants in heteroaromatics assume similar values.

Proton spin-spin coupling constants of some common systems are given in Table 5.3.

Table 5.3. Proton spin-spin coupling constants

Type	J_{ab} (Hz)	Type	J_{ab} (Hz)
>C(H$_a$)(H$_b$) (geminal)	10–16	>C=C(H$_a$)(H$_b$)	0–3
—CH$_a$—CH$_b$— (free rotation)	6–8	H$_a$(C=C)H$_b$ (cis)	6–12
—CH$_a$—C—CH$_b$—	~0	H$_a$(C=C)H$_b$ (trans)	12–18
Cyclohexane (H$_{ax}$, H$_{eq}$)		H$_a$—C—C=C—H$_b$	0–3
ax–ax	8–10	—C(H$_a$)—C(H$_b$)=C<	4–10
ax–eq	2–3		
eq–eq	2–3		
		H$_a$—C—C=C—C—H$_b$	0–2
		—C=C(H$_a$)—C(H$_b$)=C—	10–13
		H$_a$—C—C≡C—H$_b$	2–3
		H$_a$—C—C≡C—C—H$_b$	2–3
		Benzene H$_a$, H$_b$: *ortho*	6–10
		meta	1–3
		para	0–1

5.12 Analysis (Interpretation) of NMR Spectra

To obtain structurally useful information, we must analyze the NMR spectrum and correlate the NMR parameters with structure. The process of deriving the NMR parameters δ and J from multiplets is called *analysis of the NMR spectrum.*

(i) First Order Spectra

When the chemical shifts are large compared to the coupling constants ($\Delta\nu/J$* is greater than about 10), δ and J values may be measured directly from the spectrum, and spectra of this type are known as *first order spectra.* Nuclei (e.g. protons) which give rise to such spectra are said to be *weakly coupled.* First order spectra can usually be interpreted by using the following splitting rules (Sections 5.10 and 5.11) which are features of these spectra:

1. The number of component peaks in a multiplet is given by $n + 1$, where n is the number of equally coupled protons causing the splitting. The general formula which covers all the nuclei is $2nI + 1$, where n is the number of the coupling nuclei with spin I.
2. The relative intensities of the component peaks of a multiplet are given by coefficients of the terms in binomial expansion of $(x + 1)^n$ for nuclei with $I = \frac{1}{2}$.
3. Center of the multiplet gives the resonance position of the nucleus, and hence its chemical shift.
4. In the case of only two different groups of coupling nuclei, the separation between the component peaks of the multiplet are equal and correspond to the coupling constant.

A large number of PMR spectra are first order spectra, and can be analyzed by inspection and direct measurement in terms of the above rules. The chemical shift separation $\Delta\nu$ (in Hz) increases as the strength of applied magnetic field increases, but the value of J remains constant, and thus $\Delta\nu/J$ ratio is increased. Hence, a large proportion of PMR spectra become first order spectra at high applied magnetic fields. It should be noted that NMR instruments operating at high magnetic fields (i.e. at high radio frequencies) give better resolution and relatively easily interpretable spectra.

As an example of the analysis of first order NMR spectra, let us analyze the PMR spectrum of ethyl chloride shown in Fig. 5.8. The signal at δ 0.00 is due to the internal reference TMS. The downfield quartet

$$\overset{a}{C}H_3\overset{b}{C}H_2Cl$$

centered at δ 3.57 is due to methylene protons as they are more deshielded than the methyl protons by the chlorine. The methyl protons appear as a triplet centered at δ 1.48, i.e. upfield. The integration curve show relative peak areas of 2 : 3 corresponding to the number of protons causing the peak. The spacing between the component peaks of both the multiplets are equal and have the value ~9 Hz, i.e. J_{ab} = ~9 Hz. The chemical shift separation is 3.57–1.48 = 2.09 δ, i.e. $\Delta\nu$ = 2.09 × 60 = 125.4 Hz (the spectrum has been recorded at 60 MHz, hence δ 1 = 60 Hz). Thus $\Delta\nu/J$ is about 14, a large enough ratio for first order analysis. The system is A_3X_2 (Section 5.13). The leaning of the two coupled signals towards each other even at such a high $\Delta\nu/J$ ratio may be noticed, which shows that the

---

*$\Delta\nu$ is the difference in chemical shifts (in Hz) between two groups of coupled protons.

multiplets are not perfectly symmetrical, i.e. there is no exact intensity ratio of 1 : 2 : 1 for the triplet and 1 : 3 : 3 : 1 for the quartet. It should be remembered that such minor deviations from ideality are almost always apparent.

(ii) Second Order (More Complex) Spectra

When the chemical shifts of coupled protons are of approximately the same magnitude as the coupling constants, the NMR spectra cannot be analyzed by inspection and direct measurement in terms of the simple splitting rules summarized above. Such spectra are more complex and are known as *second order spectra.* Nuclei which give rise to second order spectra are said to be *strongly coupled.* Second order spectra may be recognized by the following features:

1. Often more lines are present than are predicted by $(2nI + 1)$ rule used for first order spectra.
2. Even in the case of only two different groups of coupling nuclei, the lines of a particular multiplet are not equally separated.
3. The relative intensities of the peaks of a multiplet are not given by coefficients of the terms in binomial expansion of $(x + 1)^n$.
4. The chemical shifts and coupling constant both cannot be measured directly from the spectrum. However, in certain cases one of these parameters may be obtained by inspection.

In second order spectra, mathematical analysis of the spectral data is required to obtain values for the chemical shifts and the coupling constants. In principle, any NMR spectrum, however complicated, can be analyzed by quantum mechanical calculations performed by a computer.

(a) Distortion of Multiplets

As $\Delta\nu/J$ becomes smaller the coupled multiplets approach one another, the inner peaks increase in intensity and the outer peaks decrease. Thus, there is a leaning of coupled multiplets towards one another (Fig. 5.20). Generally, a multiplet points upward towards the signal of the protons which causes the splitting. Now the center of the multiplet does not give the resonance position, i.e. the chemical shift of the protons(s) causing that multiplet. In such cases the chemical shift position is at the 'center of gravity'. Fig. 5.21 shows this type of distortion of two doublets. The chemical shift position can be calculated by the following formula, where the peak positions 1, 2, 3 and 4 from left to right (Fig. 5.22), are in Hz from reference

$$(1 - 3) = (2 - 4) = \sqrt{(\Delta\nu)^2 + J^2}$$

The chemical shift positions (in Hz) of each kind of protons are $\pm\frac{1}{2}\Delta\nu$ from the mid-point of the pattern.

5.13 Nomenclature of Spin Systems

A spin system, which is a group of coupled protons, may not include a whole molecule. For example, the ethyl protons in ethyl isopropyl ether constitute one spin system and the isopropyl protons another. By convention, protons of a spin

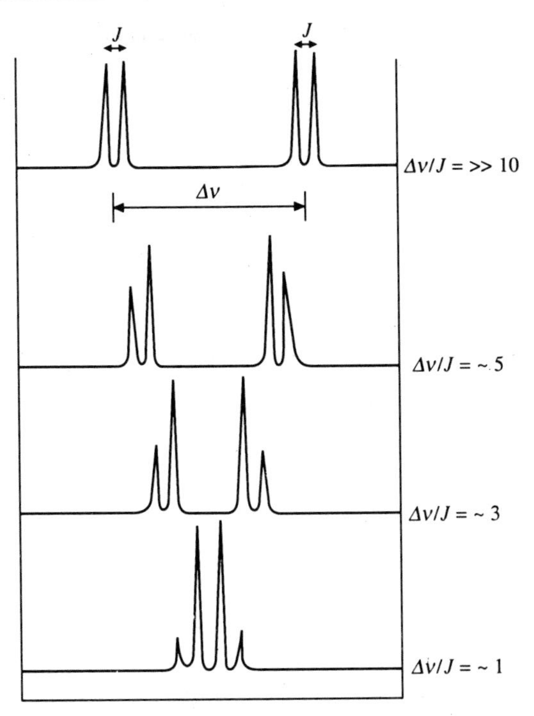

Fig. 5.21 **As $\Delta v/J$ becomes smaller, the doublets approach one another, the inner two peaks increase in intensity and the outer two peaks decrease**

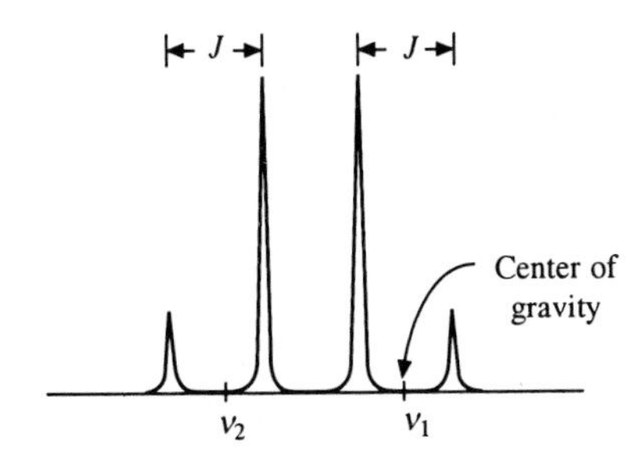

Fig. 5.22 **When $\Delta v/J$ ratio is low, centers of gravity (v_2 and v_1) are chemical shift positions instead of linear mid-points**

system which are separated by small chemical shifts are denoted by A, B and C (usually in order of decreasing δ value), and those far away in chemical shift from these by X, Y and Z, whereas those intermediate in chemical shift by M, N and O. In brief, the protons widely differing in chemical shift by ($\Delta v/J = 6$) are assigned letters widely separated in the alphabet, e.g. A, M and X. The protons with about the same chemical shifts are assigned letters adjacent to one another in the alphabet. The number of protons of each kind is denoted by a subscript. For example, A_2B system means the spin system has two kinds (A and B) of protons; there are two protons of kind A, and the protons of this spin system are separated by small chemical shift. A_2B denotes a strongly coupled 3-spin system.

The nuclei which are chemically equivalent but magnetically nonequivalent are differentiated by primes, e.g. $AA'XX'$ is a 4-spin system, where A and A' (as well as X and X') are the protons which are chemically equivalent but magnetically nonequivalent. The above nomenclature of spin systems is illustrated by the following examples:

H_3C—CH_2I (B, A)
A_2B_3 system
Ethyl Iodide

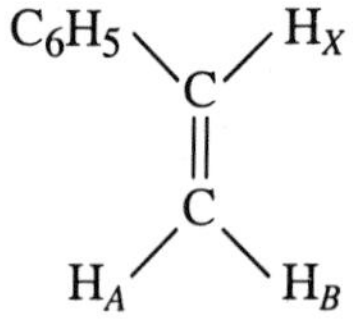

ABX system
Styrene

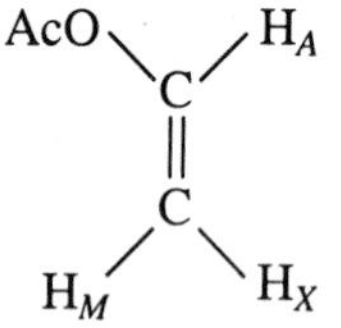

AMX system
Vinyl acetate

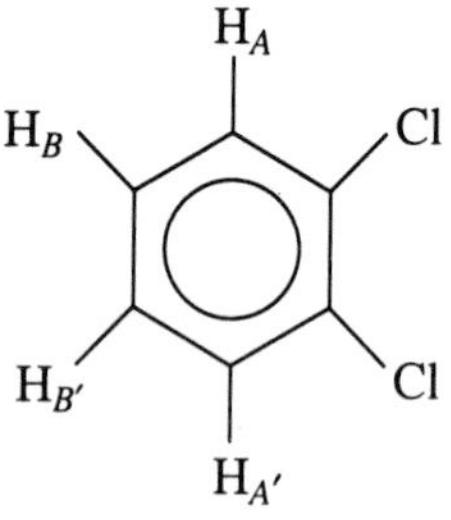

AA'BB' system
o-Dichlorobenzene

5.14 Magnetic Equivalence

Magnetically equivalent nuclei (e.g. protons) have the same chemical shift and the same coupling constant J to every other nucleus in the spin system. All the magnetically equivalent nuclei are chemically equivalent but the reverse is not always true. For example, protons H_A and $H_{A'}$ in *p*-chloroanisole are chemical shift equivalent. Protons H_A and $H_{A'}$ are coupled to proton H_B (or $H_{B'}$) with different geometry (through different bond angles and bond distances). Hence

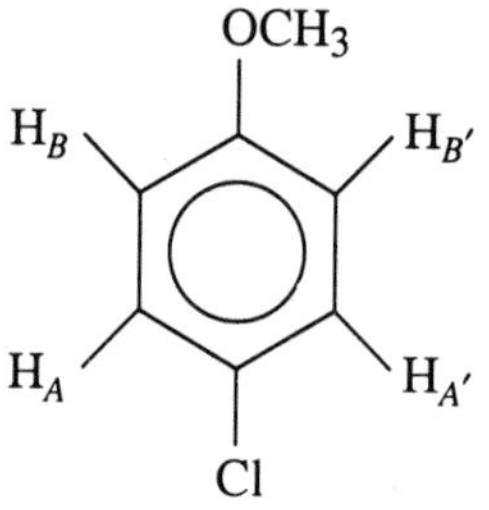

have different coupling constants. Thus, protons H_A and $H_{A'}$ are magnetically nonequivalent, protons H_B and $H_{B'}$, when treated in the same way, are also found to be magnetically nonequivalent. The system is $AA'BB'$. Spin systems which contain groups of chemically equivalent protons which are magnetically nonequivalent cannot be analyzed by first order method.

5.15 Spin-Spin Coupling of Protons with Other Nuclei

As shown in Section 5.2, any nuclei which has $I > 0$ is capable of exhibiting a NMR spectrum. Different nuclei (e.g. ^1H, D, ^{13}C, ^{19}F etc.) require different oscillator frequencies for exhibiting NMR in a given magnetic field. For example, in a magnetic field of 14,092 gauss, ^1H, D, ^{13}C and ^{19}F nuclei resonate at 60.000, 9.211, 15.085 and 56.446 MHz, respectively. Thus, under a given set of conditions for NMR of a particular nucleus, signals due to other nuclei are not observed in the spectrum. Proton may couple with any nucleus (having $I > 0$) to which it is covalently bonded.

Nuclei with $I \geq 1$ have an electric quadrupole moment, the magnitude of which is a measure of the nonspherical nature of the electric charge distribution within the nucleus. The NMR signals for protons attached to a nucleus which has an electric quadrupole moment are broadened, i.e. splitting is not observed. The greater the magnitude of the electric quadrupole moment, the more is the broadening of the signal. Thus, the signals of protons coupled with deuterons, which have only small quadrupole moment, are not appreciably broadened. The coupling constant for a proton with deuteron (J_{HD}) in the situation —CHD is ~2 Hz and in the situation —CHCD—, it is less than 1 Hz. The signals of protons coupled with nitrogen nucleus are almost always broadened because of the intermediate value for its electric quadrupole moment. Protons do not couple with halogen atoms (except fluorine) on adjacent atoms or on the same atom, because the very large quadrupole moments of the halogen atoms cause spin-spin decoupling of adjacent protons.

Fluorine (^{19}F) nucleus has $I = \frac{1}{2}$ and does not have electric quadrupole moment. Thus, ^{19}F nuclei can efficiently couple with each other as well as with protons. Hence, the splitting of their signals occurs (J_{HCF} = ~ 60 Hz; J_{HCCF} = ~ 20 Hz). For example, in 1,2-dichloro-1,1-difluoroethane, the coupling of two equivalent protons with fluorine nuclei gives a triplet in the spectrum

```
      F   H
      |   |
 Cl—C—C—Cl
      |   |
      F   H
```

1,2-dichloro-1,1-difluoroethane

We know that spin-spin splitting is a reciprocal affair. Hence, if a spectrum contains a multiplet then it must be accompanied by at least one more multiplet. Thus, appearance of only one triplet (one multiplet only) in the PMR spectrum of 1,2-dichloro-1,1-difluoroethane looks rather surprising. However, it is easily understandable because different nuclei require different radio frequency for exhibiting NMR in a given magnetic field. Thus, under the PMR conditions another triplet due to two fluorine atoms is not observed in the PMR spectrum of 1,2-dichloro1,1-difluoroethane. Similarly, in the ^{19}F NMR spectrum of this compound, only a triplet due to the two fluorine atoms will be observed and the triplet due to the two protons will not be observed. ^{13}C gives rise to observable ^{13}C—H coupling ($J\,^{13}C$—H) = ~100-250 Hz; $J\,^{13}CCH$ = ~40 Hz), especially in spectra recorded at high amplitude. Peaks resulting from coupling with ^{13}C are called ^{13}C satellite peaks. These weak peaks appear symmetrically on either side of the much stronger signals of ^{12}C—H groups.

Unlike PMR spectra, NMR spectra of nuclei which have large electric quadrupole moments have very broad bands rather than sharp peaks. The spread in resonance frequencies of other nuclei and the magnitude of the coupling constants are very large when compared with the corresponding values for PMR spectra. Similar to the PMR spectroscopy, NMR spectra of other nuclei also provide structural information about compounds.

5.16 Protons on Heteroatoms: Proton Exchange Reactions

Protons on a heteroatom differ from protons on a carbon atom as they are:

(i) exchangeable.
(ii) subject to hydrogen bonding.
(iii) subject to partial or complete decoupling by electrical quadrupole effects of some heteroatoms (Section 5.15).

We have already discussed the effect of hydrogen bonding on chemical shift in Section 5.7 (iii). Now we shall discuss the effect of proton exchangeability on PMR signals.

Let us take the example of ethanol. Under ordinary conditions, ethanol (neat, acidified) shows a triplet at δ 1.17 due to methyl protons, a quartet at δ 3.62 due to methylene protons and a singlet at δ 5.37 due to hydroxylic proton (Fig. 5.23). Spectrum of pure anhydrous ethanol (Fig. 5.24) exhibits

(i) a triplet for $—CH_3$ protons at δ 1.17.
(ii) a multiplet consisting of eight lines for $—CH_2—$ protons at δ 3.62. The $—CH_2—$ protons are under the influence of two kinds of protons ($—CH_3$ and —OH). Thus, the multiplet for $—CH_2—$ protons consists of $(n + 1)(n' + 1) = (3 + 1)(1 + 1) = 8$ lines.
(iii) a triplet for —OH proton at δ 5.28. The —OH appears as a triplet because of its coupling with $—CH_2—$ protons.

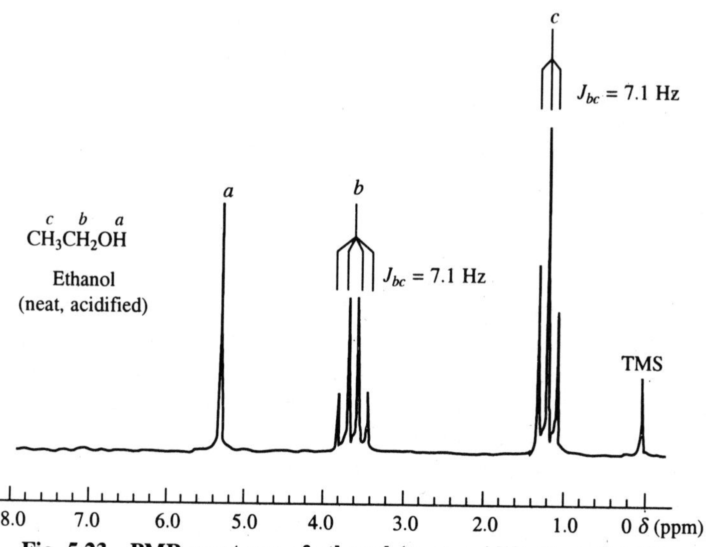

Fig. 5.23 PMR spectrum of ethanol (neat, acidified) at 60 MHz

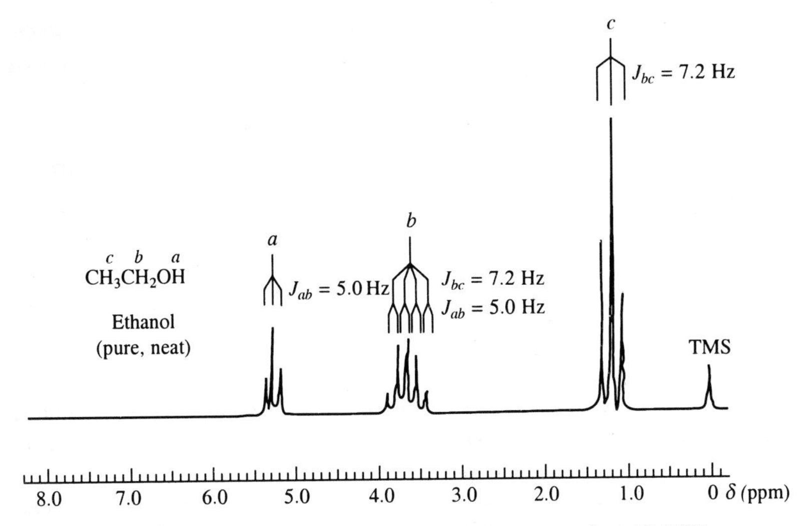

Fig. 5.24 PMR spectrum of ethanol (pure, neat) at 60 MHz

The above observations can be explained on the basis that the *proton exchange reaction* (chemical exchange) becomes faster in the presence of water or acidic or basic impurity

$$RO\overset{a}{H} + HOH{=}ROH + HO\overset{a}{H}$$

Similarly, the exchange of —OH protons among ethanol molecules also occurs. The rate of proton exchange reactions increases with increasing temperature. Proton exchange in the presence of water or at high temperature or in acidic or basic medium is faster than the NMR transition time. Thus, a particular proton does not reside on a particular oxygen atom long enough to 'see' the three spin states of methylene protons or to show its two spin states to the methylene protons. Thus, the expected spin-spin coupling is not observed. Rapid chemical exchange causes spin-spin decoupling.

When the rate of chemical exchange is made very slow by the removal of water, acidic or basic impurities, the expected couplings are observed, i.e. —CH_2— protons appear as a multiplet consisting of 8 lines and —OH protons as a triplet. The rate of proton exchange can also be made slower through strong solvation by using highly polar solvents like dimethyl sulphoxide (DMSO, CH_3SOCH_3) or acetone.

Since the rate of chemical exchange increases with increasing temperature, spin-spin decoupling can sometimes be observed by raising the temperature of the sample. On the other hand, spin-spin coupling can sometimes be observed by lowering the temperature of the sample. For example, the PMR spectrum of methanol at very low temperature (–40°C) shows a quartet for hydroxyl proton and a doublet for methyl protons. This indicates that the chemical exchange is very slow at – 40° C as compared to the NMR transition time. When the temperature is raised to +31°C, the signals for —CH_3 and —OH protons appear as sharp singlets. This indicates that the rate of chemical exchange at +31°C is faster than the NMR transition time. At −4°C, the component peaks of both the multiplets coalesce to give relatively broad singlets. This shows that at this temperature, the rate of chemical exchange is intermediate.

Fast chemical exchange will usually occur when two hydroxylic species are present, like in solutions of ethanol-water or acetic acid-water. In such solutions, only one PMR signal is observed for hydroxylic protons of both the species present although individually they have different chemical shifts. The signal appears at an average concentration dependent position according to the following formula:

$$\delta_s = N_a\delta_a + N_b\delta_b$$

where δ_s is the chemical shift of hydroxylic protons in the solution, N_a the mole fraction of the hydroxylic proton *a*, N_b the mole fraction of the hydroxylic proton *b*, δ_a the chemical shift of unexchanged hydroxylic proton *a* and δ_b chemical shift of unexchanged hydroxylic proton *b*.

It helps in the quantitative analysis of mixtures like ethanol-water; acetic acid-water etc. If a single compound contains both carboxyl and hydroxyl groups, proton exchange usually causes these groups to appear as a single signal.

Protons attached to other heteroatoms, such as N, S etc., also behave like hydroxylic protons. Protons of —OH, —NH_2, —SH etc. groups have no characteristic chemical shift ranges, as their chemical shifts depend on concentration, solvent and temperature. However, such groups are identified by exchange with D_2O which causes their signals to disappear from the spectrum (Section 5.17(iii)).

It should be noted that even when exchange is very slow, the signal due to N—H proton is broadened by quadrupolar interaction with nitrogen. However, the N—H proton splits the signal of the proton on an adjacent carbon. For example, in the spectrum of ethyl N-methylcarbamate ($CH_3NHCOOCH_2CH_3$), the N—H proton shows a broad signal centered at δ~5.16, and the signal of N—CH_3 protons at δ2.78 (J = ~ 5 Hz) is split into a doublet by the N—H proton. The ethoxy (—OCH_2CH_3) protons give the usual triplet at δ 1.23 and quartet at δ 4.14.

5.17 Simplification of Complex NMR Spectra

The complete analysis of a NMR spectrum becomes difficult when signals overlap and thus, useful information is buried due to complexity of the spectrum. For example, if several closely related methylene groups are present in a molecule, their signals may overlap and may not be clearly recognized. Sometimes, an intense, broad and unresolved signal due to several methylene groups is observed at about δ 1-2 which is called as the *methylene envelope*. When there is not much difference between the chemical shifts and coupling constants, more complex (second order) spectra are obtained. The important methods for simplifying a NMR spectrum to get the maximum information are discussed as follows.

(i) High Field Strengths

We have noted that the chemical shift in Hz is directly proportional to the applied magnetic field, whereas the value of coupling constant in Hz remains constant in different applied magnetic fields. On increasing the field strength the chemical shift separation $\Delta\nu$ (Hz) increases, but the value of J (Hz) remains constant. Thus, the multiplets which are overlapped at lower field are expected to separate out at high field strengths. In this way, the NMR spectrometers operating at high magnetic fields (i.e. at high radio frequencies) give better resolution and relatively easily interpretable spectra.

(ii) Spin-Spin Decoupling (Double Irradiation or Double Resonance)

Spin-spin coupling between neighboring nuclei splits their signals into multiplets and the analysis of the splitting patterns is useful for structure determination of compounds. However, in certain cases, the splitting patterns and spectra are so complex that for the simplification of spectra, spin-spin decoupling is desired.

Irradiation of a nucleus or a group of equivalent nuclei at their resonance using a second strong radio frequency oscillator results in the removal of all couplings arising from the irradiated nuclei called *spin-spin decoupling* (spin decoupling or double irradiation or double resonance) because an additional radio frequency is used. Such an irradiation imparts extra energy which causes rapid transitions between the different spin states of the irradiated group of nuclei. Thus, neighboring nuclei (e.g. protons) cannot see different spin states (but they can see only an average view of spin states) of the irradiated nuclei and consequently spin-spin couplings are effectively removed. This is very similar to spin-spin decoupling through rapid proton exchange reaction (Section 5.16).

Let us take the example of PMR spectrum of ethanol (Fig. 5.25) where methyl protons appear as a triplet due to the spin-spin splitting by the methylene protons. If the methylene protons are irradiated strongly with an additional radio frequency at their resonance, then they change their spin states very rapidly. Thus, the methyl protons can see only an average of the possible spin orientations of the methylene protons and the coupling will be removed. Consequently, the triplet resulting from the methyl protons collapses to a singlet and appears at its

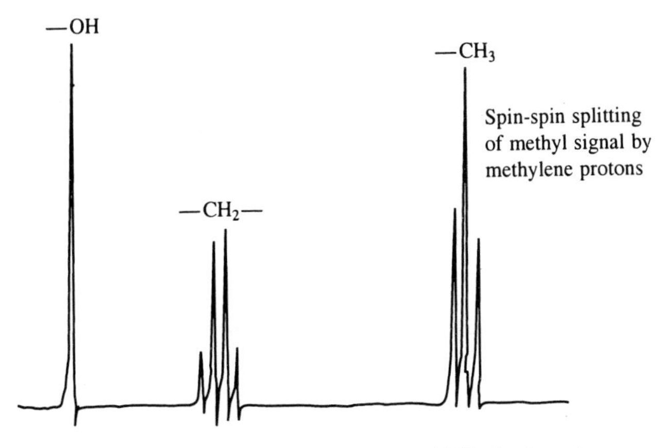

Fig. 5.25 PMR spectrum of acidified ethanol

usual position and the methylene absorption is eliminated (Fig. 5.26). The disappearance of the methylene absorption is due to saturation (Section 5.2). Similarly, when the methyl protons are irradiated, the quartet resulting from the methylene protons collapses to a singlet and the methyl absorption is eliminated.

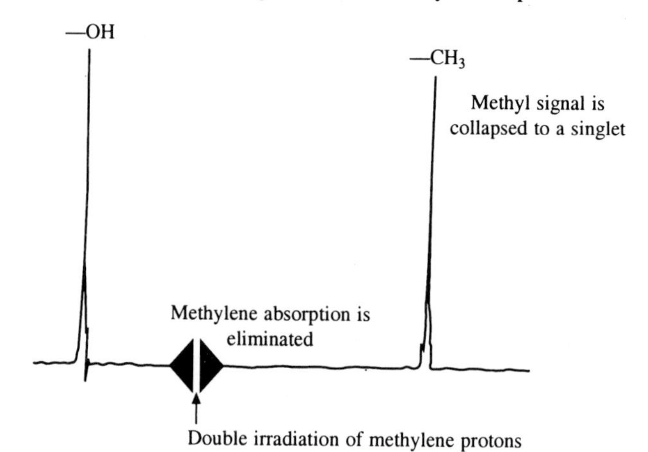

Fig. 5.26 PMR spectrum of acidified ethanol on double irradiation of methylene protons

Since strong irradiation is used for decoupling, this technique is not suitable when the chemical shifts positions for the coupling multiplets are closer than ~20 Hz at 100 MHz. In such cases, another technique *spin tickling* is applicable. Spin tickling involves irradiation of a nucleus with a much less intense radiation than required for spin-spin decoupling. This results in an increase in the number of lines in the coupled multiplets.

(iii) Deuteration-Deuterium Exchange and Deuterium Labelling

The protons of —OH, —NH_2, —SH etc. groups are exchangeable with D_2O. Thus, such groups

$$ROH + D_2O \rightleftharpoons ROD + HOD$$

are identified by exchange with D_2O which causes their signal to disappear from the spectrum. This exchange reaction is similar to proton exchange reactions (Section 5.16), and is called *deuterium exchange reaction.* For detecting the protons exchangeable with D_2O, either the PMR spectrum is recorded in D_2O or a few drops of D_2O are added to the sample. For example, in the case of an alcohol on D_2O exchange, the proton signal due to the alcoholic —OH will disappear and instead, a signal due to HOD proton will appear in the PMR spectrum.

Deuterium is easily introduced into a molecule and its presence in a molecule is not detected in the PMR spectrum because it absorbs at different field strengths (Section 5.14). Deuterium couples only slightly with the proton, hence it does not split its signal. Deuterium labelling also simplifies PMR spectra as illustrated by the example of ethyl chloride.

protons and various conformations of a molecule can be distinguished on the basis of different values of their coupling constants, chemical shifts and peak areas.

(v) Detection of Partial Double Bond Character

In certain cases, it can be detected by PMR spectroscopy whether a particular single bond in a molecule has acquired partial double bond character. One of the most thoroughly studied example is the hindered rotation about the C—N bond in simple amides, e.g. N,N-dimethylformamide (DMF). There is hindered rotation about C—N bond because it has acquired partial double bond character through resonance as shown below:

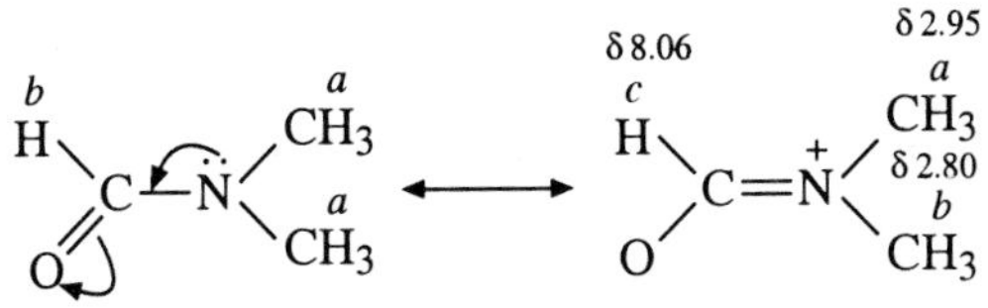

$$\underset{\text{(XI)}}{H^{b}-C(=O)-\ddot{N}(CH_3^{a})_2} \longleftrightarrow \underset{\text{(XII)}}{\overset{\delta\,8.06}{H^{c}}-C(O^{-})=\overset{+}{N}(\overset{\delta\,2.95}{CH_3^{a}})(\overset{\delta\,2.80}{CH_3^{b}})}$$

The hindered rotation, i.e. the partial double bond character of C—N bond in DMF is demonstrated by the presence of two doublets in its PMR spectrum at δ 2.80 (J = 0.6 Hz) and 2.95 (J = 0.3 Hz) due to the two methyl groups at room temperature. This is because the methyl groups have become nonequivalent in structure (XII) due to the presence of C—N double bond. At elevated temperatures, the rapid rotation about the C—N bond makes both the methyl groups equivalent and only one signal (doublet) is observed for both the methyl groups.

(vi) Quantitative Analysis

The fact that areas under the peaks are directly proportional to the number of protons causing the respective peaks is the basis for the quantitative analysis by NMR spectroscopy. For quantitative analysis, the components of the mixture must be known and each component must give at least one signal which is well separated from the other signals in the spectrum. Impure samples may be determined by the addition of a known pure compound as an internal standard. If the reactants and products are known, then the rate of the reaction may be determined.

Automatic integration of NMR signals afford an easy and rapid quantitative means for determining the ratio of compounds in a mixture provided that at least one signal from each constituent is free from overlap by other signal(s). The estimation of the keto-enol ratio in acetyl acetone will illustrate the quantitative analysis of a mixture.

$$\underset{\text{Keto form}}{H_3C—CO—CH_2—CO—CH_3} \rightleftharpoons \underset{\text{Enol form}}{H_3C—\underset{\substack{|\\OH}}{C}=CH—CO—CH_3}$$

Acetylacetone

In the PMR spectrum of acetylacetone, the height of the integration curve at the methylene (—CH_2—) signal was found to be 10 mm and that at the methine

(=CH—) signal was 22 mm. Let us calculate the % of keto and enol forms in the sample.

The methylene group of the keto form ≡ 2H ≡ 10 mm
The methine group of the enol form ≡ 1H ≡ 22 mm
Therefore, 2H ≡ 44 mm

$$\% \text{ of the keto form} = \frac{10}{44 + 10} \times 100 = 18.5\%$$

$$\% \text{ of the enol form} = \frac{44}{44 + 10} \times 100 = 81.5\%$$

Quantitative analysis of the mixtures like ethanol-water; acetic acid-water etc. has already been discussed in Section 5.16. PMR spectroscopy has also been used for the quantitative analysis of the mixtures of diastereomers as well as for determining the enantiomeric excess (ee), i.e. optical purity.

5.20 Continuous Wave (CW) and Fourier Transform (FT) NMR Spectroscopy

The common method for obtaining NMR spectra is to irradiate the sample with a constant radio frequency while changing (sweeping) the applied magnetic field (field sweep). Alternatively, NMR spectrometers operate at a constant magnetic field while the radio frequency is varied (frequency sweep). Both the methods give the same NMR spectrum. This commonly used technique is called *continuous wave (CW) NMR spectroscopy.*

In a recent method for obtaining NMR spectra, the sample is irradiated with an intense pulse of all radio frequencies in the desired range (e.g. covering all ^1H frequencies) at once while keeping the magnetic field constant. All the nuclei under study absorb at their individual frequencies and are flipped to their higher energy spin states. This results in an interferogram (called *free induction decay, FID* or *time-domain* spectrum) which cannot be interpreted directly. The time-domain spectrum is converted into ordinary frequency-domain spectrum (showing the intensity of absorption against frequency) by performing a mathematical operation known as *Fourier transformation*. This technique is called *pulsed-Fourier transform nuclear magnetic resonance (FT-NMR) spectroscopy*. It gives good spectra even with very small quantities of samples (less than a milligram). The principal advantage of FT-NMR spectroscopy is a great increase in sensitivity per unit time of experiment. It is the increase in sensitivity brought about by the introduction of FT-NMR spectroscopy which has allowed the routine observation of ^{13}C NMR spectra.

5.21 Some Solved Problems

Problem 1. Give the relative positions of the PMR signals and their multiplicity in each of the following compounds:

(i) $CH_3CH_2COCH_3$ (ii) CH_3CH_2CHO
(iii) $CH_3CH_2OOCCH_2CH_2COOCH_2CH_3$ (iv) $(CH_3)_2CHCOOH$

Solution

(i) $\overset{a}{C}H_3\overset{b}{C}H_2CO\overset{c}{C}H_3$

On moving downfield, the sequence of signals is protons *a*, then *c* and *b*. *Multiplicity of signals:* Protons *a* (triplet), *b* (quartet) and *c* (singlet).

(ii) $\overset{a}{C}H_3\overset{b}{C}H_2\overset{c}{C}HO$

On moving downfield, the sequence of signals is proton *a*, then *b* and *c*. *Multiplicity of signals:* Protons *a* (triplet), *b* (multiplet) consisting of eight lines [(3 + 1) (1 + 1) = 8] and *c* (triplet).

(iii) $\overset{a}{C}H_3\overset{b}{C}H_2OOC\overset{c}{C}H_2\overset{c}{C}H_2COO\overset{b}{C}H_2\overset{a}{C}H_3$

On moving downfield, the sequence of signals is protons *a*, then *c* and *b*.

Multiplicity of signals: Protons *a* (triplet), *b* (quartet) and *c* (singlet).

(iv) 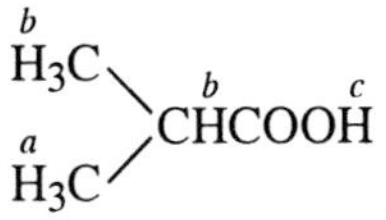 $(\overset{a}{H_3C})_2\overset{b}{C}HCOO\overset{c}{H}$

Multiplicity of signals: Protons *a* (doublet), *b* (septet) and *c* (singlet).

Problem 2. Comment on the number of signals and their splitting, if any, in the PMR spectra of the following compounds:

(i)  $C_6H_5-CH_2COCH_2-C_6H_5$

(ii) 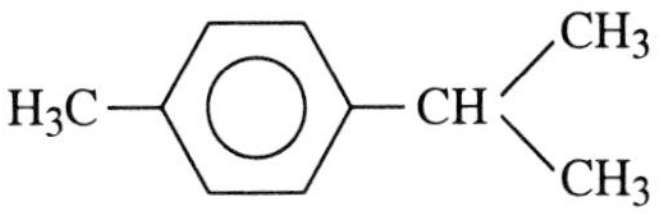 $H_3C-C_6H_4-CH(CH_3)_2$

(iii) 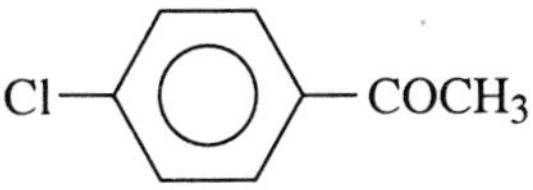 $Cl-C_6H_4-COCH_3$

Solution

(i) 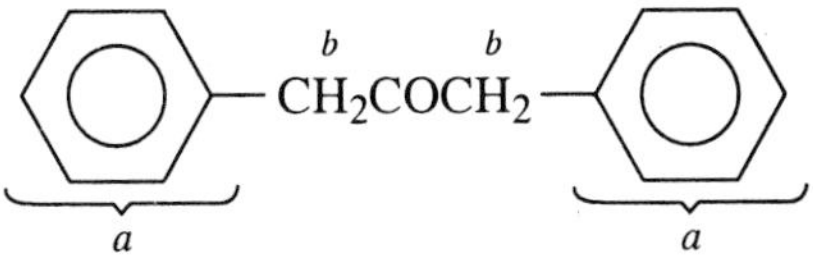 $\underbrace{C_6H_5}_{a}-\overset{b}{C}H_2CO\overset{b}{C}H_2-\underbrace{C_6H_5}_{a}$

In certain cases, environments of chemically nonequivalent protons are not different enough for the signals to be noticeably separated, and in such cases we may see fewer signals than we predict. For example, in some (but not all) aromatic compounds the *ortho*, *meta* and *para* protons have nearly the same chemical shifts, and hence for NMR purposes they are nearly equivalent, i.e. exhibit only one signal (singlet). This is the situation in the present case. Thus, all the five phenyl protons *a* are nearly equivalent and appear as a singlet. This compound shows two singlets—one due to the phenyl proton *a* and the other due to the methylene proton *b*.

(ii) $\overset{a}{CH_3}$—C_6H_4 (b)—$\overset{c}{CH}(\overset{d}{CH_3})_2$

This compound contains four kinds of protons, hence will exhibit four PMR signals. Proton *a*, a singlet; proton *b*, a singlet; protons *c*, a septet because it has six neighboring protons (6 + 1 = 7); protons *d* have one neighboring proton, hence will appear as a doublet (1 + 1 = 2).

(iii) Cl—C_6H_4—$\overset{c}{COCH_3}$ (ring protons: H^a, H^a ortho to Cl; H^b, H^b ortho to $COCH_3$)

In this compound, protons *b* which are *ortho* to the carbonyl group will have quite different chemical shift compared to protons *a* which are *ortho* to the chloro group. Thus, this compound has three kinds of protons and will exhibit three signals. Protons *a* have one neighboring proton, hence will appear as a doublet (1 + 1 = 2). Similarly, protons *b* will also appear as a doublet. The methyl proton *c* has no neighboring proton, hence will appear as a singlet.

Problem 3. In an organic compound, three kinds of protons appear at 60, 100 and 180 Hz when the spectrum is recorded at 60 MHz NMR spectrometer. What will be their relative positions (in Hz) when 90 MHz spectrometer is used?

Solution. The chemical shift in Hz is directly proportional to the strength of the applied magnetic field (and, therefore, to the applied frequency). Thus

(i) $\frac{60}{60} \times 90 = 90 \text{ Hz}$

(ii) $\frac{100}{60} \times 90 = 150 \text{ Hz}$

(iii) $\frac{180}{60} \times 90 = 270 \text{ Hz}$

Problem 4. Propose the structure for the compounds that fit the following ^1H NMR data:

(i) $C_5H_{10}O$

δ 0.95, 6H, doublet
δ 2.10, 3H, singlet
δ 2.43, 1H, multiplet

(ii) C_4H_7BrO

δ 2.11, 3H, singlet
δ 3.52, 2H, triplet, $J = 6$ Hz
δ 4.40, 2H, triplet, $J = 6$ Hz

Solution (i) The compound with molecular formula $C_5H_{10}O$ shows a doublet due to six protons. This indicates that these six protons are equivalent and the carbons bearing them are attached to a —CH— group, i.e. the molecule has $(CH_3)_2CH$— group; the —CH— proton appears as a multiplet (septet). One singlet due to three protons indicates that the compound contains a methyl group which has no proton on the atom to which it is attached. Thus, the structure for the compound which fits the above data is

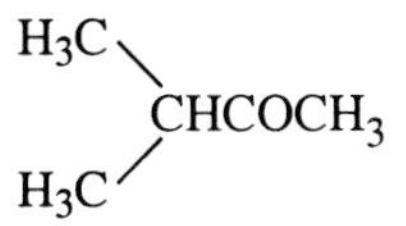

$$(H_3C)_2CHCOCH_3$$

(ii) The compound C_4H_7BrO shows two triplets with the same coupling constant ($J = 6$ Hz) showing that it contains two nonequivalent adjacent methylene groups coupled with each other (—CH_2—CH_2—). The presence of a three proton singlet indicates the presence of a methyl group which has no proton on the atom to which it is attached. Thus, the structure for the compound is

$$BrCH_2CH_2COCH_3$$

Problem 5. A compound $C_6H_{10}O_2$ shows a significant IR bond at 1770 cm^{-1}, and three ^1H NMR signals at τ 5.8, 7.5 and 9.1 with relative intensity 1 : 1 : 3, respectively. Deduce the structure of the compound.

Solution. The presence of an IR bond at 1770 cm^{-1} indicates that the compound is a lactone (cyclic ester). It shows three ^1H NMR signals. Hence, there are three kinds of protons in the compound. Since the total number of protons is 10 and the intensity ratio is 1 : 1 : 3, the number of each kind of protons is 2, 2 and 6, i.e. the compound has two nonequivalent CH_2 and two equivalent CH_3 groups. Thus, the structure of the compound is

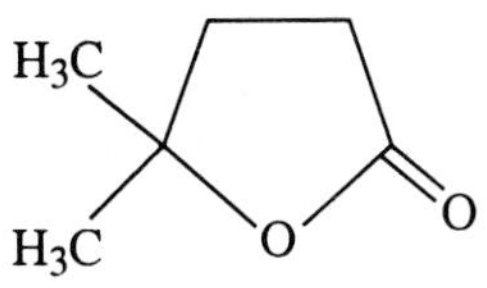

H₃C, H₃C, O, O

Problem 6. Draw the structure of a compound with each of the following molecular formulae that will show only one PMR signal:

(i) $C_3H_6Cl_2$ (ii) C_5H_{12} (iii) C_4H_6 (iv) C_5H_{10}

Solution

(i) $H_3C—C(Cl)_2—CH_3$

Cl
|
H₃C—C—CH₃
|
Cl

(ii) $H_3C—C(CH_3)_2—CH_3$

(iii) $H_3C—C{=}C—CH_3$

(iv) 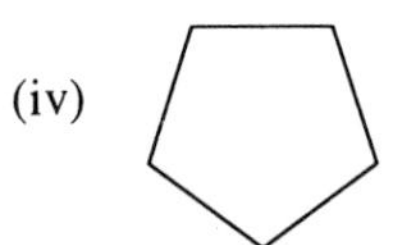

Problem 7. An organic compound has molecular formula C_3H_7Br. Its PMR spectrum is shown in Fig. 5.28. Deduce the structure of the compound.

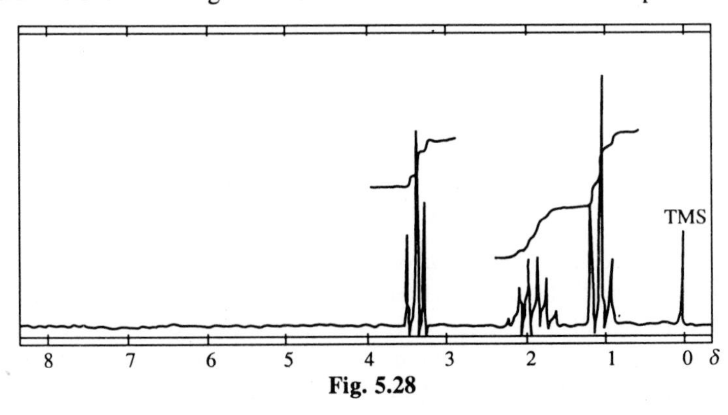

Fig. 5.28

Solution. The PMR spectrum of the compound shows three signals, viz. two triplets and one sextet, hence it contains three kinds of protons. On moving upfield the successive heights of the integration curves at the signals are 8 mm, 8 mm and 12 mm, i.e. the ratio of the number of each kind of protons is 1 : 1 : 1.5. Since the molecular formula of the compound is C_3H_7Br, the number of each kind of protons is 2H, 2H and 3H. This indicates that the compound has $CH_3CH_2CH_2$— group and thus, its structure is $\overset{a}{C}H_3\overset{b}{C}H_2\overset{c}{C}H_2Br$.

The proton *a* appear as an upfield triplet, while protons *c* as *a* downfield triplet. The protons *b* appear as a sextet. However, the protons *b* might be expected to exhibit twelve lines, i.e. (3 + 1) (2 + 1) = 12 because they are coupled with two equivalent groups of three protons *a* and two protons *c*. In practice, J_{ab} is approximately equal to J_{bc} and therefore overlapping of lines occurs as shown in Fig. 5.28 and thus, a sextet is observed. Since $J_{ab} = J_{bc}$, we can say that the *b* protons have five equivalent neighboring protons, hence appear as a sextet, i.e. 5 + 1 = 6. Thus, the structure of the given compound is:

$$CH_3CH_2CH_2Br$$

Problem 8. The PMR spectrum of an organic compound C_8H_9Br shows a quartet at 5.5 δ, a doublet at 2.0 δ and an unsymmetrical multiplet at ~7.4 δ in the intensity ratio 1: 3: 5, respectively. Deduce the structure of the compound.

Solution. The molecular formula of the compound is C_8H_9Br and the intensity ratio of different kinds of protons present is 1 : 3 : 5. Therefore, the number of each kind of protons will be 1H, 3H and 5H. The presence of an unsymmetrical multiplet at ~7.4 δ due to 5H shows the presence of a phenyl group (C_6H_5—), and the presence of a quartet due to 1H and a doublet due to 3H indicates the presence of —CH—CH$_3$ group. Thus, the structure of the given compound is

$$C_6H_5—\underset{\displaystyle Br}{\underset{|}{C}}H—CH_3$$

Problem 9. Using PMR spectroscopy, how will you distinguish the following pairs?

(i) 1,2-dimethoxyethane and 1,1-dimethoxylethane
(ii) *cis*-1-chloropropene and *trans*-1-chloropropene
(iii) Acetone and methyl acetate.

Solution (i) 1,2-dimethoxyethane ($\overset{a}{C}H_3O\overset{b}{C}H_2\overset{b}{C}H_2O\overset{d}{C}H_3$) will exhibit two singlets due to protons *a* and *b*, while 1,1-dimethoxyethane ($(\overset{a}{C}H_3O)_2\overset{b}{C}H\overset{c}{C}H_3$) will exhibit one singlet due to proton *a*, one quartet due to proton *b* and one doublet due to protons *c*.

(ii) *cis*-1-chloropropene and *trans*-1-chloropropene can be distinguished on the basis of their coupling constants. The *cis* isomer will have lower coupling constant (J_{cis} = 6-12 Hz) than the *trans* isomer (J_{trans} = 12-18 Hz).

(iii) Both the methyl group in acetone (CH_3COCH_3) are equivalent, hence it will show only one singlet. In methyl acetate (CH_3COOCH_3) the two methyl groups are nonequivalent, hence it will show two singlets.

Problem 10. An organic compound having molecular formula $C_5H_{11}Cl$ gave the following ^1H NMR data:

$$\delta\, 1.0\ (t,\ 3H),\ 1.5\ (s,\ 6H)\ \text{and}\ 1.8\ (q,\ 3H)$$

Deduce the structure of compound.

Solution. The ^1H NMR spectrum of the compound exhibits a three proton triplet and a two proton quartet which indicate the presence of the CH_3CH_2— group. The appearance of a six proton singlet shows the presence of two equivalent CH_3— groups which must be attached to a carbon containing no hydrogen. Thus, the structure of the compound having molecular formula $C_5H_{11}Cl$ is

$$H_3C—CH_2—\overset{\displaystyle CH_3}{\underset{\displaystyle Cl}{C}}—CH_3$$

Problem 11. Fig. 5.29 shows the PMR spectrum of a compound having molecular formula $C_8H_{10}O$. Deduce the structure of the compound.

Solution. The PMR spectrum of the compound shows four signals, viz. two singlets, one doublet and one quartet. Hence, it contains four kinds of protons.

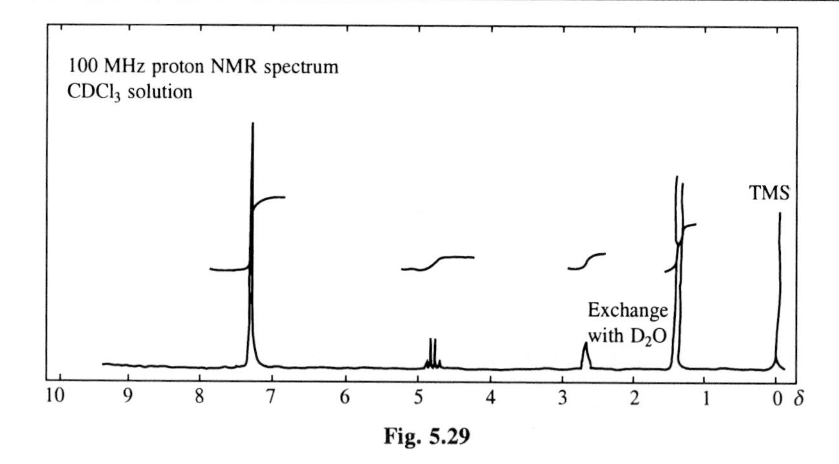

Fig. 5.29

On moving upfield, the successive heights of the integration curves at the signals are 15 mm, 3 mm, 3 mm, and 9 mm, i.e. the ratio of the number of each kind of protons is 5 : 1 : 1 : 3. Since the molecular formula of the compound is $C_8H_{10}O$, the number of each kind of protons is 5H, 1H, 1H and 3H.

The appearance of a five proton singlet at ~7.2 δ indicates the presence of a phenyl (C_6H_5—) group. The presence of a three proton doublet and one proton quartet shows the presence of a CH_3CH group. The proton (1H) causing a singlet is exchangeable with D_2O which shows the presence of a hydroxyl group. Thus, the structure of the compound consistent with the above observations is

$$H_5C_6—\underset{\displaystyle OH}{\underset{|}{CH}}—CH_3$$

PROBLEMS

1. For which of the following isotopes NMR spectroscopy is possible and why?

   ^{12}C, ^{14}N, ^{2}H, ^{35}Cl, ^{32}S, ^{16}O and ^{31}P

2. Discuss the process of absorption of energy during the nuclear magnetic transitions.
3. Predict the number of signals and their relative intensities in the low resolution PMR spectra of the following compounds:

   (a) Toluene (b) Propanal (d) Propionamide

4. What is chemical shift? Giving examples, discuss the factors which affect the magnitude of the chemical shift.
5. Explain spin-spin coupling and splitting of signals with examples.
6. Predict the number of signals and their multiplicity in the PMR spectra of the following compounds:

   (a) CH_3CH_2OOC—C_6H_4—$COOCH_2CH_3$

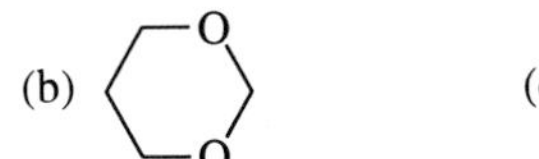
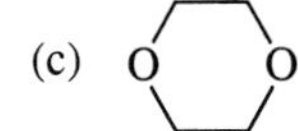

(d) Cl— —Cl

7. Write notes on:

(a) Shielding and deshielding (b) Relaxation processes (c) Coupling constant

8. (a) Why are the NMR absorption positions expressed relative to a reference compound?

(b) Why is TMS a good reference compound in NMR spectroscopy?

9. In PMR spectroscopy, what information can be obtained from the following:

(a) Number of signals (b) Chemical shifts (c) Areas under peaks
(d) Splitting of signals (e) Coupling constants

10. In an *AX* spectrum, the four lines were observed at δ 5.8, 5.7, 1.1 and 1.0 (measured from TMS with an instrument operating at 100 MHz). What are the chemical shift positions (in δ) of the *A* and *X* nuclei, and the coupling constant (in Hz) between them?

11. Give a structure consistent with each of the following sets of NMR data:

(a) $C_3H_5Cl_3$: δ 2.20 singlet, 3H; δ 4.02, singlet 2H
(b) $C_{10}H_{14}$: δ 1.30, singlet, 9H; δ7.28, singlet 5H
(c) $C_{10}H_{14}$: δ 0.88, doublet, 6H; δ 1.86, multiplet, 1H; δ 2.45; doublet, 2H; δ 7.12, singlet, 5H.

12. Predict the number of signals and their relative intensities in the PMR spectra of the following isomers:

(a) Acetone and propanal
(b) Ethylbenzene and *p*-xylene
(c) 2-pentanone and 3-pentanone

13. What is the cause of different chemical shifts for various hydrogens in NMR spectroscopy? Why are chemical shifts generally expressed in δ or τ values instead of in cps?

14. A compound $C_{10}H_{13}Cl$ gave the following NMR data:

δ 1.57, singlet, 6H; δ 3.07, singlet, 2H; δ 7.27, singlet, 5H

Deduce the structure of compound.

15. Write explanatory notes on :

(a) Shift reagents
(b) Spin-spin decoupling
(c) Nuclear Overhauser Effect (NOE)

16. Using PMR spectroscopy, how will you distinguish the following pairs:

(i) Maleic acid and fumaric acid
(ii) 1-chloropropane and 2-chloropropane
(iii) Intermolecular hydrogen bonding and intramolecular hydrogen bonding

17. Explain the following:

(a) Acetylenic protons absorb at higher field than olefinic protons.

(b) The hydroxylic proton of ethanol does not split the PMR signal of its methylene protons in the presence of a trace of acid.

(c) No signal for deuterium is observed in the PMR spectrum of a compound, e.g. $CD_3CH_2CH_3$.

18. An organic compound has molecular formula $C_3H_6Br_2$. Its PMR spectrum is given in Fig. P5.1. Interpret the spectrum and assign the structure to the compound.

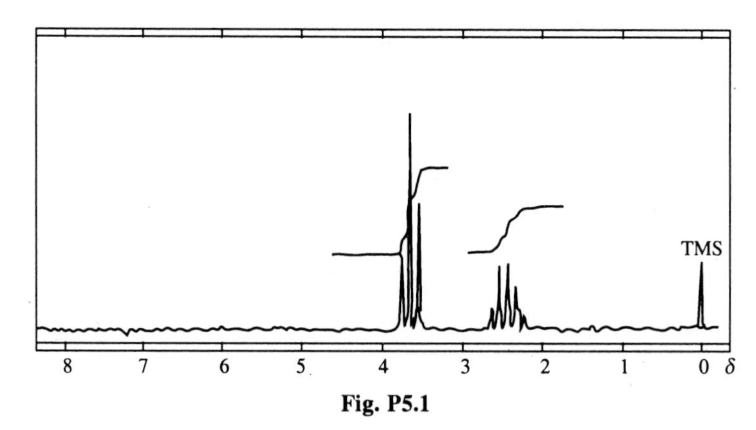

Fig. P5.1

19. Discuss the characteristic features of the first order PMR spectra. How can the more complex (second order) PMR spectra be simplified for obtaining more information?

20. Write notes on:

(a) Chemical and magnetic equivalence of protons.

(b) Factors affecting coupling constants.

21. The following ^1H NMR absorptions were recorded and are listed in Hz from TMS standard. Convert the absorption values into δ units.

(i) 451 Hz at 60 MHz spectrometer

(ii) 430 Hz at 90 MHz spectrometer

(iii) 543 Hz at 100 MHz spectrometer

22. A compound has molecular formula C_3H_8O. Its IR spectrum shows a strong absorption band at 3380 cm^{-1} with no other characteristic band. The PMR spectrum of the compound displayed signals δ 1.2 (*d*, 6H), 3.8 (*s*, 1H) and 4.9 (*s*, 1H). Deduce the structure of this compound.

23. (a) Give an account of vicinal and geminal couplings in PMR spectroscopy.

(b) Write a note on chemical exchange and spin-spin decoupling.

24. Using PMR spectroscopy, how will you distinguish the following isomeric compounds:

(i) 1,4-dioxane and 1,3-dioxane

(ii) *t*-butyl bromide and 1-bromo-2-methyl propane

(iii) 1-butyne and 2-butyne

25. Propose a structure consistent with the ^1H NMR data of each of the following compounds:

(a) $C_4H_{10}O$: δ 1.28 (*s*, 9H), 1.35 (*s*, 1H)

(b) $C_4H_{10}O_2$: δ 3.25 (*s*, 6H), 3.45 (*s*, 4H)

(c) $C_{10}H_{13}Cl$: δ1.57 (*s*, 6H), 3.07 (*s*, 2H), 7.27(*s*, 5H)

26. If the observed chemical shift of a proton is 315 Hz form TMS at a 90 MHz NMR spectrometer, what is the chemical shift in terms of δ? Express it in τ value also.
27. The PMR spectrum of a compound C_8H_{10} is given in Fig. P5.2. Analyse the spectrum and assign the structure to the compound.

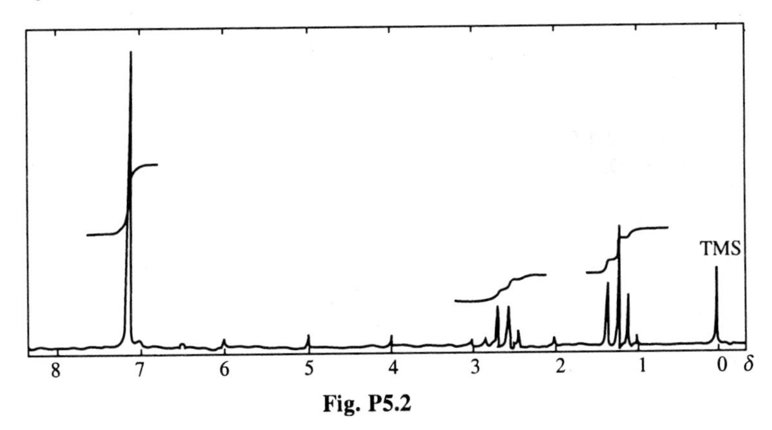

Fig. P5.2

28. A compound containing C, H, O and Cl shows a strong IR absorption band near 1710 cm^{-1} and a broad band near 2800 cm^{-1}. Its PMR spectrum displays two triplets at δ 2.8 and 3.8 and a singlet at about δ 12 in the intensity ratio 2 : 2 : 1, respectively. Deduce the structure of the compound.
29. A pale yellow organic compound with molecular formula $C_6H_5NO_3$ exhibited an unsymmetrical multiplet in the region 1.8-2.9 τ (4H) and a singlet at 0.1 τ (1H) in its NMR spectrum. Deduce the structure of the compound.
30. (a) How is PMR spectroscopy useful in the detection of aromaticity?
    (b) Discuss the use of deuterium exchange and deuterium labelling in PMR spectroscopy.
31. Fig. P5.3 shows PMR spectrum of a compound $C_8H_{10}O$. Interpret the spectrum and assign the structure to the compound.

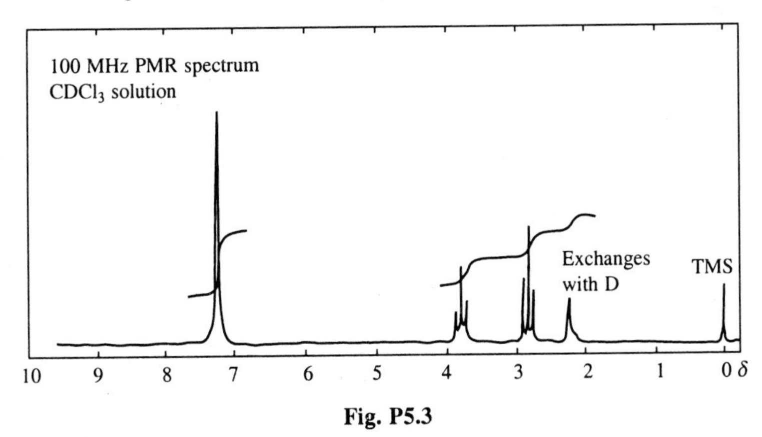

Fig. P5.3

32. Cyclohexane gives only one PMR signal (singlet) at the room temperature, whereas at –100°C it gives two sharp singlets. Explain this observation.
    (Hint: At room temperature the interconversions of the two equivalent chair conformations is so fast that the PMR spectrometer sees protons in their average

environment and records them as a singlet. At –100°C the time between interconversions is long enough for the spectrometer to record the PMR of the molecule as one conformation or the other, and thus one singlet for the six axial and the other for the six equatorial protons are observed.)

References

1. A. Ault and G.O. Dudey, An Introduction to Nuclear Magnetic Spectroscopy, Holden-Day, San Francisco, 1978.
2. A.E. Derome, Modern NMR Techniques for Chemistry Research, Pergamon, Oxford, 1987.
3. D. Neuhaus and M. Williamson, The Nuclear Overhauser Effect in Structural and Conformational Analysis, VCH Publishers Inc., New York, 1989.
4. D.H. Williams and I. Fleming, Spectroscopic Methods in Organic Chemistry, McGraw-Hill, New York, 1966.
5. E.D. Becker, High Resolution NMR, Academic Press, New York, 1969.
6. F.A. Bovey, NMR Spectrometry, Academic Press, New York, 1969.
7. H. Booth, Tetrahedron Letters, 1965, 411.
8. J.D. Roberts, Nuclear Magnetic Resonance Applications to Organic Chemistry, McGraw-Hill, New York, 1959.
9. J.D. Roberts, An Introduction to the Analysis of Spin-Spin Splitting in High-Resolution Nuclear Magnetic Resonance Spectra, McGraw-Hill, New York, 1962.
10. J.R. Dyer, Applications of Absorption Spectroscopy of Organic Compounds, Prentice-Hall, Englewood Cliffs, N.J., 1965.
11. K. Nakanishi, V. Woods and L.H. Durham, A Guide Book to the Interpretation of NMR Spectra, Holden-Day, San Francisco, 1967.
12. L.M. Jackman and S. Sternhell, Applications of NMR Spectroscopy in Organic Chemistry, 2nd Ed., Pergamon, New York, 1969.
13. R.H., Jr., Bible, Interpretation of NMR Spectra, Plenum Press, New York, 1965.
14. R.J. Abraham, J. Fisher and P. Loftus, Introduction to NMR Spectroscopy, 2nd Ed., Wiley, London-New York, 1989.
15. R.M. Silverstein, G.C. Bassler and T.C. Morrill, Spectrometric Identification of Organic Compounds, 5th Ed., Wiley, London-New York, 1991.
16. S. Sternhell and J.R. Kalman, Organic Structures from Spectra, Wiley, Chichester-New York, 1986.
17. T.C. Farrar and E.D. Becker, Pulse and Fourier Transform NMR, Academic Press, New York, 1971.
18. W.W. Paudler, Nucler Magnetic Resonance, Wiley, New York, 1987.

6

^{13}C NMR Spectroscopy

6.1 Introduction and Theory

The possibility of carbon NMR appears surprising at a first glance because

(i) the most abundant isotope of carbon ^{12}C (natural abundance 98.9%) has no net nuclear spin (spin number I is zero). Hence, it does not exhibit NMR phenomenon.

(ii) the far less abundant carbon isotope ^{13}C (natural abundance only 1.1%) has (like ^{1}H) a nuclear spin of $\frac{1}{2}$ and is detectable by NMR.

The too low natural abundance of ^{13}C had been a major obstacle for the advent of ^{13}NMR (carbon-13 NMR or CMR) spectroscopy. Since the natural abundance of ^{13}C is only 1.1%, and its sensitivity is only about 1.6% that of ^{1}H, the overall sensitivity of ^{13}C absorption compared with ^{1}H (natural abundance 99.9844%) is about 1/5700. Because of such a low sensitivity, ^{13}C gives rise to extremely weak signals. Thus, the conventional continuous wave (CW), slow-scan technique (Section 5.20) requires a very large sample and prohibitively long time to obtain a CMR spectrum. This was the main reason why ^{13}C NMR was not popular before 1970. However, the availability of improved electronics, sophisticated computers and the advent of Fourier transform (FT) NMR technique (which permits simultaneous irradiation of all ^{13}C nuclei; see Section 5.20) has now made CMR spectroscopy a routine tool for structure determination. Direct observation of carbon skeletons and carbon-containing functional groups by CMR on practical basis has been available only since the early 1970s.

The principles governing ^{13}C NMR spectroscopy are exactly similar to those of ^{1}H NMR spectroscopy. The gyromagnetic ratio of ^{13}C is about one-fourth that of the proton, consequently the resonance frequency of ^{13}C is also around one-fourth. Thus, nuclear magnetic resonance of ^{13}C can be observed at 15.1 MHz in a magnetic field of 14,092 gauss, while that of a proton is observed at 60 MHz in the same field. Similarly, a spectrometer operating with a 23,486 gauss magnet records ^{13}C NMR at 25.2 MHz compared to ^{1}H NMR at 100 MHz.

Similar to that in PMR spectroscopy, tetramethylsilane (TMS) is also used in CMR spectroscopy as the common reference compound, and the chemical shifts are usually expressed in dimensionless δ units (ppm). ^{13}C chemical shifts range from 0 to about 250 δ (compared to 0-15 δ for PMR).

6.2 Sample Handling

A routine sample (MW ~300) on a 300-MHz instrument consists of about 100-200 mg in about 0.4 ml of deuterated solvent in a glass tube with 5 mm outside diameter. Such a sample usually requires about 1 min of instrument time. It is possible to record CMR spectra of samples consisting of several hundred micrograms in highly purified solvents. In such cases, sufficient instrument time (several hours) is required. The same deuterated solvents are used in CMR spectroscopy which are used in PMR spectroscopy, e.g. $CDCl_3$, CD_3COCD_3, C_6D_6, CD_3SOCD_3 etc. (Section 5.4).

6.3 Common Modes of Recording ^{13}C Spectra

The common modes of spectrometer operation which provide different kinds of ^{13}C NMR spectra containing different structural information are:

(i) Proton-noise decoupling
(ii) Off-resonance decoupling
(iii) Gated decoupling

The exact meaning of the phrases, proton-noise decoupled, off-resonance decoupled and gated-decoupled is not important. However, these are related to the conditions during signal acquisition. What is important is what really happens when spectra are recorded in these different operating modes.

(i) Proton-Noise Decoupling

The proton-noise decoupling (also called noise or proton or broadband decoupling) mode of operation is the most common mode. In this mode *all* the protons in the molecule are decoupled from the carbons. This is done by irradiation of the sample with a noise decoupler at the ^1H frequency (e.g. 100 MHz, in a field of 23,486 gauss), while observing the spectrum at the ^{13}C frequency (25.2 MHz in the same field). If the radiation is strong and sufficiently broadband to cover all the proton frequencies in the sample (i.e. ~1000 Hz broad), then the protons change their spin states too rapidly, and are effectively decoupled from the carbons. This is an example of heteronuclear decoupling. Proton-noise decoupling simplifies the ^{13}C spectrum and increases the intensities of signals. The increase in peak intensities is due to Nuclear Overhauser Effect (NOE, Section 5.18). Thus, the carbon atoms bearing no proton exhibit low intensity peaks. A proton-noise decoupled CMR spectrum exhibits a single sharp peak (singlet) for each kind of chemically nonequivalent* carbon atom present in a molecule (Fig. 6.1). Another important feature of this spectrum is that the intensities of peaks (heights or areas) are not proportional to the number of carbon atoms causing them. Some peaks appear larger than the others even though each may be due to a single carbon (Fig. 6.1) because of following two reasons:

(a) Carbon atoms with large spin-lattice relaxation times T_1 may not completely

*The equivalence or nonequivalence of carbon atoms is judged in the same way as has been discussed with protons in Section 5.8.

return to a Boltzmann's distribution between pulses, and the resulting signals will be considerably weaker than expected from the number of carbon atoms causing those signals.

(b) Some carbon atoms give rise to exceptionally small signals due to weak NOE enhancement.

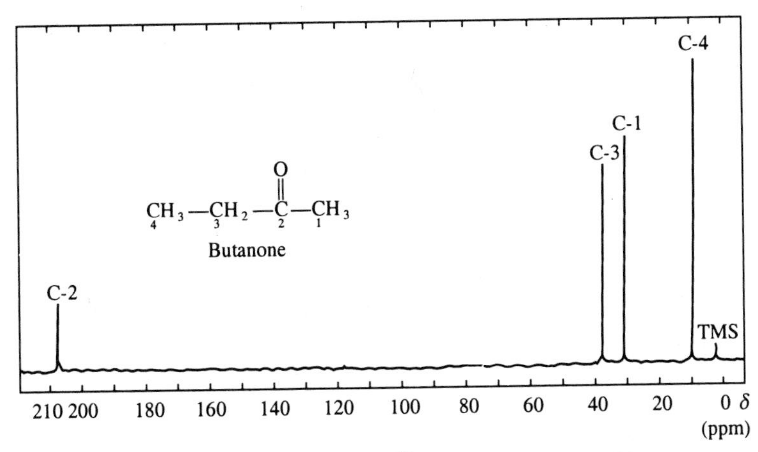

Fig. 6.1 Proton-noise decoupled ^{13}C NMR spectrum of butanone

In brief, wrong relative intensities of the signals depend on the mode of spectrometer operation.

As compared to J values 0–20 Hz for PMR, J values are large for $^{13}C—H$ (~110-320 Hz) and appreciable for $^{13}C—C—H$ (about –5-60 Hz) and $^{13}C—C—C—H$ (about –5-25 Hz) couplings. Thus, nondecoupled (completely proton coupled) CMR spectra usually show complex overlapping multiples, and are difficult to interpret. Thus, the simplification of CMR spectra is done by the proton-noise decoupling. Due to too low natural abundance of ^{13}C, the probability of the presence of two adjacent ^{13}C atoms in molecule is very small. Thus, there is no complication due to $^{13}C—^{13}C$ coupling.

(ii) Off-Resonance Decoupling

In off-resonance decoupling mode, couplings due to protons directly attached to ^{13}C atoms ($^{13}C—H$ couplings) are observed but other couplings (like $^{13}C—C—H$, $^{13}C—C—C—H$ etc.) are removed. Thus, in this mode ^{13}C signals are split into a multiplet consisting of $n + 1$ component peaks, where n is the number of protons directly attached to the ^{13}C atom. Hence, methyl carbon atoms appear as a quartets (3 + 1 = 4), methylene carbon atoms as triplets (2 + 1 = 3) (or as pair of doublets if the protons are not equivalent and their coupling constants are sufficiently different), methine carbons as doublets, and the quaternary carbon atoms as singlets. A typical off-resonance decoupled ^{13}C NMR spectrum is given in Fig. 6.2.

By knowing the number of CH_3—, —CH_2— and —CH— groups present in a molecule, we can count protons. But the number of protons obtained from the off-resonance decoupled CMR does not tally with the molecular formula in case of compounds containing two or more equivalent carbons bearing protons or heteroatoms bearing protons.

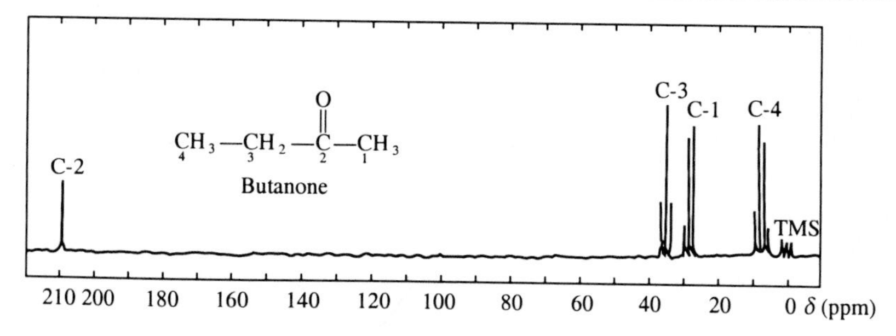

Fig. 6.2 Off-resonance decoupled ^{13}C NMR spectrum of butanone

Off-resonance decoupling is achieved by irradiating the sample at a frequency close to but not coinciding with the resonance frequency of protons. Thus, we irradiate about 1000-2000 Hz upfield or about 2000-3000 Hz downfield from the proton frequency of TMS, i.e. we irradiate upfield or downfield of the usual (1000 Hz sweepwidth) PMR spectrum. As a result residual couplings due to protons directly bonded to ^{13}C atoms are retained, while couplings beyond one bond are usually removed. The multiplets resulting from the retained couplings become narrow, hence there is no complexity due to overlapping.

(iii) Gated Decoupling

When the NMR spectrometer is operated in the gated-decoupling mode, the area under each peak (the intensity of the peak) is directly proportional to the number of carbon atoms causing that peak (Fig. 6.3). By electronically integrating the area under each peak, the relative number of carbon atoms represented by each peak can be determined in the same way as in PMR spectroscopy (Section 5.9). Gated-decoupled spectra contain more information than normal proton noise-decoupled spectra because the former can be integrated. However, this information

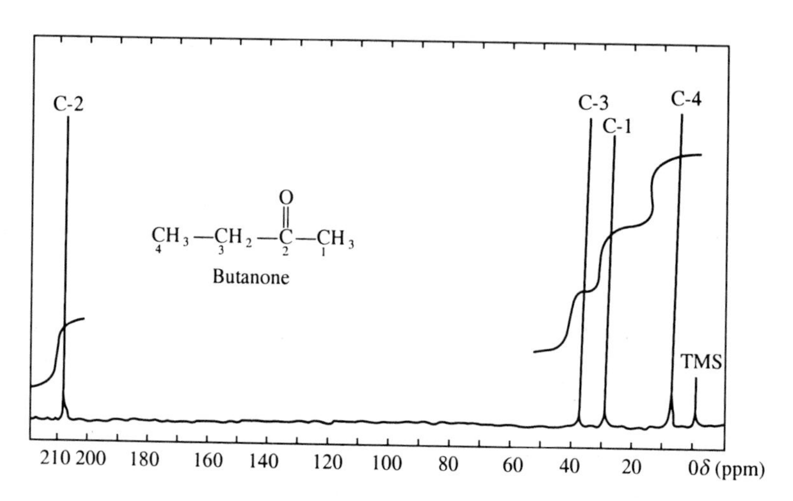

Fig. 6.3 Gated-decoupled ^{13}C NMR spectrum of butanone

comes at a price because the NMR spectrometer is two or three times less sensitive in the gated-decoupling mode than in the proton noise-decoupling mode, and so more time and larger samples are required to obtain the spectrum. Unless a particular ambiguity exists about the number of carbon atoms in a sample, the price is not worth paying, hence gated-decoupled spectra are rarely obtained in practice. Fig. 6.3 shows a typical gated-decoupled spectrum.

Some carbon atoms give rise to exceptionally small signals due to weak NOE enhancement which is one of the reasons for wrong relative intensities of the signals. In gated-decoupling mode, the noise decoupler is gated on during the pulse at the early part of free induction decay (FID), and then gated off during the pulse delay. Thus, NOE enhancement is minimized for all carbon atoms. This happens because the free induction signal decays quickly in an exponential manner, whereas the NOE factor slowly builds up in an exponential manner. Due to the removal of the NOE factor, longer signals acquisition time is required which is a demerit of the gated-decoupling mode.

The above discussion on different kinds of ^{13}C NMR spectra recorded in different modes of spectrometer operation is summarized in Table 6.1.

Table 6.1 Types of ^{13}C NMR spectra and structural information obtained from them

^{13}C NMR spectrum	Structural information obtained
(a) Proton-noise decoupled	Exhibits a single line (singlet) for each kind of chemically nonequivalent carbon present in the molecule. This allows us to count the number of kinds of carbons and deduce their environments from chemical shifts.
(b) Off-resonance decoupled	Spin-spin coupling causes splitting of ^{13}C signals into multiplets. A carbon bonded to n protons gives a signal that is split into $n + 1$ peaks. This shows the number of protons attached to each kind of carbon.
(c) Gated-decoupled	Area under each peak is directly proportional to the number of carbon atoms causing that peak. Thus, the relative number of carbon atoms represented by each peak can be determined by integrating the area under each peak, but is rarely done in practice.

6.4 Chemical Shift Equivalence

The discussion on chemical shift equivalence for protons also applies to carbon atoms and the chemical equivalence or non-equivalence of carbon atoms is judged in the same way as discussed for protons in Section 5.8. For example, carbon atoms of all the three methyl groups in *t*-butyl alcohol $(CH_3)_3COH$ are equivalent in the same sense in which the protons of a methyl group are. Thus, the CMR spectrum of *t*-butyl alcohol shows only two signals. Let us recognize various kinds of carbon atoms, and thus predict the number of CMR signals for molecules. The multiplicity of each signal can also be predicted (Section 6.3(ii)).

Example 1. Indicate the types of carbon atoms and number of signals in the proton-noise decoupled ^{13}C NMR spectra of the following compounds:

(a) Diethyl ether
(b) Ethyl propionate
(c) 2-methyl-2-butene
(d) 1,3-dichloropropane

Solution (a) Diethyl ether has two kinds of carbon atoms labelled as *a* and *b*, hence it will exhibit two CMR signals

$$\overset{a}{C}H_3\overset{b}{C}H_2O\overset{b}{C}H_2\overset{a}{C}H_3$$

Two types of carbon atoms; two CMR signals.

(b) Ethyl propionate has five types of carbon atoms, hence it will exhibit five CMR signals

$$\overset{a}{C}H_3\overset{b}{C}H_2\overset{c}{C}OO\overset{d}{C}H_2\overset{e}{C}H_3$$

Five types of carbon atoms; five CMR signals.

(c) 2-methyl-2-butene has four kinds of carbon atoms, hence it will exhibit four CMR signals.

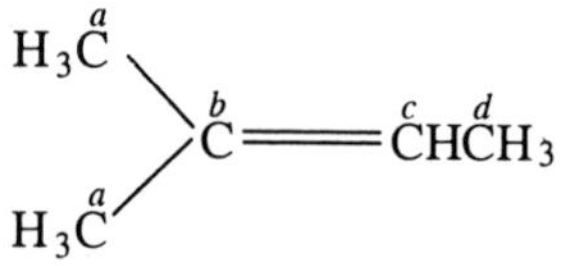

$$(H_3\overset{a}{C})_2\overset{b}{C}{=}\overset{c}{C}H\overset{d}{C}H_3$$

Four types of carbon atoms; four CMR signals.

(d) 1,3-dichloropropane has two types of carbon atoms, hence it will exhibit two CMR signals.

$$Cl\overset{a}{C}H_2\overset{b}{C}H_2\overset{a}{C}H_2Cl$$

Two types of carbon atoms; two CMR signals.

Example 2. Indicate the number of peaks and their multiplicity in the off-resonance decoupled ^{13}C NMR spectra of the following compounds:

(a) 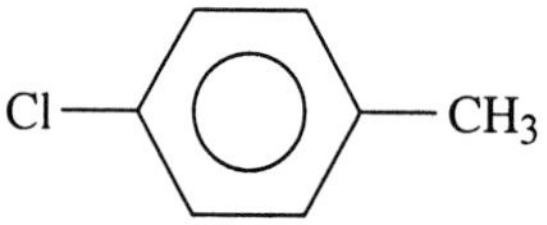 Cl—C_6H_4—CH_3 (para)
(b) $(CH_3)_3CCH_2COOCH_3$

(c) $CH_3CHOHCH_2CH_3$

Solution

(a) 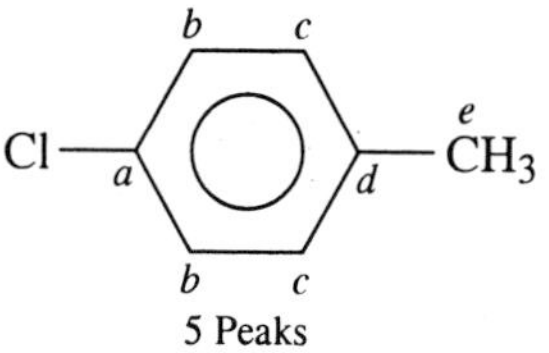 Cl—C_6H_4—$\overset{e}{C}H_3$ (ring carbons labelled *a*, *b*, *c*, *d*, *b*, *c*)

5 Peaks

Carbon *a*, singlet; *b*, doublet; *c*, doublet; *d*, singlet and *e*, quartet

(b)
$$\overset{a}{CH_3}-\underset{\underset{\overset{}{CH_3}\,(a)}{|}}{\overset{\overset{CH_3\,(a)}{|}}{C^b}}-\overset{c}{CH_2}\overset{d}{C}OO\overset{e}{C}H_3$$

5 Peaks

Carbon *a*, quartet; *b*, singlet; *c*, triplet; *d*, singlet and *e* quartet.

(c)
$$\overset{a}{C}H_3\underset{\underset{OH}{|}}{\overset{b}{C}}H\overset{c}{C}H_2\overset{d}{C}H_3$$

4 Peaks

Carbon *a*, quartet; *b*, doublet; *c*, triplet and *d* quartet.

Example 3. Using proton-decoupled ^{13}C NMR spectroscopy, distinguish the following isomeric compounds:

(a) 2-pentanone and 3-pentanone
(b) 1-propanol and 2-propanol
(c) Ethanol and dimethyl ether

Solution (a) There are five types of carbon atoms in 2-pentanone. Hence, it will show five CMR signals. 3-pentanone has only three types of carbon, and thus will exhibit only three CMR signals

$$\overset{a}{C}H_3\overset{b}{C}O\overset{c}{C}H_2\overset{d}{C}H_2\overset{e}{C}H_3$$

2-pentanone
5 types of carbon atoms
5 CMR signals

$$\overset{a}{C}H_3\overset{b}{C}H_2\overset{c}{C}O\overset{b}{C}H_2\overset{a}{C}H_3$$

3-pentanone
3 types of carbon atoms
3 CMR signals

(b) 1-propanol has three kinds of carbons, hence it will show three CMR signals, whereas 2-propanol has only two types of carbons and will show only two CMR signals

$$\overset{a}{C}H_3\overset{b}{C}H_2\overset{c}{C}H_2OH$$

1-propanol
3 types of carbon
3 CMR signals

$$\overset{a}{C}H_3\overset{b}{C}HOH\overset{a}{C}H_3$$

2-propanol
2 types of carbon
2 CMR signals

(c) Ethanol has two types of carbons, hence it will show two signals. On the other hand, dimethyl either has only one type of carbon, and thus will exhibit only one CMR signal

$$\overset{a}{C}H_3\overset{b}{C}H_2OH$$

Ethanol
2 types of carbon
2 CMR signals

$$\overset{a}{C}H_3OH\overset{a}{C}H_3$$

Dimethyl ether
1 type of carbon
1 CMR signal

Example 4. How many peaks do you expect in the proton-noise decoupled ^{13}C NMR spectra of the following compounds?

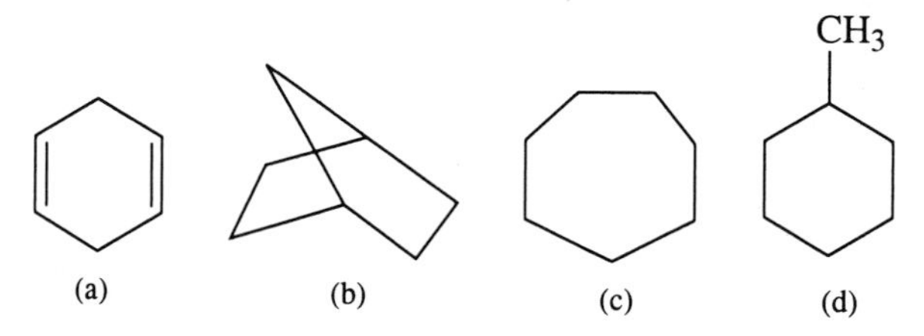

Solution (a) 2 peaks; (b) 3 peaks; (c) 1 peak and (d) 5 peaks.

Example 5. Predict the number of signals and their multiplicity in the off-resonance proton decoupled CMR spectra of the following compounds:

(i) Methyl succinate (ii) Methylcyclopentane

Solution

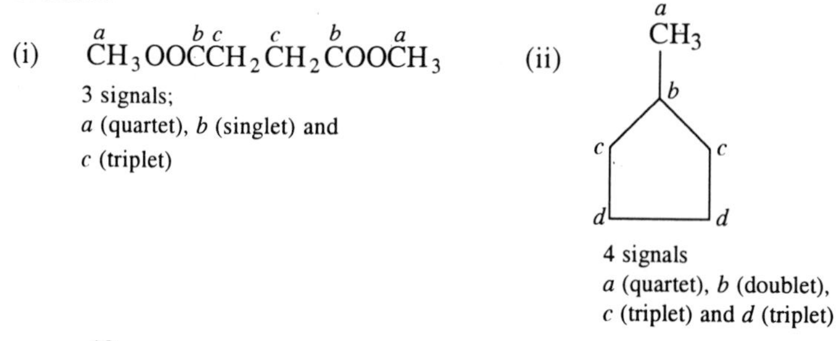

6.5 ^{13}C Chemical Shifts

As mentioned earlier (Section 6.1), ^{13}C chemical shifts range from 0 to about 250 δ (compared to δ 0-15 for PMR) with respect to TMS which is used as the common reference compound. This range is about 20 times that of routine PMR spectra (~12 δ). Due to this large spread of chemical shifts, relatively fewer peaks overlap in ^{13}C NMR spectra. At the operating frequency of 25.2 MHz, the routine spectral width (sweepwidth) is 5000 MHz (198.4 δ). Several cations absorb at ~335 δ downfield and CI_4 has been recorded at approximately –290 δ upfield from TMS.

Trends in chemical shifts of ^{13}C are to some extent parallel to those of ^{1}H. The ^{13}C chemical shifts are mainly related to the hybrid state of the carbon and electronegativities of substituents. ^{13}C chemical shifts are affected by substituents as far removed as the δ position. ^{13}C as well as ^{1}H chemical shifts are also affected by solvents and hydrogen bonding. A general correlation chart for ^{13}C chemical shifts of important chemical classes is given in Fig. 6.4.

In general, on moving downfield from TMS, the sequence alkanes, substituted alkanes, alkynes, olefins, aromatics and aldehydes for ^{13}C chemical shifts are similar to that for ^{1}H. Approximate chemical shift ranges of some ^{13}C resonances

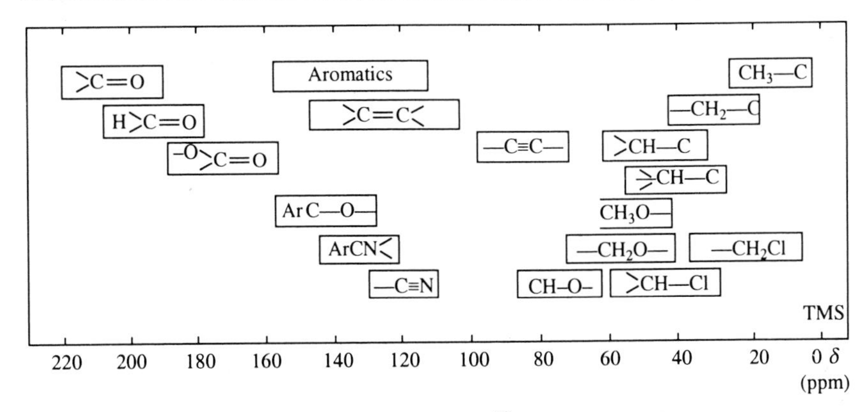

Fig. 6.4 General regions of ^{13}C chemical shifts

are given in Table 6.2.

Table 6.2 General ranges of ^{13}C chemical shifts

Type of carbon	Chemical shift δ	Type of carbon	Chemical shift δ
$—CH_3$	8–30	—C≡C—	65–90
$—CH_2—$	15–55	>C=C<	100–150
>CH—	20–60	—C(<)—O—	40–80
—C(<)—I	0–40	>C=O	150–220
—C(<)—Br	25–65	C of aromatic ring	110–160
—C(<)—Cl	35–80	—C(<)—N<	30–65

6.6 Factors Affecting ^{13}C Chemical Shifts

^{13}C chemical shifts are basically affected in the same way as ^{1}H chemical shifts. Important factors affecting ^{13}C chemical shifts are discussed as follows.

(i) α-, β- and γ-Effects

^{13}C chemical shifts are significantly affected by the presence and number of substituents at α-, β- or γ-position with respect to the carbon under study. Thus, there is an increase in chemical shifts (δ values) on going from primary to secondary to tertiary to quaternary carbon atoms and the average increase for per hydrogen atom replaced is in the range of 7 to 10 δ. This is illustrated by moving from methane to neopentane as follows:

	CH_4	$CH_3—CH_3$	$H_3C—CH_2—CH_3$	$H_3C—CH(CH_3)—CH_3$	$H_3C—C(CH_3)_2—CH_3$
δ_C^{TMS}	−2.5	5.7	16.3	25.4	28.1
	Methane	Ethane	Propane	Isobutane	Neopentane

Substituents (except iodine) at α- and β-positions enhance the δ value, i.e. the ^{13}C absorption position is shifted downfield, whereas substituents at the γ-position shift the absorption position upfield, i.e. the δ value is decreased. In other words, the presence of substituents at α- and β-positions exhibits deshielding effects, whereas that at γ-position shows shielding effect. Such effects are called α-, β- and *γ-effects*. For example, in case of alkanes the value of α-effect is $\delta + 9.1$ (downfield), β-effect is $\delta + 9.4$ (downfield), γ-effect is $\delta - 2.5$ (upfield), δ-effect is $\delta + 0.3$ (downfield) and ε-effect is $\delta + 0.1$ (downfield). The γ-effect is important in conformational analysis and stereostructure of alkenes, and is discussed in Section 6.6(v).

Based on these observations, we have the following equation for calculating chemical shifts for carbon atoms in alkanes within reasonable limits

$$\delta_C^{TMS} = 1 + 7n_1 + 8n_2 - 2n_3 \qquad (6.1)$$

where δ_C^{TMS} is the predicted chemical shift for a carbon with reference to TMS, n_1, n_2 and n_3 are the number of carbon atoms one, two and three bonds, respectively, away from the carbon atom whose chemical shift is being calculated. The calculated and observed values of chemical shifts match within ±4 ppm. Using Eq. (6.1) let us calculate the chemical shift of each carbon atom in 3-methylpentane. Here, for C-1, $n_1 = 1$, $n_2 = 1$ and $n_3 = 2$. Thus, the chemical shift is

$$1 + (7 \times 1) + (8 \times 1) - (2 \times 2) = 12\,\delta$$

Similarly, for C-2, $n_1 = 2$, $n_2 = 2$ and $n_3 = 1$. Thus, the chemical shift is

$$1 + (7 \times 2) + (8 \times 2) - (2 \times 1) = 29\ \delta, \text{ and so on}$$

$\overset{1}{C}H_3—\overset{2}{C}H_2—\overset{3}{C}H(\overset{6}{C}H_3)—\overset{4}{C}H_2—\overset{5}{C}H_3$

δ_C^{TMS}		C-1	C-2	C-3	C-6
	Calculated	12	29	38	20
	Observed	11.5	29.5	36.9	18.8

3-Methylpentane

An alternative formula for calculating ^{13}C chemical shifts is

$$\delta_C^{TMS} = -2.5 + 9.1n\alpha + 9.4n\beta - 2.5n\gamma + 0.3n\delta \qquad (6.2)$$

where δ_C^{TMS} is the predicted chemical shift for a carbon with reference to TMS. $n\alpha$, $n\beta$, $n\gamma$ and $n\delta$ are the number of carbon atoms one, two, three and four bonds, respectively, away from the carbon atom whose chemical shift is being calculated. The constant $-2.5\ \delta$ is the chemical shift of ^{13}C of methane. Using Eq. (6.2), let us calculate the chemical shift of each carbon atom in *n*-hexane. Here, for C-1, $n\alpha = 1$, $n\beta = 1$.

		$\overset{1}{C}H_3$—	$\overset{2}{C}H_2$—	$\overset{3}{C}H_2$—	$\overset{4}{C}H_2$—	$\overset{5}{C}H_2$—	$\overset{6}{C}H_3$
δ_C^{TMS}	Calculated	13.8	22.9	32.0			
	Observed	14.1	23.1	32.2			

n-Hexane

$n\gamma = 1$ and $n\delta = 1$. Thus the chemical shift is

$$-2.5 + (9.1 \times 1) + (9.4 \times 1) - (2.5 \times 1) + (0.3 \times 1) = 13.8\ \delta$$

Similarly, for C-2, $n\alpha = 2$, $n\beta = 1$, $n\gamma = 1$ and $n\delta = 1$. Thus the chemical shift is

$$-2.5 + (9.1 \times 2) + (9.4 \times 1) - (2.5 \times 1) + (0.3 \times 1) = 22.9\ \delta,$$ and so on

There is very good agreement between the calculated and observed values.

The ^{13}C chemical shifts for some linear and branched-chain alkanes are given in Table 6.3.

Table 6.3 ^{13}C chemical shifts for some linear and branched-chain alkanes (ppm from TMS)

Alkane	C-1	C-2	C-3	C-4	C-5
Methane	–2.5				
Ethane	5.7				
Propane	15.8	16.3	15.8		
Butane	13.4	25.2	25.2		
Pentane	13.9	22.8	34.7	22.8	13.9
Hexane	14.1	23.1	32.2	32.2	23.1
Heptane	14.1	23.2	32.6	29.7	32.6
Octane	14.2	23.2	32.6	29.9	29.9
Isobutane	24.5	25.4			
Isopentane	22.2	31.1	32.0	11.7	
Neopentane	31.7	28.1			
2,3-dimethylbutane	19.5	34.3			
2,2,3-trimethylbutane	27.4	33.1	38.3	16.1	
3-methylpentane	11.5	29.5	36.9	(18.8, 3-CH_3)	

(ii) Effect of Substituents on Alkanes

The α-effects exerted by polar substituents (Table 6.4) are quite large compared to that exerted by the CH_3 group. For example, the α-effect of the OH group causes a large downfield shift of the carbon absorption to which it is attached (Table 6.4). However, β- and γ-effects are nearly the same as for the methyl group. It should be noted that simple alkyl substitution has much larger effect on ^{13}C chemical shifts as compared to that on PMR shifts. The approximate ^{13}C chemical shifts for the carbon atoms of alkane derivatives may be calculated by properly applying the increments given in Table 6.4. During calculation, the required chemical shift values for different carbon atoms may be taken from Table 6.3 or calculated from Eqs. (6.1) and (6.2). For example, the approximate chemical shifts for carbon atoms of 3-pentanol may be calculated from the values for pentane in Table 6.3 and the increments for the functional group (OH), in Table 6.4, as follows:

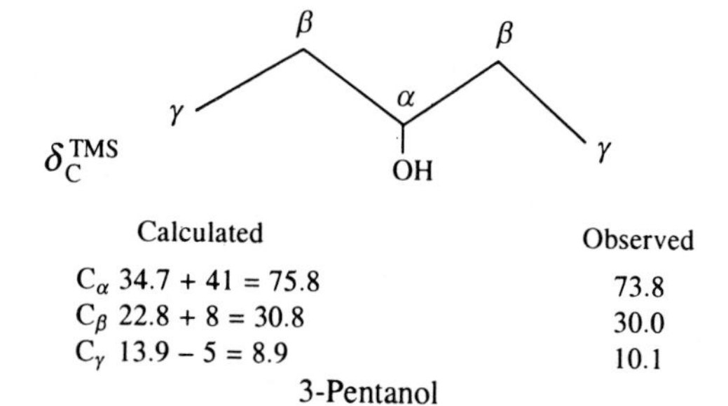

	Calculated	Observed
C_α	34.7 + 41 = 75.8	73.8
C_β	22.8 + 8 = 30.8	30.0
C_γ	13.9 − 5 = 8.9	10.1

3-Pentanol

Table 6.4 Increments* (ppm) for substituents *Y* on replacement of H by *Y* in alkanes

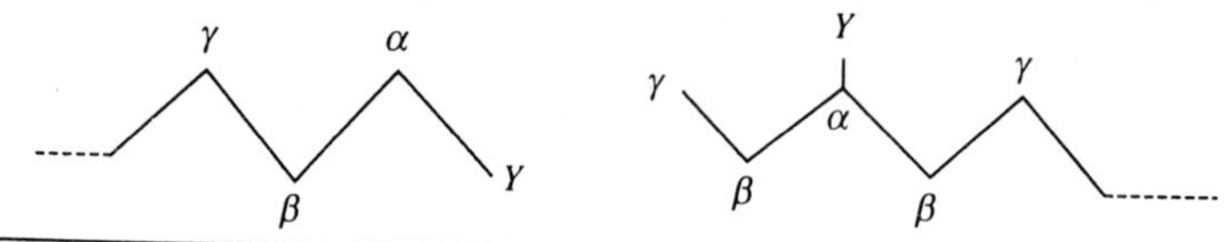

Substituent	α		β		γ
Y	Terminal	Internal	Terminal	Internal	
CH_3	+9	+6	+10	+8	−2
$HC{=}CH_2$	+20		+6		−0.5
$C{\equiv}CH$	+4.5		+5.5	+2	−3.5
COOH	+21	+16	+3	+2	−2
COO^-	+25	+20	+5	+3	−2
COOR	+20	+17	+3	+2	−2
COCl	+33	+28		+2	
$CONH_2$	+22		+2.5		−0.5
COR	+30	+24	+1	+1	−2
CHO	+31		0		−2
Phenyl	+23	+17	+9	+7	−2
OH	+48	+41	+10	+8	−5
OR	+58	+51	+8	+5	−4
OCOR	+51	+45	+6	+5	−3
NH_2	+29	+24	+11	+10	−5
NH_3^+	+26	+24	+8	+6	−5
NHR	+37	+31	+8	+6	−4
NR_2	+42		+6		−3
NH_3^+	+31		+5		−7
NO_2	+63	+57	+4	+4	
CN	+4	+1	+3	+3	−3
SH	+11	+11	+12	+11	−4
SR	+20		+7		−3
F	+68	+63	+9	+6	−4
Cl	+31	+32	+11	+10	−4
Br	+20	+25	+11	+10	−3
I	−6	+4	+11	+12	−1

*Add these increments to the chemical shift values of the appropriate carbon atom in Table 6.3 or to the shift value calculated from Eq. (6.1) or (6.2).

(iii) Effect of Hybridization of Carbon

The chemical shift of carbon is very significantly affected by its hybrid state. sp^3-hybridized carbon atoms unsubstituted by heteroatoms absorb 0-60 ppm downfield from TMS, sp^2 carbon atoms of alkenes unsubstituted by heteroatoms absorb in the range of ~100-150 ppm downfield from TMS and sp carbon atoms of alkynes unsubstituted by heteroatoms absorb in the range of ~65–90 ppm downfield from TMS. It is noteworthy that sp^3 hybridized carbon atoms in alkenes and benzenoid hydrocarbons absorb almost in the same general region of the ^{13}C spectrum (Table 6.1). Thus, PMR spectrum is more useful for distinguishing olefinic and aromatic C=C bonds.

The double bond only slightly affects the shift of sp^3 carbon in a molecule, e.g. the methyl signal of propene is at 18.7 ppm and that of propane at 15.8 ppm. In general, *cis* —CH=CH— signals are upfield compared to those of the corresponding *trans* groups, and the terminal $=CH_2$ groups absorbs upfield compared to an internal =CH— group.

Approximate shifts of acyclic olefinic carbons can be calculated by making the following substitution corrections to the shift value of carbon in ethylene (123.3 ppm). Here α, β and γ represent substituents on the same end of the double bond as the olefinic carbon whose shift is being calculated (indicated below by an arrow), and α', β' and γ' represent substituents on the far side.

$$\overset{\gamma}{C}—\overset{\beta}{C}—\overset{\alpha}{C}—\overset{\downarrow}{C}=C—\overset{\alpha'}{C}—\overset{\beta'}{C}—\overset{\gamma'}{C}$$

Substituent carbon C	Correction
α	+10.6
β	+7.2
γ	−1.5
α'	−7.9
β'	−1.8
γ'	−1.5
Z (*cis*) correction	−1.1

For example, the chemical shifts of olefinic carbons (C-1 and C-2) in 1-pentene can be calculated as

$$\underset{\gamma'}{\overset{5}{C}H_3}—\underset{\beta'}{\overset{4}{C}H_2}—\underset{\alpha'}{\overset{3}{C}H_2}—\overset{2}{C}H=\underset{\uparrow}{\overset{1}{C}H_2} \qquad \underset{\gamma}{\overset{5}{C}H_3}—\underset{\beta}{\overset{4}{C}H_2}—\underset{\alpha}{\overset{3}{C}H_2}—\underset{\uparrow}{\overset{2}{C}H}=\overset{1}{C}H_2$$

$$\delta_{C-1}^{TMS} = 123.3 + (1 \times -7.9) + (1 \times -1.8) + (1 \times -1.5) = 112.1$$

$$\delta_{C-2}^{TMS} = 123.3 + (1 \times 10.6) + (1 \times 7.2) + (1 \times -1.5) = 139.6$$

The observed values are C-1 = 114.3 and C-2 = 138.5 δ. Thus, there is a good agreement between the calculated and observed values.

Olefinic carbon atoms in polyenes are treated as alkane carbon substituents on one of the double bonds in the above calculations. Thus, in calculating the chemical shift of C-2 in 1,4-pentadiene, C-4 is treated as $\alpha\beta$-sp^3 carbon atom.

The sp carbon atom shifts the directly attached sp^3 carbon about 5-15 ppm upfield compared to the corresponding alkane. The terminal ≡CH absorbs upfield compared to the internal ≡C*R*.

(iv) Effect of Substituents on Olefinic, Acetylenic and Aromatic Carbons
Compared to saturated sp^3 carbon atoms, the polar effects of substituents on unsaturated (sp^2 and sp) carbons are longer-ranging because of resonance (mesomeric) effects in addition to inductive effects. The ^{13}C chemical shifts for vinyl ethers can be explained on the basis of electron density of the contributing structures:

$$H_2C{=}CH{-}\ddot{O}{-}CH_3 \leftrightarrow H_2\ddot{C}^{-}{-}CH{=}\overset{+}{O}{-}CH_3$$

δ 84.2 153.2 52.5

Thus, the olefinic or an aromatic carbon to which a methoxy group is attached is deshielded due to its negative inductive effect. On the other hand, the other carbon of the double bond as well as the carbons in the *ortho* and *para* positions to the methoxy group, e.g. in anisole, are shielded due to electron-donating ability of oxygen through mesomeric effect.

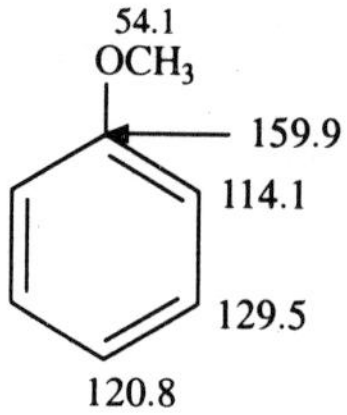

54.1
OCH_3
159.9
114.1
129.5
120.8

Carbon atoms of benzene absorb at 128.5 ppm

The ^{13}C chemical shifts of α, β-unsaturated ketones can also be explained on the basis of the same effects.

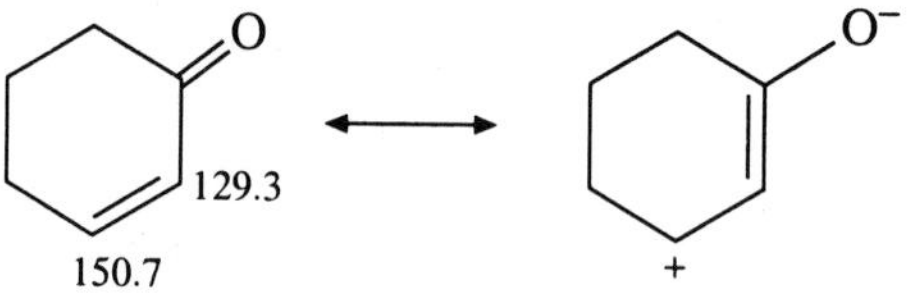

O
129.3
150.7
↔
O^-
+

The ^{13}C chemical shifts of the *sp* carbons in the following ethers can be explained in the same way as discussed above for vinyl ethers.

$HC{\equiv}C{-}OCH_2CH_3$ (23.2, 89.4) $CH_3C{\equiv}C{-}OCH_3$ (28.0, 88.4)

The same explanation applies to the proton chemical shifts in such compounds.

Benzene carbon atoms absorb at 128.5 ppm. As stated earlier (Section 6.6(iii)), sp^2 carbons in alkenes and benzenoid hydrocarbons absorb in the same region of the CMR spectrum. For example, it is shown by the comparison of the ^{13}C chemical shifts in benzene and cyclohexene.

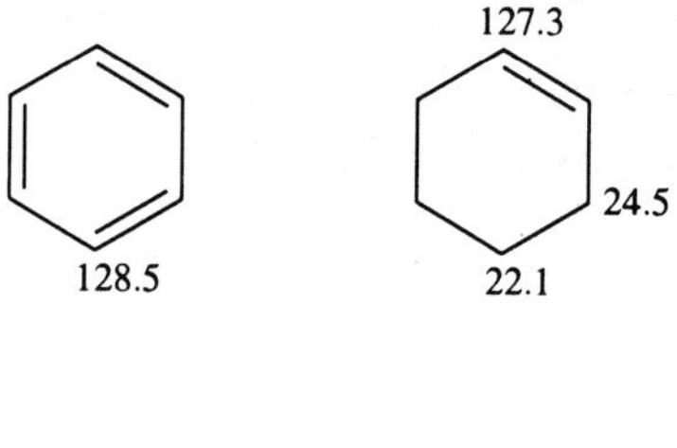

128.5   127.3
24.5
22.1

The effect of substituents on the substituted carbon (C-1) and on the *ortho*, *meta* and *para* carbons in benzene derivatives is given in Table 6.5.

Table 6.5 Increments in the shifts of the aromatic carbon atoms of monosubstituted benzenes (ppm from benzene at 128.5 ppm)

Substituent	C-1	C-2 (*ortho*)	C-3 (*meta*)	C-4 (*para*)
H	0.0	0.0	0.0	0.0
CH_3	+9.3	+0.7	–0.1	–2.9
CH_3CH_2	+15.6	–0.5	0.0	–2.6
$CH(CH_3)_2$	+20.1	–2.0	0.0	–2.5
$HC{=}CH_2$	+9.1	–2.4	+0.2	–0.5
C_6H_5	+12.1	–1.8	–0.1	–1.6
F	+35.1	–14.3	+0.9	–4.5
Cl	+6.4	+0.2	+1.0	–2.0
Br	–5.4	+3.4	+2.2	–1.0
I	–32.2	+9.9	+2.6	–7.3
OH	+26.6	–12.7	+1.6	–7.3
OCH_3	+31.4	–14.4	+1.0	–7.7
CHO	+8.2	+1.2	+0.6	+5.8
$COCH_3$	+7.8	–0.4	–0.4	+2.8
COOH	+2.9	+1.3	+0.4	+4.3
CN	–16.0	+3.6	+0.6	+4.3
NO_2	+19.6	–5.3	+0.9	+6.0
NH_2	+19.2	–12.4	+1.3	–9.5
$NHCOCH_3$	+11.1	–9.9	+0.2	–5.6
$COOCH_3$	+2.0	+1.2	–0.1	+4.8

The chemical shifts for ring carbon atoms in polysubstituted benzenes can be approximately calculated by applying the principle of substituent additivity. For example, the chemical shifts of ring carbon atoms in 4-chlorobenzonitrile can be calculated as follows by using the increments given in Table 6.5:

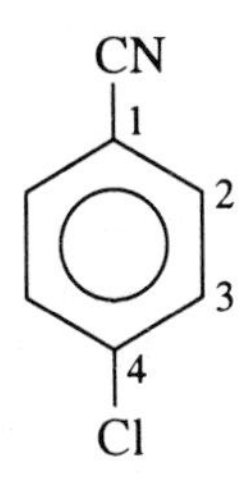

$\delta_{\text{C-1}}^{\text{TMS}}$ = 128.5 (chemical shift of benzene C) – 16 (attached CN group) – 2(*p*-Cl)
= 110.5 (observed value = 111.9 δ)

$\delta_{\text{C-2}}^{\text{TMS}}$ = 128.5 + 3.6 (*o*-CN) + 1.0 (*m*-Cl) = 133.1 (observed value = 133.6 δ)

δ_{C-3}^{TMS} = 128.5 + 0.2 (*o*-Cl) + 0.6 (*m*-CN) = 129.3 (observed value = 129.8 δ)

δ_{C-4}^{TMS} = 128.5 + 6.4 (attached Cl) + 4.3 (*p*-CN)
= 139.2 (observed value = 139.3 δ)

The substituted aromatic carbon atom can be distinguished from the unsubstituted aromatic carbon atom by its decreased peak height. This is because the substituted aromatic carbon lacks a proton and thus, suffers from a longer relaxation time T_1 and a diminished NOE.

(v) γ-Effect

In substituted alkanes (Tables 6.3 and 6.4), we saw the ^{13}C upfield γ shift due to the γ carbon or γ substituent because of steric compression of a Gauche interaction.

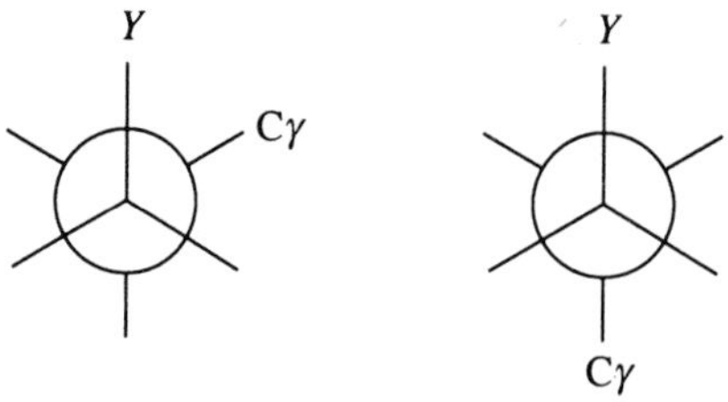

Y Y
Cγ
Cγ
Gauche Anti

However, it has no counterpart in ^{1}H spectra. In some cases, the upfield shift with the substituent *Y* in the anti-conformation is attributed to hyperconjugation. The γ-Gauche steric compression (γ-effect) accounts for the upfield ^{13}C shift of an axial methyl substituent compared to an equatorial methyl, and for the upfield shift of the γ carbon atoms of the ring (with respect to the methyl group). In case of alkenes, the γ-effect mainly depends on their stereostructures, i.e. the spatial relationship between the carbon being observed and the substituent. For example, because of the γ effect the ^{13}C shift of the methyl groups in *cis*-2-butene are at 12.1 ppm and that in *trans*-2-butene are at 17.6 ppm. Thus, the methyl groups in *cis*-2-butene are exerting a γ effect of –5.5 ppm on each other. It is comparable with the carbon shifts of the methyl groups in butane which appear at 13.4 ppm due to the γ-effect. Since propene has no γ carbon, there is no γ-effect in this molecule. Thus, the carbon of its methyl group appears at 18.7 ppm which is comparable with the ^{13}C shifts (17.6 ppm) of the methyl groups in trans-2-butene. It indicates that there is negligible γ-effect in *trans*-2-butene.

The appearance of olefinic protons around δ 5 in the PMR spectrum and olefinic carbons around δ 100-150 in the CMR spectrum is used to detect the presence of a double bond. Similarly, the magnitudes of the couping constants in the PMR spectrum and the γ-effects in the CMR spectrum are highly useful for the determination of the stereostructure of alkenes.

(vi) Carbonyl Carbons

Carbonyl carbons absorb in the range of ~150-220 ppm downfield from TMS.

The absorption position of a particular carbonyl carbon depends on the electron-donating ability of the attached atoms or groups. General regions of chemical shifts of various carbonyl carbons are given in Table 6.6.

Table 6.6 **^{13}C chemical shifts of various carbonyl carbons**

Compound	^{13}C chemical shift (ppm from TMS)
Ketone	205–220
Aldehyde	200–210
Carboxylic acid	175–185
Primary amide	170–180
Ester and anhydride	165–175
Secondary amide and imide	160–170

6.7 ^{13}C Chemical Shifts (ppm from TMS) of Some Compounds

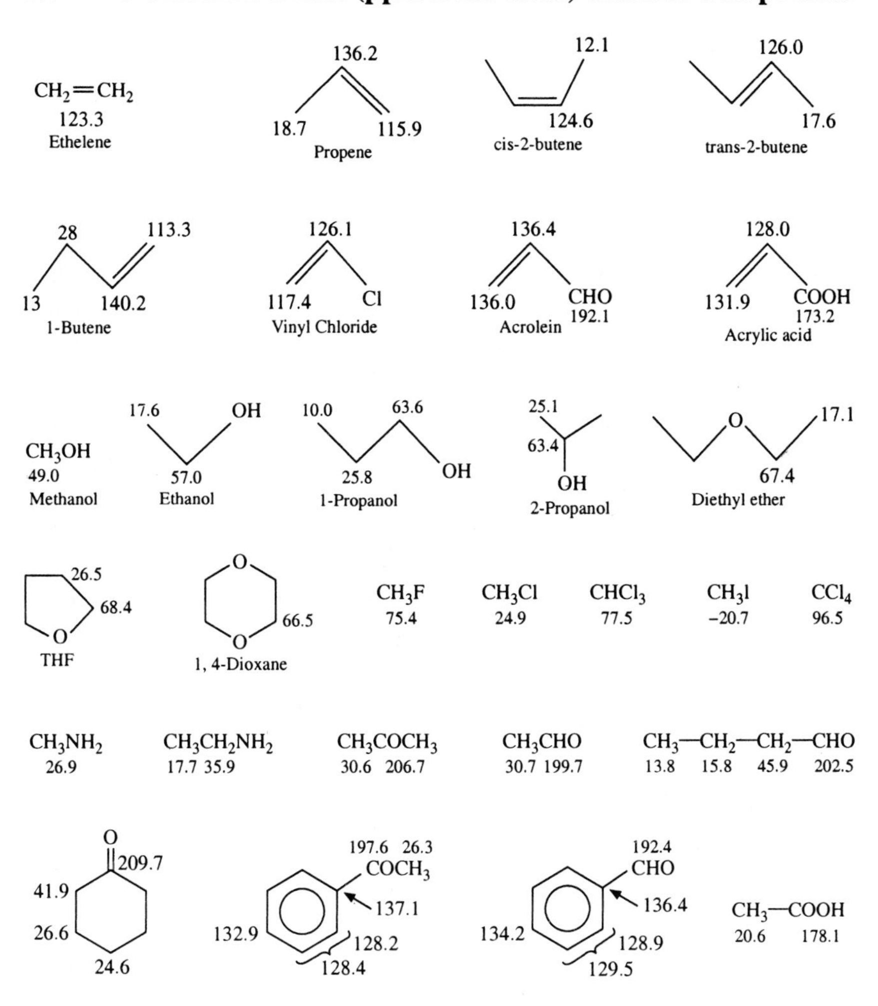

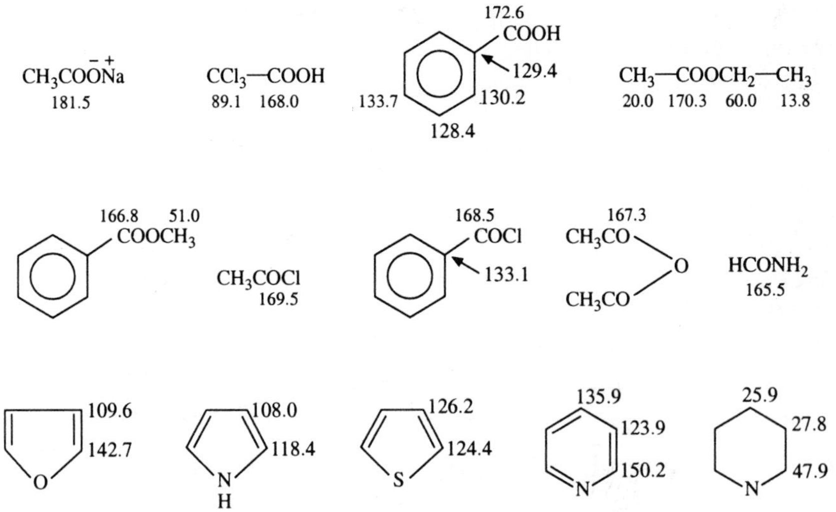

$CH_3COO^-Na^+$ 181.5

CCl_3—COOH 89.1 168.0

C_6H_5COOH: 172.6 (COOH), 129.4, 130.2, 128.4, 133.7

CH_3—$COOCH_2$—CH_3 20.0 170.3 60.0 13.8

$C_6H_5COOCH_3$ 166.8 51.0

CH_3COCl 169.5

C_6H_5COCl 168.5 (COCl), 133.1

$(CH_3CO)_2O$ 167.3

$HCONH_2$ 165.5

Furan: 109.6, 142.7

Pyrrole: 108.0, 118.4

Thiophene: 126.2, 124.4

Pyridine: 135.9, 123.9, 150.2

Piperidine: 25.9, 27.8, 47.9

6.8 Spin-Spin Coupling

Since routine CMR spectra are usually noise decoupled, spin-spin coupling is less important in CMR spectra than in PMR spectra. However, in off-resonance decoupling mode (Section 6.3(ii)), couplings due to protons directly attached to ^{13}C atoms (^{13}C—H couplings) are observed but other couplings (e.g. ^{13}C—C—H, ^{13}C—C—C—H etc.) are removed. For spin-spin splitting of signals see Section 6.3(ii).

The ^{13}C—^{13}C coupling is usually not observed because of the low probability of two adjacent ^{13}C atoms in a molecule.

The values for one-bond ^{13}C—H coupling constants ($^1J_{CH}$) range from about 110 to 320 Hz. The magnitude of coupling constant increases with increased *s* character of the ^{13}C—H bond, by the presence of electron-withdrawing groups on the carbon atom, and with angular distortion. Appreciable coupling of ^{13}C to ^1H has also been observed over two (^{13}C—C—H) or three (^{13}C—C—C—H) bonds (*J* values denoted as $^2J_{CH}$ and $^3J_{CH}$, respectively). Some representative $^1J_{CH}$ and $^2J_{CH}$ values are given in Table 6.7. $^2J_{CH}$ values usually range from about –5 to 60 Hz, whereas $^3J_{CH}$ values generally range from about –5 to 25 Hz. However, in aromatic rings, $^3J_{CH}$ values are characteristically larger than $^2J_{CH}$ values, e.g. in benzene, $^3J_{CH}$ = 7.4 Hz, whereas $^2J_{CH}$ = 1.0 Hz.

Coupling of ^{13}C to other nuclei, e.g. D, ^{19}F etc. may be observed in proton-noise decoupled spectra. Some representative coupling constants are:

CH_3CF_3, $^1J_{CF}$ = 271 Hz; CF_2H_2, $^1J_{CF}$ = 235 Hz; C_6H_5F, $^1J_{CF}$ = 245 Hz and $^2J_{CF}$ = 21.0 Hz; $CDCl_3$, $^1J_{CD}$ = 31.5 Hz; CD_3 $COCD_3$, $^1J_{CD}$ = 19.5 Hz; $(CD_3)_2SO$, $^1J_{CD}$ = 22.0 Hz; C_6D_6, $^1J_{CD}$ = 25.5 Hz

Table 6.7 Some $^1J_{CH}$ and $^2J_{CH}$ values

Compound	$^1J_{CH}$ (Hz)	Compound	$^1J_{CH}$ (Hz)
sp^3		*sp*^{2}	
CH_3CH_3	124.9	$CH_2{=}CH_2$	156.2
$CH_3CH_2CH_3$	119.2	CH_3CHO	172.4
$(CH_3)_3CH$	114.2	C_6H_6	159.0
CH_3NH_2	133.0	*sp*	
CH_3OH	141.0	$CH{\equiv}CH$	249.0
CH_3Cl	150.0	$C_6H_5C{\equiv}CH$	251.0
CH_2Cl_2	178.0	$HC{\equiv}N$	269.0
$CHCl_3$	209.0	*sp*^{3}	$^2J_{CH}$
		CH_3CH_3	−4.5
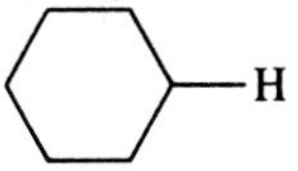 H	123.0	CH_3CCl_3	5.9
		CH_3CHO	26.7
		sp^2	
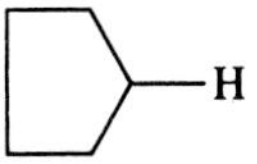 H	128.0	$CH_2{=}CH_2$	−2.4
		CH_3COCH_3	5.5
		$CH_2{=}CHCHO$	26.9
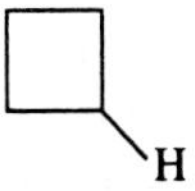 H	134.0	C_6H_6	1.0
		sp	
		$CH{\equiv}CH$	49.3
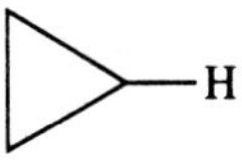 H	161.0	$C_6H_5OC{\equiv}CH$	61.0

6.9 Effect of Deuterium Substitution on CMR Signals

There is dramatic decrease in the height of the ^{13}C signal in a proton-noise decoupled spectrum when hydrogen on a carbon is substituted by deuterium. This is because

(i) deuteron-bearing carbons (^{13}C—D) have longer relaxation times than ^{13}C—H due to decreased dipole-dipole relaxation.

(ii) the enhanced line intensities caused by the NOE are lost, as there is no irradiation of deuterium during the proton-noised decoupled mode.

(iii) deuterium splits ^{13}C signal into three lines, i.e. a triplet (ratio 1 : 1 : 1). The multiplicity is calculated from the general formula $2nI + 1$. Since deuterium has $I = 1$, in molecules with one deuteron attached to each carbon (e.g. $CDCl_3$ and C_6D_6), the ^{13}C signal is $= 2 \times 1 \times 1 + 1 = 3$, i.e. a triplet (ratio 1 : 1 : 1).

The above points also explain the relatively weak signal shown by deuterated solvents. Similarly, CMR signals of carbons which do not bear protons, e.g. carbonyl carbons, phenyl C-1 and quaternary carbons, are recognized by their low intensities.

6.10 Use of Shift Reagents

The lanthanide shift reagents can also be used to spread out a ^{13}C spectrum in the same way as they are used in PMR spectroscopy (Section 5.17(iv)). The use of shift reagents may separate coincident peaks and proton shifts are also spread out, thus selective proton decoupling is facilitated.

6.11 Applications of CMR Spectroscopy

Similar to PMR spectroscopy, CMR spectroscopy also has become useful for organic chemists for structure determination. Direct observation of carbon skeleton and carbon-containing functional groups has now become possible by CMR spectroscopy. It is also useful in stereochemical studies. CMR spectroscopy is useful in the study of natural products, polymers and other complex biological molecules. Some important applications of CMR spectroscopy, besides obtaining routine structural information, are summarized as follows.

(i) Identification of Structural Isomers

Structural isomers can easily be distinguished by CMR spectroscopy. For example

(a) $CH_3CH_2CH_2Cl$ and $CH_3CHClCH_3$

1-Chloropropane ($CH_3CH_2CH_2Cl$) having three types of carbon will exhibit three signals in its proton-decoupled CMR spectrum, whereas only two signals will be observed for 2-chloropropane ($CH_3CHClCH_3$). Further, 1-chloropropane will exhibit one quartet and two triplets in its off-resonance decoupled CMR spectrum, whereas one quartet and one doublet will be observed in case of 2-chloropropane. Thus, the above isomers can be clearly distinguished.

(b) CH_3OCH_3 and CH_3CH_2OH

The proton-noise decoupled spectrum of dimethyl ether will exhibit only one signal, whereas that of ethanol will show two signals. Further, the off-resonance decoupled spectrum of dimethyl ether will exhibit a quartet, whereas that of ethanol will show a quartet and a triplet. This clearly distinguishes the above structural isomers.

(ii) Detection of a Double Bond and Distinction Between *cis* and *trans* Alkenes

The presence of an olefinic double bond is detectable by the low field resonances (~ 5 ppm) of olefinic hydrogens in the PMR spectrum and of olefinic carbons in the CMR spectrum (~110-150 ppm). The *cis* and *trans* alkenes can be distinguished by the magnitudes of the PMR coupling constants (Section 5.11(ii)) and the γ effects in the CMR spectra (Section 6.6(v)).

(iii) Distinction Between Conformers

γ-effect (Section 6.6(v)) is useful in distinguishing conformers. For example, because of the γ-Gauche steric compression (γ-effect) the carbon of an axial methyl group absorbs upfield compared to that of an equatorial methyl. Further, an axial methyl group at C-1 causes an upfield shift of several ppm at C-3 and C-5. Thus, the conformer having an axial methyl group can easily be distinguished from that having an equatorial methyl group in a rigid cyclohexane ring.

(iv) Study of Natural Products

^{13}C NMR spectroscopy is useful in the study of complex molecules, e.g. of natural products, polymers and other bioactive molecules because CMR spectra are usually less complex than PMR spectra. For example, cholesterol (a natural product with molecular formula $C_{27}H_{46}O$), shows a complex PMR spectrum due to the presence of a large number of hydrogens which cause overlap of the signals and the component peaks of the multiples. Thus, it becomes difficult to get clear insight into the structure. On the other hand, the proton, the proton-decoupled CMR spectrum of cholesterol is relatively simple and displays 26 resolved lines for the 27 carbon atoms present in the molecule.

6.12 Some Solved Problems

Problem 1. Suggest the number of signals and their multiplicity, if any, in the off-resonance ^{13}C NMR spectra of the following compounds:

(i) $CH_3COCH_2CH_3$ (iii) $(CH_3)_3COH$

(ii) $CH_3C{\equiv}CCH_3$ (iv) $CH_3OCH_2CH_3$

Solution

(i) $\overset{a}{C}H_3\overset{b}{C}O\overset{c}{C}H_2\overset{d}{C}H_3$

This compound has four kinds of carbon atoms. Hence, it will exhibit four CMR signals whose multiplicity will be

carbon *a*, quartet; *b*, singlet; *c*, triplet and *d*, quartet

(ii) $\overset{a}{C}H_3\overset{b}{C}{\equiv}\overset{b}{C}\overset{a}{C}H_3$

2-Butyne has two kinds of carbon atoms. Hence, it will exhibit two CMR signals. The signal due to carbon *a* will appear as a quartet, while that due to carbon *b* as a singlet.

(iii) $(\overset{a}{C}H_3)_3\overset{b}{C}OH$

tert-butyl alcohol has two kinds of carbons. Thus, it will exhibit two CMR signals. The signal due to carbons *a* will appear as a quartet, while that due to carbon *b* as a singlet.

(iv) $\overset{a}{C}H_3O\overset{b}{C}H_2\overset{c}{C}H_3$

Methoxyethane has three types of carbons. Hence, it will exhibit three CMR signals whose multiplicity will be

carbon *a*, quartet; *b*, triplet and *c* quartet

Problem 2. Predict the number of peaks and their multiplicity in the off-resonance ^{13}C NMR spectra of the following compounds. In each case, indicate carbons with the highest and the lowest chemical shifts in δ unit with reference to TMS.

(i) $CH_3COCH_2COOCH_3$ (ii) $(CH_3)_2CHCOOH$

(iii) $H_3C{-}O{-}\overset{\overset{O}{\|}}{C}{-}C_6H_4{-}\overset{\overset{O}{\|}}{C}{-}O{-}CH_3$

Solution

(i) $\overset{a}{C}H_3\overset{b}{C}O\overset{c}{C}H_2\overset{d}{C}OO\overset{e}{C}H_3$

This compound has five types of carbons. Hence, it will exhibit five peaks whose multiplicity will be

carbon *a*, quartet; *b*, singlet; *c*, triplet; *d*, singlet and *e*, quartet

Carbon *b* will have the highest chemical shift in δ unit, whereas carbon *a* the lowest δ value.

(ii) $(\overset{a}{C}H_3)_2\overset{b}{C}H\overset{c}{C}OOH$

This compound has three types of carbon atoms. Thus, it will exhibit three peaks whose multiplicity will be as follows:

carbon *a*, quartet; *b*, doublet and *c*, singlet

Carbon *c* will have the highest chemical shift, while carbon *a* the lowest chemical shift in δ unit.

(iii) $H_3\overset{a}{C}{-}O{-}\underset{b}{\overset{\overset{O}{\|}}{C}}{-}C_6H_4{-}\underset{b}{\overset{\overset{O}{\|}}{C}}{-}O{-}\overset{a}{C}H_3$ (ring carbons attached to C=O labelled *c*; other ring carbons labelled *d d*)

This compound has four kinds of carbon atoms. Hence, it will exhibit four peaks whose multiplicity will be

carbon *a*, quartet, *b*, singlet, *c*, singlet and *d*, doublet

Carbon *b* will have the highest and carbon *a* the lowest chemical shift in δ unit.

Problem 3. A highly symmetrical compound C_6H_8 shows two singlets in its proton-noise decoupled ^{13}C NMR spectrum. Off-resonance decoupled ^{13}C NMR spectrum of the same compound shows a triplet and a doublet. Deduce the structure of the compound.

Solution The proton-noise decoupled ^{13}C NMR of the compound shows two singlets. Thus, it has two kinds of carbon atoms. Since the off-resonance decoupled ^{13}C NMR spectrum of the compound shows a triplet and a doublet, the compound contains only CH_2 and CH groups. Thus, the structure of the compound C_6H_8 which fits the above ^{13}C NMR spectral data is

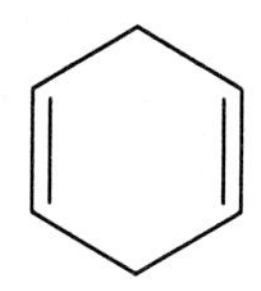

Problem 4. Acid-catalyzed dehydration of 1-methylcyclohexanol yields a mixture of two alkenes. After separating the two products, how will you distinguish between them with the help of proton-noise decoupled ^{13}C NMR spectra?

Solution. The acid-catalyzed dehydration of 1-methylcyclohexanol will yield 1-methylcyclohexene and methylenecyclohexane which can easily be distinguished from their proton-noise decoupled ^{13}C NMR spectra because 1-methylcyclohexene will exhibit seven signals in its proton-noise decoupled ^{13}C NMR whereas methylenecyclohexane will exhibit only five signals.

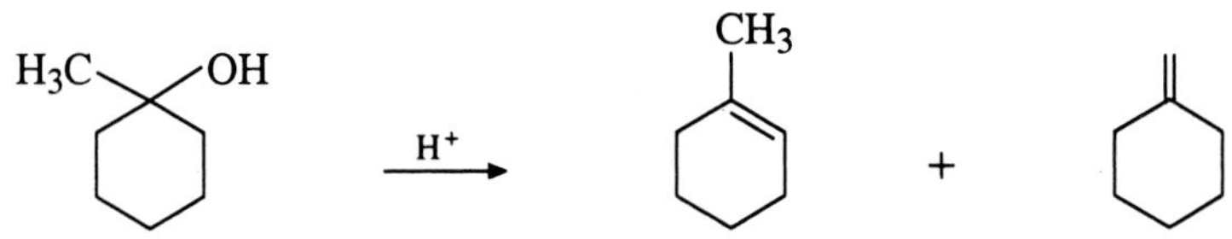

H_3C OH → (H^+) CH_3 +

1-Methylcyclohexanol — 1-Methylcyclohexene — 1-Methylenecyclohexane

Problem 5. Propose structures of compounds that fit these descriptions:

(a) A hydrocarbon with seven lines in its proton-noise decoupled ^{13}C NMR spectrum.
(b) A six-carbon hydrocarbon that shows only five resonance lines in its ^{13}C NMR spectrum.
(c) A four-carbon alcohol that shows only two resonance lines in its ^{13}C NMR spectrum.

Solution. (a) Any hydrocarbon containing seven types of carbon atom will show seven lines in its proton-noise decoupled spectrum, e.g.

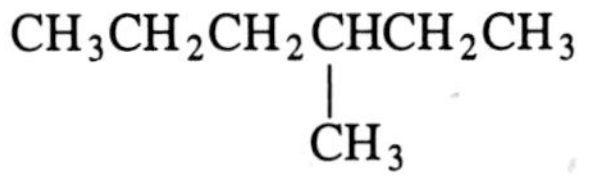

$$CH_3CH_2CH_2\underset{\displaystyle CH_3}{\underset{|}{C}H}CH_2CH_3$$

(b) This six-carbon hydrocarbon must have only five kinds of carbon atoms to show five resonance lines in its ^{13}C NMR spectrum, i.e.

$$(H_3C)_2CHCH_2CH_2CH_3$$

(c) This four-carbon alcohol must have only two types of carbon to show two resonance lines in its ^{13}C NMR spectrum. This alcohol is *t*-butyl alcohol

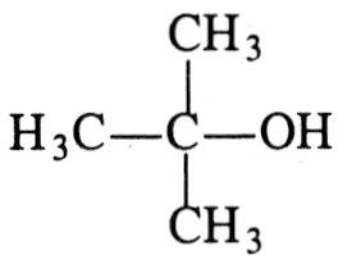

$$H_3C-\overset{\displaystyle CH_3}{\overset{|}{\underset{\displaystyle CH_3}{\underset{|}{C}}}}-OH$$

Problem 6. An organic compound $C_5H_{10}O$ shows one singlet, one triplet and one quartet at δ 211.0, 35.4 and 7.9, respectively, in its off-resonance decoupled ^{13}C NMR spectrum. Deduce the structure of the compound and explain the ^{13}C NMR spectral data.

Solution. The downfield (211.0 δ) singlet indicates the presence of a *keto* group. The appearance of a triplet and a quartet indicates the presence of CH_2 and CH_3 groups, respectively. Thus, the structure of the compound which fits its molecular formula $C_5H_{10}O$ and the ^{13}C NMR spectral data is

$$CH_3CH_2COCH_2CH_3$$

The given spectral data can be explained with this structure because it is expected to show a downfield singlet due to the *keto* group, a triplet due to CH_2 and a quartet due to the CH_3 groups.

Problem 7. On dehydration 3-hexanol gives four isomeric hexenes (I)-(IV). The proton-noise decoupled ^{13}C NMR data of these are given as follows. Analyze the CMR spectral data and correlate them to the structures of the four isomeric hexenes.

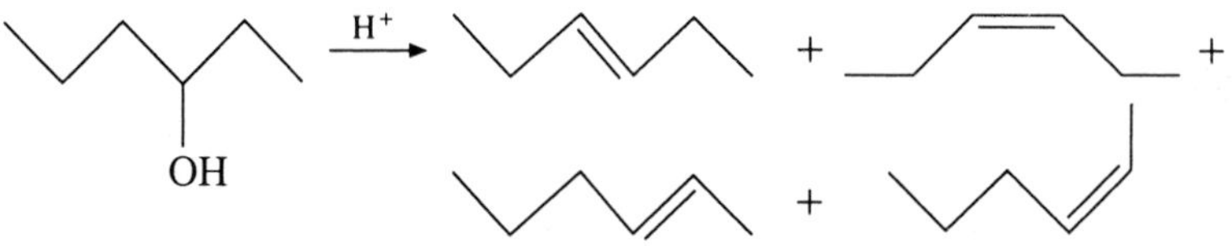

OH $\xrightarrow{H^+}$ + + +

(I) δ 12.3, 13.5, 23.0, 29.3, 123.7, 130.6 (III) δ 14.3, 20.6, 131.0
(II) δ 13.4, 17.5, 23.1, 35.1, 124.7, 131.5 (IV) δ 13.9, 25.8, 131.2

Solution. The isomers (III) and (IV) represent two isomeric 3-hexenes because each shows three signals. In these, the signals around δ 131 are due to sp^2 carbons and that around δ 14 are probably due to methyl carbons. The chemical shift of C-2 in (III) and (IV) differ by ~5 δ. This is due to γ-effect in a pair of *cis* and *trans* isomers. Thus, the isomer (III) with the upfield C-2 absorption must be the *cis* isomer, i.e. (III) is *cis*-3-hexene and (IV) is *trans*-3-hexene. On the basis of similar arguments, the isomer (I) is *cis*-2-hexene and (II) is *trans*-2-hexene because in (I) two of the signals (at δ 13.5 and 29.3) are about 5 ppm upfield compared with their counterparts (at δ 17.5 and 35.1) in the isomer (II).

Problem 8. Fig. 6.5 shows the proton-noise decoupled CMR spectrum of a compound having molecular formula C_3H_6O. The signal appearing at the lowest field is split into a doublet and the other two signals into triplets in the off-resonance spectrum of the compound. Deduce the structure of the compound.

Solution. The proton-noise decoupled spectrum of the compound C_3H_6O shows three signals. Hence, it has three kinds of carbons. Out of these, one signal is split into a doublet and two signals into triplets showing that the compound contains a CH group and two non-equivalent CH_2 groups. The doublet and one of the triplets appear in the region of olefinic carbons (δ 110-150). This indicates the presence of a $CH_2{=}CH{-}$ group. Thus, the structure which fits the given ^{13}C NMR spectra and the molecular formula C_3H_6O is

$$H_2C{=}CH{-}CH_2OH$$

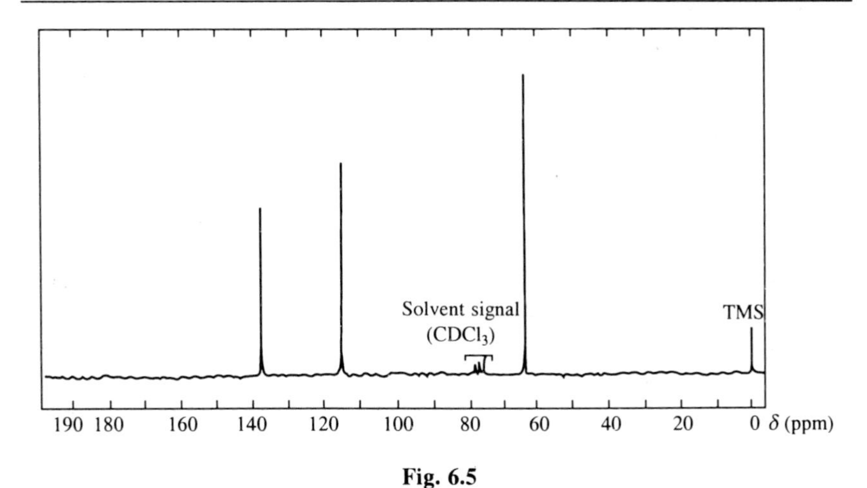

Fig. 6.5

Problem 9. Fig. 6.6 shows off-resonance decoupled and proton-noise decoupled CMR spectra of a compound $C_5H_{11}Cl$. Deduce the structure of the compound.

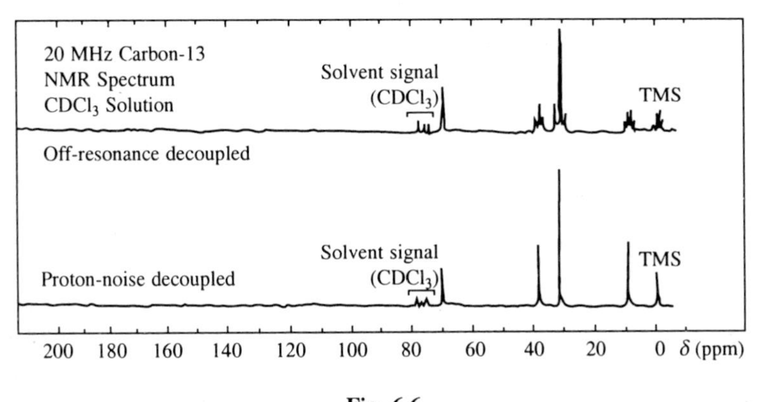

Fig. 6.6

Solution. The molecular formula $C_5H_{11}Cl$ of the compound shows that it is an open-chain saturated compound. The presence of four CMR signals in the proton-noise decoupled spectrum of the compound shows that it has four kinds of carbons. The appearance of a singlet, a triplet and two quartets in its off-resonance decoupled CMR spectrum indicates the presence of a CH_2 group, two nonequivalent CH_3 groups and a carbon attached to no hydrogen. Thus, the structure which fits the above descriptions is

$$H_3C-CH_2-\overset{\displaystyle Cl}{\underset{\displaystyle CH_3}{C}}-CH_3$$

PROBLEMS

1. Indicate the number of signals in the proton-noise decoupled spectra of the following compounds:

(a) 2-Butene (b) Toluene (c) Propane (d) Ethane

2. Discuss what happens when CMR spectra are recorded in proton-noise decoupling mode.

3. How will you explain the indicated ^{13}C chemical shifts (in δ) of the carbon atoms in each of the following compounds:

(a) $CH_2{=}CH{-}O{-}CH_3$

84.2 153.2 52.5

(b) $CH_3{-}C{\equiv}C{-}O{-}CH_3$

28.0 88.4

(c)

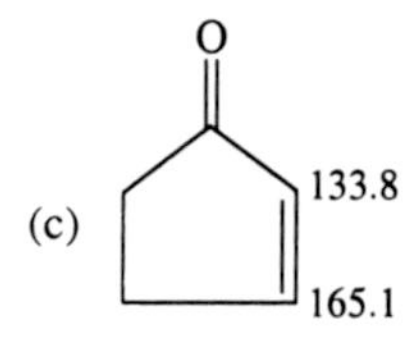

4. Predict the number of peaks and their multiplicity in the off-resonance decoupled CMR spectra of the following compounds:

(a) *p*-Dichlorobenzene (b) *p*-Xylene
(c) 1,4-Dioxane (d) Methylcyclopropane

5. Draw the structure of each of the compound which meets the following requirements:

(a) C_5H_8; one singlet, two doublets and one quartet in its off-resonance ^{13}C NMR spectrum.
(b) C_6H_{14}; four peaks in its proton-noise decoupled ^{13}C NMR spectrum.
(c) $C_3H_3Cl_5$; one singlet, one doublet and one triplet in its off-resonance CMR spectrum.

6. Write notes on the following:

(a) Off-resonance proton decoupled ^{13}C NMR spectra
(b) ^{13}C chemical shifts
(c) Effect of deuterium substitution on CMR signals

7. Using ^{13}C NMR spectroscopy, how will you distinguish the following pairs of compounds:

(a) 1-butyne and 2-butyne
(b) Phthalic acid and terephthalic acid
(c) 1,4-dioxane and 1,3-dioxane

8. A compound with molecular formula $C_5H_{10}O$ gave the following CMR spectral data:

(i) One singlet, δ 212 (iii) One quartet, δ 27
(ii) One doublet, δ 42 (iv) One quartet, δ 18

Deduce the structure of the compound consistent with the given data.

9. Predict the number of peaks and their multiplicity in the off-resonance CMR spectra of the following compounds. In each case, indicate carbons with the highest and the lowest chemical shifts in δ unit from TMS.

(a) Acetophenone (b) Ethyl malonate
(c) Ethyl acetate (d) Pentane

10. Draw the structure of each of the following compounds which meets the given requirements:

(a) C_7H_{16}; three signals in its proton-noise decoupled CMR spectrum.
(b) $C_4H_{10}O$; one doublet, one triplet and two quartets in its off-resonance proton decoupled CMR spectrum.
(c) C_5H_8; two singlets, one triplet and two quartets in its off-resonance proton decoupled CMR spectrum.

11. Give an account of the structural information obtained from the following kinds of CMR spectra:

(i) Proton-noise decoupled
(ii) Off-resonance decoupled
(iii) Gated decoupled

12. Discuss the following:

(a) α-, β-, γ-effects
(b) Effect of hybridization of carbon on its chemical shift.

13. Using proton-noise decoupled CMR spectroscopy, how will you distinguish the following:

(a) *o*-, *m*- and *p*-xylenes
(b) *t*-butyl bromide and 1-bromo-2-methyl propane
(c) Isomeric alcohols represented by the molecular formula C_3H_8O
(d) 1-butene and 2-butene

14. Indicate the number of signals and their multiplicity in the off-resonance ^{13}C NMR spectra of the following compounds:

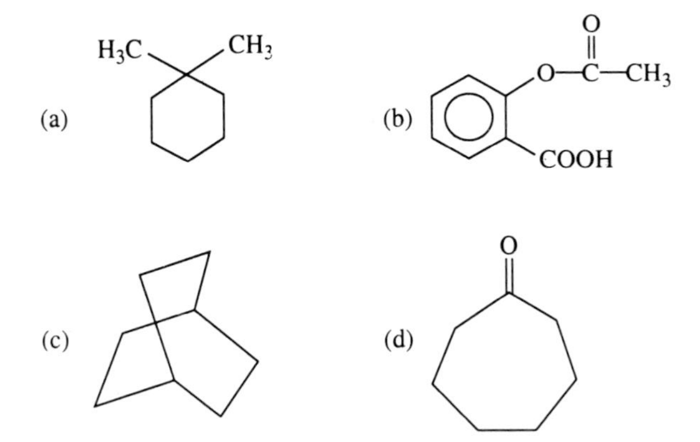

15. Explain the following:

(a) In proton-noise decoupled ^{13}C NMR spectra, some peaks are larger than the others even though they arise due to the same number of carbon atoms.
(b) The proton-decoupled CMR spectra of deuterated solvents exhibit low intensity peaks.
(c) The methyl carbons in *trans*-2 butene appear downfield (at δ 17.6) compared to those of *cis*-2-butene (at δ 12.1).

16. The reaction of 1-chloro-1-methylcyclohexane with ethanolic KOH yields a mixture of two alkenes. After separating the two alkenes, how will you distinguish between them with the help of proton-decoupled ^{13}C NMR spectra?

17. Fig. P6.1 shows the proton-noise decoupled CMR spectrum of compound $C_5H_{11}Br$. The signals are split into doublet (*d*), triplet (*t*) and quartet (*q*) as indicated in Fig. 6.7 when the spectrum was recorded in the off-resonance decoupling mode. Deduce the structure of the compound.

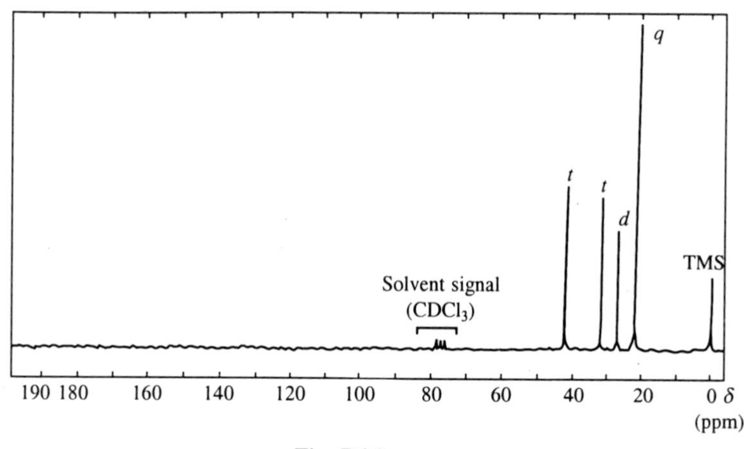

Fig. P6.1

18. Fig. P6.2 shows off-resonance decoupled and proton-noise decoupled CMR spectra of a compound having molecular formula C_8H_7OCl. Deduce the structure of the compound.

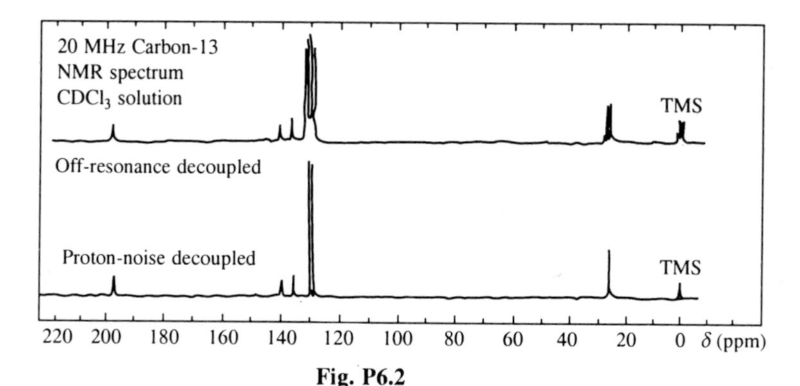

Fig. P6.2

References

1. D. Shaw, Fourier Transform NMR Spectroscopy, 2nd Ed., Elsevier, Amsterdam, 1984.
2. E. Breitmaier and W. Voelter, Carbon-13 NMR Spectroscopy, 3rd Ed., VCH Publishers, New York, 1987.
3. F.W. Wehrli, A.P. Marchand and S. Wehrli, Interpretation of Carbon-13 NMR Spectra, 2nd Ed., Wiley, New York, 1988.
4. G.C. Levy, R.L. Lichter and G.L. Nelson, Carbon-13 Nuclear Magnetic Resonance for Organic Chemists, 2nd Ed., Wiley, New York, 1980.
5. H.O. Kalinowski, S. Berger and S. Braun, Carbon-13 NMR Spectroscopy, Wiley, New York (English Ed., Translated by J.K. Becconsall, 1988).
6. J.B. Stothers, Carbon-13 NMR Spectroscopy, Academic Press, New York, 1972.
7. K. Muller and P.S. Pregosin, Fourier Transform NMR: A Practical Approach, Academic Press, New York, 1976.
8. R.M. Silverstein, G.C. Bassler and T.C. Morrill, Spectrometric Identification of Organic Compounds, 5th Ed., Wiley, London-New York, 1991.
9. S. Sternhell and J.R. Kalman, Organic Structures from Spectra, Wiley, Chichester-New York, 1986.
10. T.C. Farrar, An Introduction to Pulse NMR Spectroscopy, Farragut Press, Chicago, 1987.
11. W.W. Paudler, Nuclear Magnetic Resonance, Wiley, New York, 1987.

7

Electron Spin Resonance (ESR) Spectroscopy

7.1 Introduction

Electron spin resonance (ESR) spectroscopy, invented by Zavoiskii in 1944, is similar to NMR spectroscopy. ESR spectroscopy is an absorption spectroscopy which involves the absorption of radiation in the microwave region (10^4-10^6 MHz) by substances containing one or more unpaired electrons. This absorption of microwave radiation takes place under the influence of an applied magnetic field. The substances with one or more unpaired electrons are paramagnetic and exhibit ESR. Thus, ESR spectroscopy is also called *electron paramagnetic resonance* (EPR) spectroscopy or electron magnetic resonance spectroscopy.

Substances containing unpaired electrons, i.e. paramagnetic substances are of two types:

(i) Stable Paramagnetic Substances

These include simple molecules like NO, O_2 and NO_2, and the ions of transition metals and their complexes, e.g. Fe^{3+}, $[Fe(CN)_6]^{3-}$ etc. Such stable paramagnetic substances can be easily studied by ESR spectroscopy.

(ii) Unstable Paramagnetic Substances

These are generally called *free radicals* or *radical ions* and are formed either as intermediates in chemical reactions or by irradiation of a stable molecule with UV or X-ray radiation or with a beam of nuclear particles. If the lifetimes of such radicals is greater than 10^{-6} s, they may be studied by ESR spectroscopy. Paramagnetic substances with lifetimes shorter than 10^{-6} s, may also be studied by ESR spectroscopy if they are produced at low temperatures in the solid state, called *matrix technique*, as this increases their lifetimes. ESR spectroscopy is most useful in the study of free radicals.

7.2 Theory

The principle of ESR is similar to NMR, except that electron spin is involved in ESR instead of nuclear spin which is involved in NMR. An unpaired electron, like a proton, has a spin and this spin has an associated magnetic moment. An electron of spin $s = \frac{1}{2}$ can have the spin angular momentum quantum number values of $m_s = \pm\frac{1}{2}$. In the absence of an applied magnetic field, the two values of m_s, i.e. $+\frac{1}{2}$ and $-\frac{1}{2}$ will give rise to a doubly degenerate spin energy state.* When a magnetic field is applied, this degeneracy disappears and two non-degenerate spin energy states result. The low energy state (more stable) has the spin magnetic moment aligned with the applied magnetic field and corresponds to the quantum number $m_s = -\frac{1}{2}$, whereas the high energy state (less stable), $m_s = +\frac{1}{2}$, has its spin magnetic moment aligned against the applied field. These energy states are illustrated in Fig. 7.1. These two states will possess energies that are split up from the original state with no applied magnetic field by the amount $-\mu_e H_0$ and $+\mu_e H_0$ for the low energy and high energy states, respectively (Fig. 7.1). Here μ_e is the magnetic moment of the spinning electron and H_0 the applied magnetic field acting on the unpaired electron.

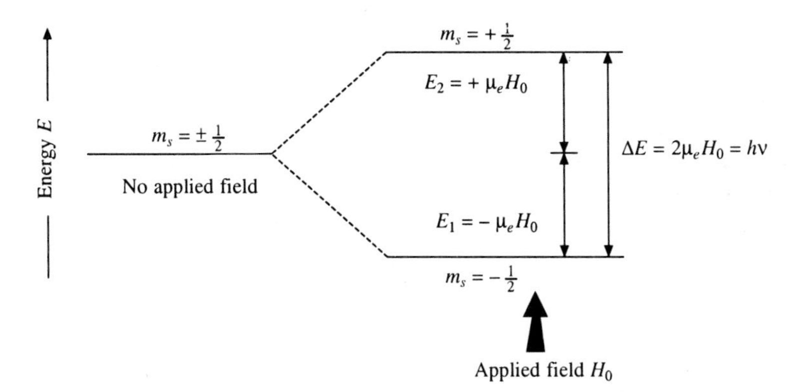

Fig. 7.1 Energy states of an unpaired electron in an applied magnetic field H_0

In ESR, a transition between the two different electron spin energy states takes place by absorption of a quantum of radiation of an appropriate frequency in the microwave region. When the absorption takes place, the following relation holds good:

$$2\mu_e H_0 = h\nu = \Delta E \tag{7.1}$$

where $2\mu_e H_0$ is the difference between the two electron spin energy states (Fig. 7.1), i.e.

$$\mu_e H_0 - (-\mu_e H_0) = 2\mu_e H_0$$

Strictly speaking, the relation given in Eq. (7.1) holds good for a free electron.

The energy of transition ΔE in substances containing an unpaired electron is more accurately given by the relation

$$\Delta E = h\nu = g\beta H_0 \tag{7.2}$$

where h is the Planck's constant, ν the frequency in cycles per sec, β the Bohr

*Two spin energy states having the same energy.

magneton which is a factor for converting angular momentum into magnetic moment and g the spectroscopic splitting factor or Lande splitting factor. The value of g (g factor) is not constant but it depends on the environment of the unpaired electron. For a free electron, the value of g is 2.0023 at 8388.255 MHz in a field of 0.30 T (3000 gauss). Virtually all free radicals and some ionic crystals have almost the same g value, i.e. 2.0023 with a variation of ± 0.003 from this value. The reason for this is essentially that in free radicals the electron behaves in very much the same way as an electron in free space.

The value of β is defined as

$$\beta = \frac{eh}{4\pi mc} = 9.273 \times 10^{-24}\ \mathrm{JT}^{-1*} \tag{7.3}$$

where e is the electronic charge, m the mass of electron and c the velocity of light.

The value of g depends on the orientation of the molecule containing the unpaired electron with respect to the applied magnetic field. It also depends on the physical state of the sample, i.e. gas, liquid or solid. In gas and liquid states, the molecules have free motion and the value of g is averaged over all orientations. In case of a paramagnetic ion or radical situated in a perfectly cubic crystal site, the value of g does not depend on the orientation of the crystal, i.e. it is the same in all directions. However, in case of a paramagnetic ion or radical situated in a crystal of low symmetry, the value of g depends on the orientation of the crystal. The values of g along x, y and z axes may vary with orientation of the crystal with respect to the applied magnetic field.

The electron spin magnetic moment is about 1000 times greater than that of protons. Thus, the frequency of absorption of electrons is about 1000 times that of protons in the same magnetic field. For example, in a field of 3200 gausses, where NMR absorption would occur at about 14 MHz (in the radio-frequency region), ESR absorption occurs at 9000 MHz, i.e. at a much higher frequency which falls in the microwave region.

Similar to that in NMR spectroscopy, the ESR spectra are most commonly recorded by varying the strength of the applied magnetic field H_0 and keeping the frequency constant. However, it can also be done by varying the frequency and keeping H_0 constant.

7.3 ESR Absorption Positions: The *g* Factor

Eq. (7.2) shows that an ESR absorption will occur at a frequency $\nu = \Delta E/h$ Hz. Thus, the position of an ESR absorption may be expressed in terms of absorption frequency. It is clear from Eq. (7.2) that the absorption frequency, i.e. the absorption position, varies with the applied field H_0. Since different ESR spectrometers operate at different magnetic fields, it is desirable to express ESR absorption positions in same form independent of the field strength. Thus, the ESR absorption

*10,000 G (gauss) = 1 T (tesla).

positions are more conveniently expressed in terms of the observed g values. Rearranging Eq. (7.2), we get

$$g = \frac{\Delta E}{\beta H_0} = \frac{h\nu}{\beta H_0} \tag{7.4}$$

If, for example, an ESR absorption were observed at 8388.255 MHz in a field of 0.30 T, the ESR absorption position would be reported at a g value of 2.0023.

For measuring the g values of free radicals it is convenient to measure the field separation between the center of the spectrum of the unknown sample and that of a reference substance whose g value is accurately known. The most widely used reference is 1,1-diphenyl-2-picrylhydrazyl free radical (DPPH) which is completely in free radical state and its g value is 2.0036. The reference substance is placed along with the unknown in the same dual resonant cavity.

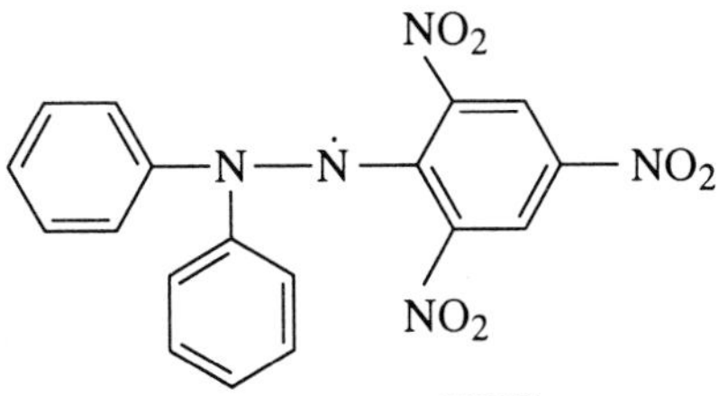

DPPH

Two signals will be observed simultaneously with a field separation of ΔH_0. The g factor for unknown sample is given by

$$g = g_{\text{ref}}\left(1 - \frac{\Delta H_0}{H}\right) \tag{7.5}$$

where g_{ref} is the g factor for the reference and H the resonance frequency. ΔH_0 is positive if the unknown has its center at a higher field than the reference.

Chemical shifts can be compared with g-value shifts. The reference in ESR is the 'free' electron (g = 2.0023). In ESR, it is not necessary to use a reference because calibration is such that g values can be estimated directly from the applied microwave frequency and the magnetic field at which the resonance absorption occurs (cf. Chemical shift, Section 5.5).

7.4 Instrumentation

The energies required to bring about transitions in NMR and ESR are different. In NMR, transitions occur in the radio-frequency region, whereas in ESR, transitions occur at frequencies in the microwave region. Thus, the instrumentation for the NMR and ESR spectroscopy must be different.

Fig. 7.2 shows a schematic diagram of an ESR spectrometer. The description of various components of the instrument is given as follows.

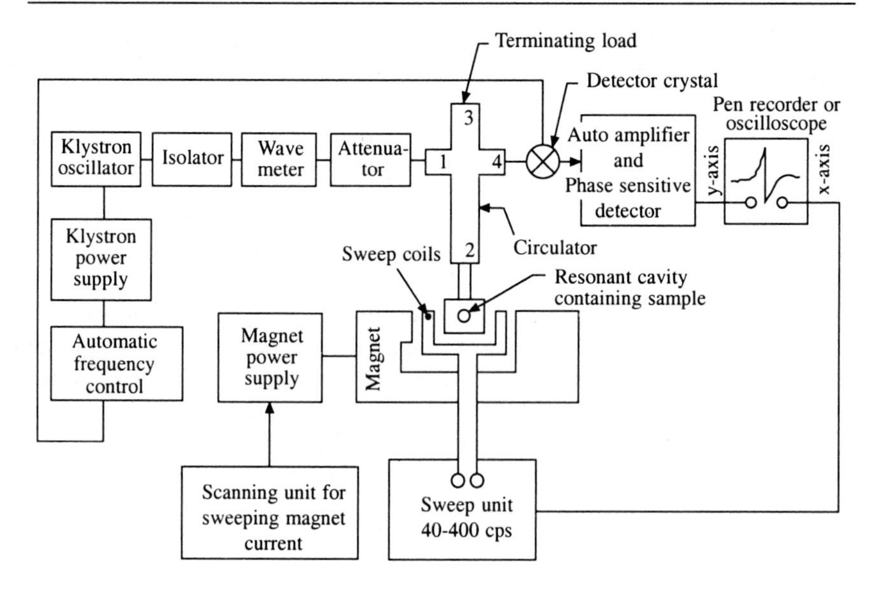

Fig. 7.2 Schematic diagram of an ESR spectrometer

(i) Source of the Microwave Radiation

It consists of the following:

(a) Klystron valve is a powerful source of microwave radiation of a small frequency range. For free radical studies, klystron is normally operated in the microwave region of 3 cm wavelength. The frequency of the monochromatic radiation is determined by the voltage applied to klystron. A klystron oscillator is generally operated at 9500 MHz.

(b) Isolator is a strip of ferrite material. It minimizes variations in the frequency of microwaves produced by klystron. The variations in frequency are caused by the backward reflections in the regions between the klystron and the circulator.

(c) Wavemeter is put in-between the isolator and attenuator to know the frequency of microwaves produced by the klystron. The wavemeter is usually calibrated in frequency units (MHz) instead of wavelength.

(d) Attenuator is put in-between the wavemeter and circulator. It has an absorption element and corresponds to a neutral filter in light absorption measurements. It adjusts the level of the microwave power incident upon the sample.

(ii) Circulator or Magic-T

From the attenuator, the microwave radiations enter the microwave circulator (bridge). Fig. 7.3 indicates the operation of a four-port circulator which works as a balanced bridge with all the advantages of null method in electrical circuits. The microwave radiations enter arm 1 and arm 2 is attached to resonant cavity which contains the sample. Arm 3, which has a balancing load, absorbs any power reflected from the detector arm 4 which is connected to the detector. The microwave, circulator does not allow the microwave power to pass in a straight line from one arm to the opposite arm.

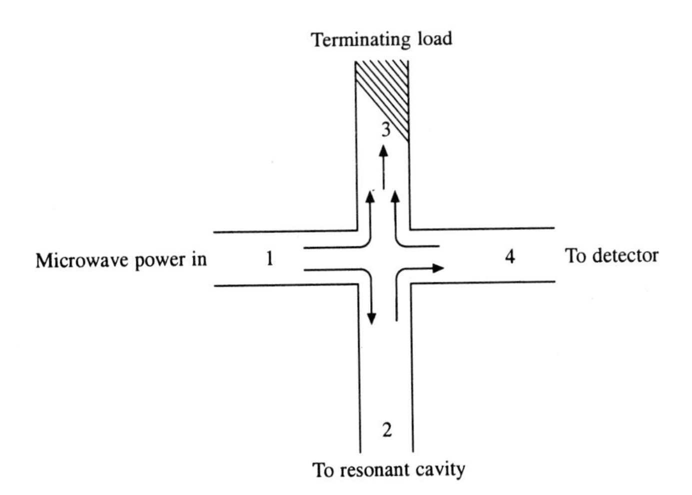

Fig. 7.3 A four-port microwave circulator showing the directions of microwave transmission among the four arms

(iii) Resonant Cavity

The resonant cavity containing the şample is the heart of an ESR spectrometer. The cavity system is constructed in such a way that the applied magnetic field along the sample dimension is minimized. There is provision to introduce the sample at the point of maximum microwave power. The sample may be in the form of a single crystal, solid powder, liquid or solution, and it is usually contained in a tube of about 5 mm diameter. For studying anisotropic effects in single crystals and solid samples, rotable sample cavities are generally used. For simultaneous observation of a sample and a reference material, in most of the ESR spectrometers dual resonant cavities are generally used. By using a reference material, the sources of error are compensated by comparing relative signal heights.

(iv) Magnet System

It consists of an electromagnet which provides a homogeneous magnetic field. The strength of the magnetic field can be varied (swept) over a small range (zero to 500 gauss) by varying the current in a pair of sweep coils. The resonant cavity is placed between the pole pieces of the electromagnet. The magnetic field should be uniform and stable over the sample volume. The stability of the field is achieved by energizing the magnet with a highly regulated power supply. The stability of 1 part in 10^6 gives satisfactory resolution of ESR spectra of samples having g factor ranging from 1.5 to 6. However, for paramagnetic ions and free radicals in solid matrices, the stability of the field might be as low as 1 part in 10^3 to give satisfactory resolution of ESR spectra. An electromagnet which is capable of producing steady fields from 50 to 5500 gauss is needed for studying samples whose g factor ranges from 1.5 to 6.

(v) Crystal Detectors

The most commonly used detector is a semiconducting silicon-tungsten crystal which acts as a microwave rectifier and converts the microwave power into a direct current output.

(vi) Autoamplifier and Phase Sensitive Detector

The signal, in the form of direct current, received from the detector undergoes narrow band amplification by the operation of the autoamplifier. This amplified signal contains a lot of noise. The operation of the phase sensitive detector reduces noise by rejecting all the noise components except those in a very narrow band.

(vii) Recorder

The signal from the phase sensitive detector and sweep unit is recorded by an oscilloscope or a pen recorder.

7.5 Working of an ESR Spectrometer

The sample whose ESR spectrum is to be recorded is placed in the resonant cavity. The cavity serves as a very long path-length cell in which the waves are reflected to-and-fro for thousands of times.

The microwaves produced by the klystron oscillator pass through the isolator, wavemeter and attenuator, and then they are received by the circulator through arm 1. The microwave power entering arm 1 is divided between arms 2 and 3. Generally, arm 3 has a balancing load. If the impedances of arms 2 and 3 are the same, then the circulator (bridge) is balanced and no power will be received by the crystal detector through arm 4. If the impedance of arm 2 (connected to the sample cavity) changes because of some ESR absorption by a sample, then bridge becomes unbalanced and some microwave power will enter into the crystal detector through arm 4. The detector acts as a rectifier and converts the microwave power into direct current. If the strength of the magnetic field around the resonant cavity containing the sample is changed to the value required for resonance, then the recorder will show an absorption curve, i.e. an absorption ESR spectrum and the absorption curve is obtained by plotting intensity against the strength of the magnetic field (Fig. 7.4), and has no fine structure. When the

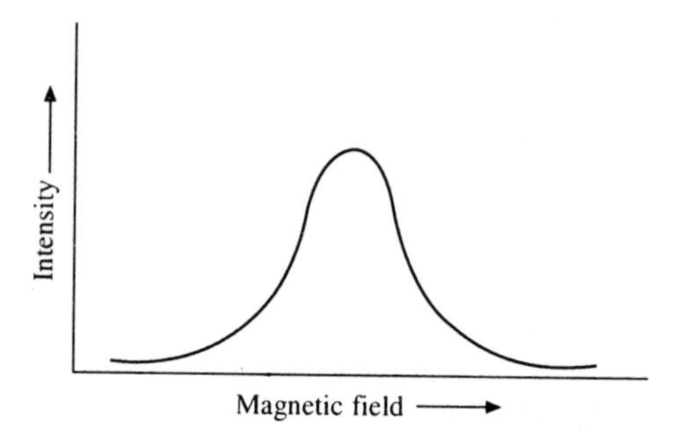

Fig. 7.4 An absorption ESR spectrum

main field is swept slowly over a period of several minutes, the recorder will show the first derivative (the slope) of the absorption curves plotted against the strength of the magnetic field. This is the first derivative or dispersion ESR spectrum (Fig. 7.5).

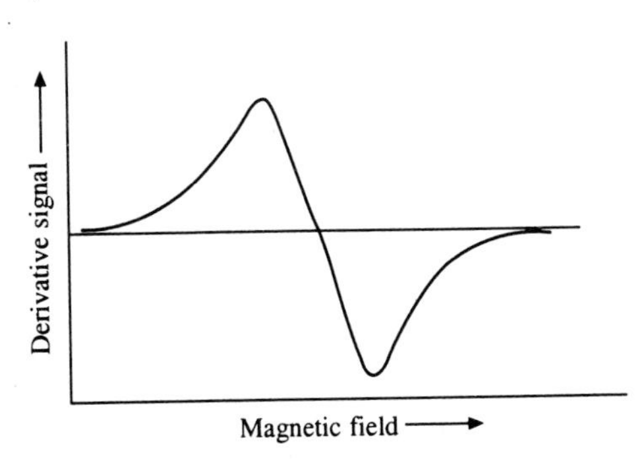

Fig. 7.5 A first derivative or dispersion ESR spectrum

Because of instrumental considerations associated with the signal-to-noise ratio, ESR spectra are generally recorded as first derivative spectra.

For low-frequency modulation (400 Hz or less), the coils can be mounted outside the cavity and even on the magnet pole pieces. Since higher modulation frequencies cannot penetrate metal effectively, the modulation coils must be mounted inside the sample cavity in case of high-frequency modulation.

7.6 Sample Handling

The sample may be in the form of a single crystal, solid powder, liquid, solution, or frozen solutions and is usually contained in a tube of about 5 mm diameter. A sample volume of 0.15 to 0.5 ml can be used for samples which do not have a high dielectric constant. For the samples with high dielectric constant, flat cells with a thickness of 0.25 mm and sample volume of 0.05 ml are generally used.

Water, alcohol and other solvents with high dielectric constant are not generally used in ESR studies because they strongly absorb microwave power. However, where there is no alternative choice of solvents other than water, alcohol and other solvents with high dielectric constant, then these solvents can be used if the sample has strong absorbance and contained in a specially designer cell (a very narrow sample tube).

When a sample is used in the form of a frozen solution, the best results are obtained if the solvent freezes to form a glass. It is observed that symmetrical molecules do not form good glasses, e.g. methylcyclohexane forms a good glass, while cyclohexane does not. Some solvents and mixtures of solvents which form good glasses are given in Table 7.1.

In case of a sample having very small dielectric loss, the minimum concentration should be of the order of 10^{-9} M. For aqueous solutions, the minimum concentration should be 10^{-7} M. For carrying out structure determination and quantitative analysis, the concentration should be about 10^{-6} M.

Table 7.1 Solvents and mixtures of solvents which form good glasses

Solvents	Mixtures of solvents
Methylcyclohexane	Propane + propene
Isooctane	3-methylpentane + isopentane
Nujol	Toluene + chloroform
Toluene	Toluene + acetone
Glycerol	Ethanol + methanol
Triethanolamine	Ethanol + diethyl ether
Sulphuric acid	Propanol + diethyl ether

7.7 Sensitivity of an ESR Spectrometer

The sensitivity of an ESR spectrometer is generally expressed as

$$N_{\text{min}} = 1 \times 10^{11} \frac{\Delta H}{\sqrt{\tau}} \tag{7.6}$$

where N_{min} is the minimum number of detectable spins per gauss, ΔH the width between deflection points on the derivative absorption curve and τ the time constant of the detecting system which is inversely proportional to the band width of the detection circuit.

The sensitivity of ESR spectrometers increases on working at higher magnetic field strengths. For example, an ESR spectrometer working at 35,000 MHz (K-band spectrometer) has twenty times greater sensitivity than that working at 9500 MHz (X-band spectrometer).

7.8 Multiplet Structures in ESR Spectroscopy

There are two types of multiplet structures in ESR spectroscopy, viz. (i) fine and (ii) hyperfine.

(i) Fine Structure

It occurs only in crystals containing more than one unpaired electronic spins. Let us consider the case of a crystal which contains molecules or ions with two parallel rather than unpaired electron spins. In this case there would be a total spin of 1, i.e. $S = 1$ and $2S + 1 = 3$, thus it is a triplet state. Molecular triplet states are generally unstable and they revert to the ground state with paired spins. For example, when naphthalene is irradiated with UV light, its individual molecules undergo excitation to the triplet state which reverts quite rapidly to the ground state with paired spins. However, by cooling the crystal to low temperatures, or by diluting the naphthalene in a solid, inert lattice, the triplet state can be maintained and examined by ESR spectroscopy. The ESR spectrum of naphthalene, recorded in its excited state (triplet state), consists of two fine-structure lines. Transition metal ions are quite stable in their triplet states and can be easily studied by ESR spectroscopy at room temperature.

In general, if there are n parallel electron spins, there will be n equally spaced resonance lines in the ESR spectrum.

(ii) Hyperfine Structure

Hyperfine structure results from hyperfine coupling which involves coupling of the unpaired electronic spin with neighboring nuclear spins. This is similar to the nuclear spin-spin coupling discussed in Section 5.10. The hyperfine coupling splits ESR signals into multiplet structures which are called hyperfine structures.

When an unpaired electronic spin couples with a nucleus with spin I, the absorption signal of the electron is split into a multiplet with $2I + 1$ lines of equal intensity. In general, when the absorption signal of an unpaired electron is split by n equivalent nuclei with equal spin I_i, the number of lines is given by $2nI_i + 1$. When the splitting is caused by both, a set of n equivalent nuclei with spin I_i, and a set of m equivalent nuclei with spin I_j, the number of lines is given by $(2nI_i + 1)(2mI_j + 1)$, and so on.

The separation between the lines is usually of the order of 10^{-3}-10^{-4} T (about 50 MHz) which is approximately 10^6 times larger than nucleus-nucleus coupling because an electron can approach a nucleus much more closer than another nucleus, and thus will interact more strongly with it. The biggest factor influencing the magnitude of electron-nucleus coupling is the electron density at the coupled nucleus, i.e. the amount of time which the electron spends in the vicinity of the coupled nucleus.

Let us illustrate the hyperfine structure resulting from the splitting of an ESR signal by considering an example of hydrogen atom having one proton and one electron. The two energy levels of a free electron in an applied magnetic field are shown in Fig. 7.6 (a) with $m_s = -\frac{1}{2}$ aligned with the field and $m_s = +\frac{1}{2}$ aligned opposing the field. Thus, the ESR spectrum of a free electron would consist of a single peak (Fig. 7.7) corresponding to a transition between these energy levels.

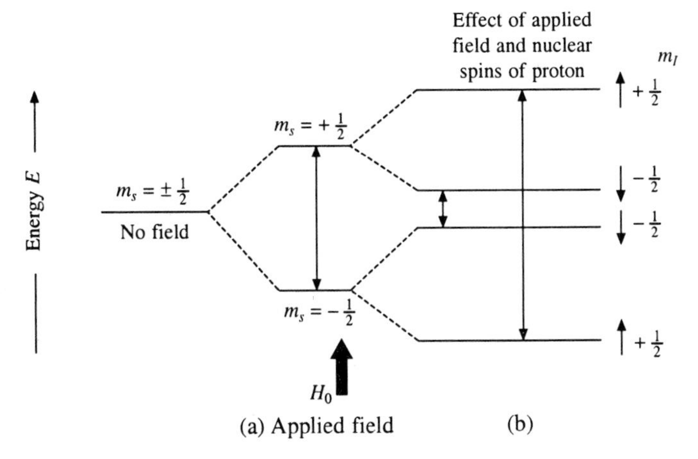

Fig. 7.6 Effect of: (a) applied field on the spin energy states $m_s = \pm\frac{1}{2}$ of an electron and (b) applied field and nuclear spins $m_I = \pm\frac{1}{2}$ of proton on $m_s = -\frac{1}{2}$ and $m_s = +\frac{1}{2}$ energy states of an electron

Each of the two energy levels of the electron in hydrogen atom is split into two energy levels by the interaction with the nuclear spins of proton $m_I = \pm\frac{1}{2}$,

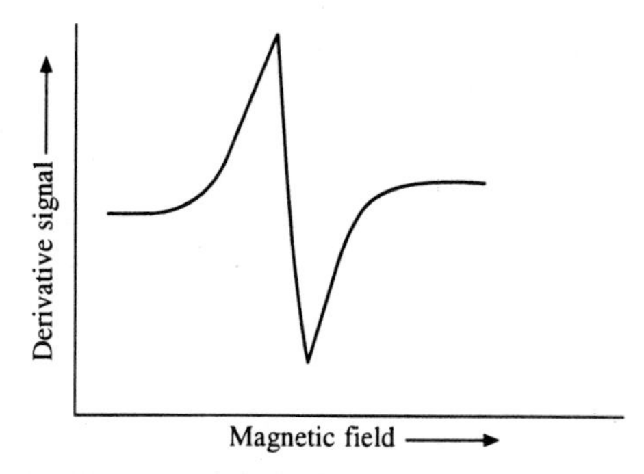

Fig. 7.7 First derivative ESR spectrum of a free electron

where m_I is the nuclear spin angular momentum quantum number. Thus, corresponding to the two energy states $m_s = -\frac{1}{2}$ and $m_s = +\frac{1}{2}$, four different energy levels are obtained as shown in Fig. 7.6(b). This is why ESR spectrum of hydrogen atom consists of two peaks of equal intensity (Fig. 7.8) corresponding to two transitions shown by two arrows in Fig. 7.6(b) rather than one corresponding to the arrow in Fig 7.6(a). The two peaks are of equal intensity because the probability of orientations of the nuclear spin of hydrogen atom causing different energy levels is equal. The selection rules in ESR are $\Delta m_I = 0$ and $\Delta m_s = \pm 1$ which allow us to decide between which levels transitions will give rise to spectral lines.

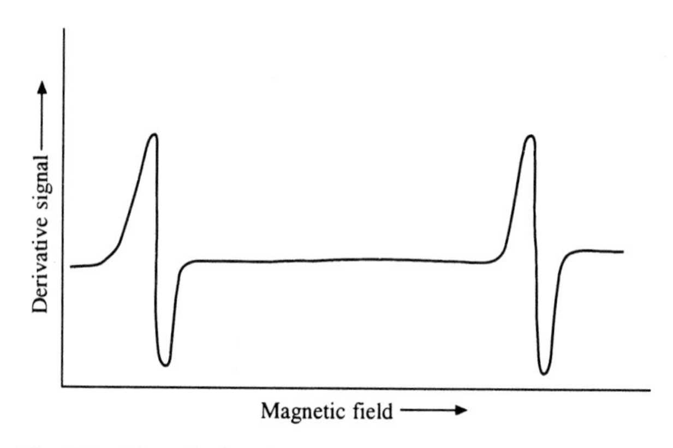

Fig. 7.8 First derivative ESR spectrum of hydrogen atom

Similarly, in the case of methyl radical, each of the two energy levels of the single unpaired electron on the carbon atom is split into four energy levels by the interaction with the nuclear spins of the three hydrogen nuclei (m_I for a single hydrogen nucleus is $\pm\frac{1}{2}$ and that for the three hydrogen nuclei $\pm 3 \times \frac{1}{2} = \pm\frac{3}{2}$). Thus, corresponding to the two spin energy states of the unpaired electron, eight different energy levels are obtained as shown in Fig. 7.9. Four transitions are possible according to the ESR selection rules $\Delta m_I = 0$ and $\Delta m_s = \pm 1$. Thus, the ESR spectrum of methyl radical consists of four peaks (Fig. 7.10). The number of component peaks in the ESR signal of methyl radical can also be determined

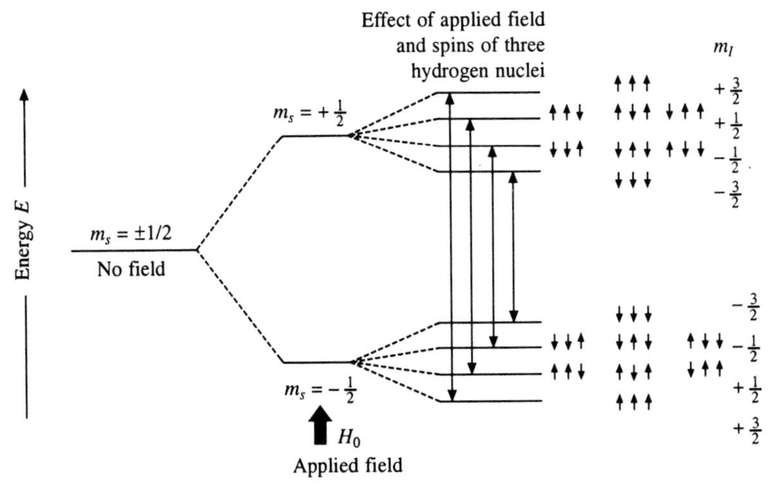

Fig. 7.9 Energy level diagram illustrating coupling between the spins of the unpaired electron and hydrogen nuclei in methyl radical

by the formula $2nI + 1$ which also indicates $(2 \times 3 \times \frac{1}{2}) + 1 = 4$ peaks. The observed relative intensities of the four component peaks are in the ratio 1 : 3 : 3 : 1. The relative intensities of the component peaks of a multiplet are directly proportional to the number of nuclear spin orientations of equivalent energy causing different energy levels (Fig. 7.9) and are given by coefficient of terms in the binomial expansion of $(x + 1)^n$ exactly in the same way as discussed in Section 5.10.

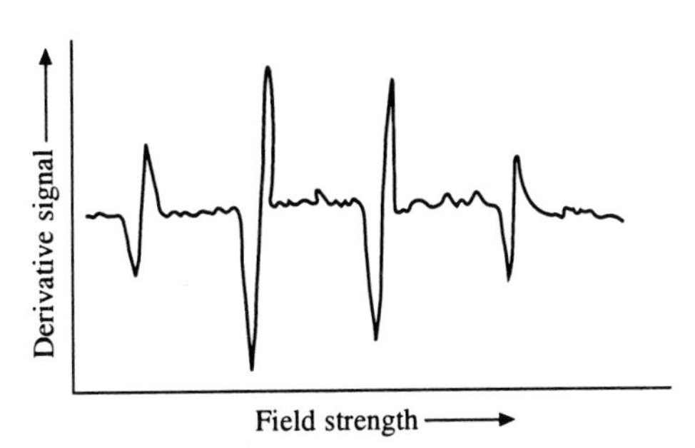

Fig. 7.10 First derivative ESR spectrum of methyl radical

Let us now discuss the ESR spectrum of 1,4-benzosemiquinone shown in Fig. 7.11. It exhibits five peaks with their intensities in the ratio 1 : 4 : 6 : 4 : 1. In 1,4-benzosemiquinone the unpaired electron is delocalized over all the carbon and oxygen atoms. Thus, all the four protons are equivalent and the unpaired electron is coupled with four equivalent protons (for proton $I = \frac{1}{2}$) resulting in the splitting of its ESR signal into five peaks according to the formula $2nI + 1$. The ratio of relative intensities of these peaks is given by coefficients of the binomial expansion of $(x + 1)^n$. In this case $n = 4$ (four equivalent protons are coupled with the unpaired electron). Hence, the intensity ratio comes to be 1 : 4 : 6 : 4 : 1.

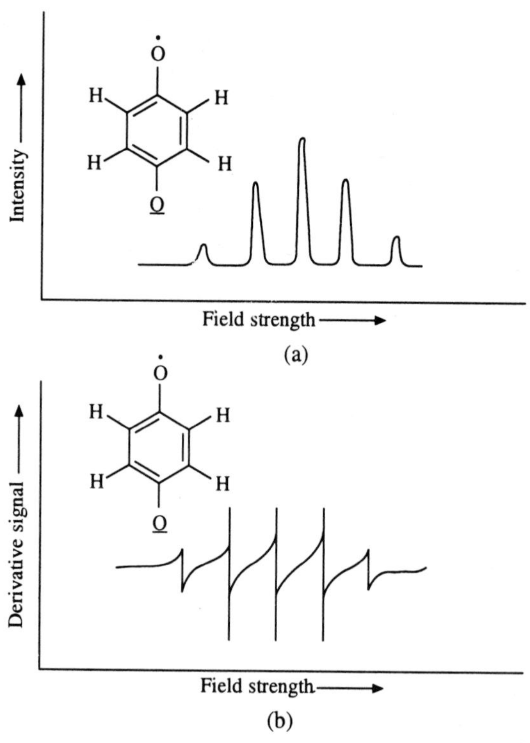

Fig. 7.11 (a) Absorption ESR spectrum of 1,4-benzosemiquinone and (b) first derivative ESR spectrum of 1,4-benzosemiquinone

The ESR spectrum of 1,4-benzosemiquinone can also be explained graphically as shown in Fig. 7.12.

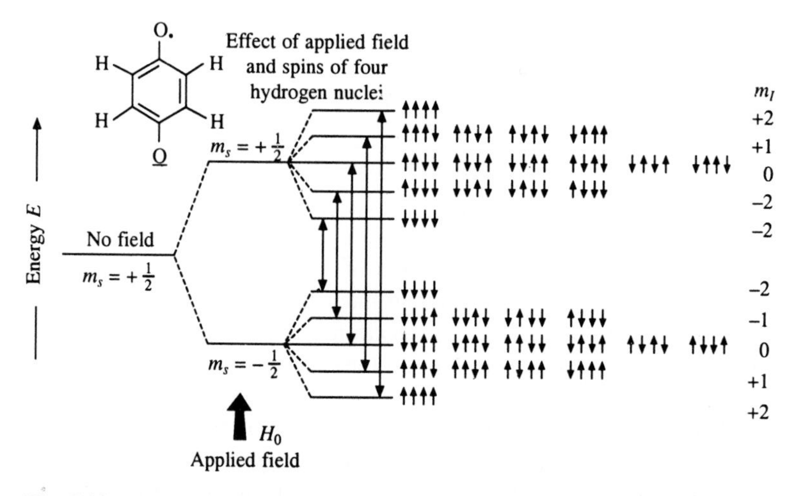

Fig. 7.12 Energy level diagram illustrating coupling between the unpaired electron and the four protons in 1,4-benzosemiquinone

The deuterium atom ($^2_1$H) is a simple example of a system having $I = 1$. In this case, there are three values of m_I, i.e. $m_I = +1$, 0 and -1 corresponding to $m_s = +\frac{1}{2}$, and three values of m_I corresponding to $m_s = -\frac{1}{2}$, i.e. $m_I = -1$, 0 and + 1. Thus, each of the two energy levels of the electron in deuterium is split into three energy levels by coupling with the nuclear spins of deuterium resulting in six different energy levels as shown in Fig. 7.13. According to the ESR selection rules ($\Delta m_s = \pm 1$ and $\Delta m_I = 0$), there are three allowed transitions. Thus. ESR spectrum of deuterium atom consists of three peaks of equal intensity (Fig. 7.14) corresponding to three transitions shown by arrows in Fig. 7.13. The three peaks are of equal intensity because the probability of the orientations of the nuclear spin of deuterium atom causing different energy levels is equal.

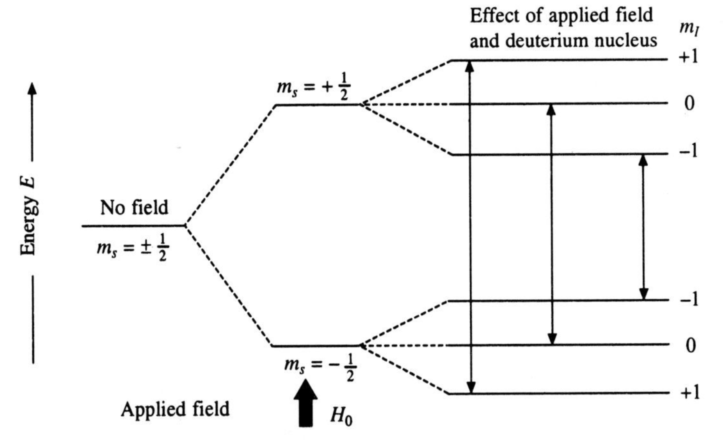

Fig. 7.13 Energy level diagram illustrating coupling between the unpaired electron and deuterium nucleus

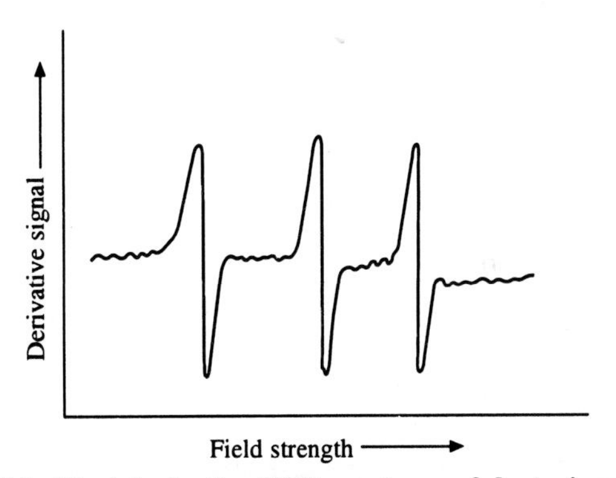

Fig. 7.14 First derivative ESR spectrum of deuterium atom

7.9 Interpretation of ESR Spectra

As in other forms of spectroscopy, four properties of spectral lines, viz. their position, intensity, multiplet structure and width are of importance in interpretation of ESR spectra.

The ESR absorption positions are generally expressed in terms of g values which depend on the environment of the unpaired electron (Section 7.3). Since all free radicals have almost the same g value, the ESR absorption position does not give much information regarding the structure of free radicals. Thus, g values are not so important as the chemical shifts in the NMR spectroscopy.

The intensity of an ESR absorption (the total area covered by either the absorption or derivative curve) is proportional to the number of unpaired electrons present in the sample, i.e. it is proportional to the concentration of the free radical or paramagnetic material. Thus, ESR method is useful for estimating the amount of free radical present. This method is very sensitive because up to about 10^{-12} mole of free radical can be detected under favourable conditions. The number of unpaired electrons in an unknown sample can be calculated by comparison with a standard sample having a known number of unpaired electrons and possessing the same line shape (Gaussian or Lorentzian). The most widely used standard is 1,1-diphenyl-2-picrylhydrazyl (DPPH) free radical which contains 1.53×10^{21} unpaired electrons per gram. However, it cannot be used as internal reference because there is only slight difference in the g values; so the absorption due to the unknown substance cannot be distinguished from that of the reference. A trace of Cr (III) entrapped in a tiny clip of ruby crystal cemented permanently to the sample cell is used as an internal reference for free radicals. It shows a strong absorption at its g value 1.4. The integral of the ESR curve must be used for the quantitative measurements.

Multiplet structures have already been discussed in Section 7.8. The fine structure lines give information about the number of electrons with unpaired electronic spins in a crystal. If there are n unpaired electronic spins, there will be n equally spaced resonance lines in the ESR spectrum. Hyperfine structures resulting from the *hyperfine coupling* give information about the neighboring magnetic nuclei coupled with the unpaired electron in a free radical. This has already been discussed with several examples in Section 7.8(ii).

As in NMR spectroscopy (Section 5.2), the width of an ESR spectral line depends on the relaxation time of the spin state under study. For most of the samples, a typical relaxation time is 10^{-7} s. Using this value in the Heisenberg uncertainty relation

$$\delta v \approx \frac{1}{2\pi \delta t}$$

we can calculate a frequency uncertainty (line width) which is ≈1 MHz. A shorter relaxation time increases this width often up to ≈10 MHz. This is a much wider spectral line than the NMR spectral line which has normally a width of 0.1 Hz for a liquid.

The wider lines in ESR spectra possess advantages and disadvantages. It is advantageous because in ESR the homogeneity of the applied magnetic field is far less critical than that in NMR. Thus, for NMR, it is essential to use a magnetic field homogeneous to 1 in 10^8 over the sample, whereas for ESR, a figure of 1 in 10^5 is adequate. The major disadvantage of wider lines is that these are more difficult to observe and measure than sharp lines, and they cause overlapping of ESR signals. For this reason, ESR spectra are generally recorded as first derivative spectra.

Increase in the width of spectral lines is termed as line broadening. The line broadening can be overcome by working at low temperatures and low concentrations of the sample.

The results represented by a first derivative spectrum can be more readily interpreted in comparison to an absorption ESR spectrum because of the following two main reasons:

1. The point of maximum absorption is difficult to measure accurately with a broad absorption curve in an absorption ESR spectrum, but it can be measured with much greater precision in a first derivative spectrum. Every crossing of the derivative axis with a negative slope indicates a true maximum, whereas a crossing with positive slope shows a minimum. The number of peaks or shoulders in the absorption curve can be determined from the number of maxima or minima of negative slopes in the derivative curve.
2. The intensity of a signal can be estimated more accurately in a derivative ESR spectrum than that in the corresponding broad absorption ESR spectrum.

In many cases the recorded spectrum may not contain all the expected lines because the g factor and coupling values can be such that two lines come very close to each other and are not resolved. In cases where many equivalent nuclei couple with the unpaired electronic spin, relative peak heights become very large and it is difficult to see the smaller peaks.

7.10 Double Resonance (or Double Irradiation) in ESR Spectroscopy

The concept of double irradiation in ESR spectroscopy is exactly parallel to that in NMR, i.e. observation of a spectrum at one frequency while simultaneously irradiating at another (Section 5.17(ii)). The two possibilities for double resonance in ESR spectroscopy are (i) ENDOR and (ii) ELDOR.

(i) ENDOR (Electron Nuclear Double Resonance)

In ENDOR technique, a microwave frequency suitable for electron spin resonance and a radio frequency suitable for nuclear spin resonance are simultaneously required. In an ENDOR experiment, the sample is irradiated with a fairly intense microwave radiation to reach microwave saturation (i.e. an ESR transition $\Delta m_s = \pm 1$ is saturated) and at the same time the radio frequency is swept slowly upwards so that an NMR transition $\Delta m_I = \pm 1$ occurs. This causes partial desaturation of the ESR transition ($\Delta m_s = \pm 1$) which results in the observation of an ENDOR signal in the ESR spectrum. Thus, the ENDOR can be considered as a special variety of NMR where the unpaired electron serves as the detector. The ENDOR display is ESR signal height as a function of the swept radio frequency.

The ENDOR technique is used for improving the effective resolution of an ESR spectrum. The following are the main applications of the ENDOR technique:

(a) There is broadening of ESR signals due to large variety of nuclear energy levels. Thus, the spectral structures which contain important physical and chemical information are masked. In such cases the ENDOR technique is useful for revealing the desired information. For example, nuclear couplings are much easier to observe in ENDOR spectra than that in usual ESR spectra.

(b) If hyperfine structures are resolved but more precise values of coupling constants are desired, then these are more accurately measured from the sharper spectral lines given by ENDOR.

(c) In cases where an unpaired electron is coupled with a nucleus having a spin number $I \geq 1$, the ESR signal is split into $2I + 1$ lines. This splitting gives a complex spectrum if resolved or a very broad signal if unresolved. In such cases, the ENDOR spectrum is much simpler because whatever the spin number of the coupled nucleus, its resonance is split only into a doublet by coupling with the single electron. Thus, the ENDOR method is also used for measuring quadrupole coupling constants in a system with spin number $I \geq 1$.

(ii) ELDOR (Electron-Electron Double Resonance)

In ELDOR technique, the sample is irradiated simultaneously with two microwave frequencies. One of these frequencies is used to observe an ESR signal at some point of the spectrum while the other is swept through other parts of the spectrum. Thus, the ESR signal height is displayed as a function of the difference of the two microwave frequencies.

This technique is used for resolving overlapping multiradical spectra and for studying relaxation mechanisms. The ELDOR method has been used for very precise measurements of the coupling constants of DPPH.

7.11 Applications of ESR Spectroscopy

ESR spectroscopy is a powerful tool for the study of chemical species with unpaired electrons. It gives information about the presence and number of unpaired electrons and their distribution in the molecule and hence, the structure. ESR spectroscopy is useful in the study of chemical, photochemical and electrochemical reactions which proceed through free-radical mechanism. The important applications of ESR spectroscopy are summarized as follows.

(i) Study of Free Radicals

ESR spectroscopy is used to detect the presence of free radicals and to determine their concentration. The method is very sensitive (much more sensitive than NMR) and concentrations as low as 10^{-12} M can be detected and estimated under favourable conditions (Section 7.9). For example, a signal for DPPH radical can be detected even if there is 10^{-12} g of material in the spectrometer. Information concerning the electron distribution (and hence, the structure) in free radicals can be obtained from the splitting pattern of the ESR spectrum (ESR signals are split by nearby protons and other magnetic nuclei) (see Section 7.8(ii)).

One of the most exciting uses of ESR spectroscopy is the study of free-radical intermediates in organic reactions. For example, in the oxidation of hydroquinone in alkaline solution by oxygen, the formation of the semiquinone free radical has been conclusively proved by ESR spectroscopy. The ESR spectrum (Fig. 7.11) of quinone-hydroquinone redox system contains five lines with their intensities in the ratio 1 : 4 : 6 : 4 : 1. This is due to coupling between the unpaired electron and the four equivalent protons on the ring (Section 7.8).

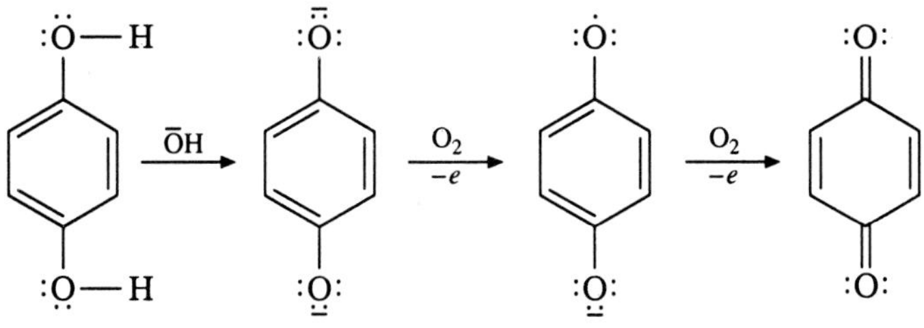

1,4-Benzosemiquinone

(ii) Reaction Velocities and Mechanisms

A large number of organic reactions proceed through free-radical mechanism. Such reactions can be studied by ESR spectroscopy. A direct measure of variation of radical concentrations with time during these reactions would be a valuable aid for determining their mechanisms. In cases of reactions which are not extremely fast, rate studies can be simply performed by following the signal intensity of a paramagnetic reactant or product as a function of time.

Very rapid electron exchange reactions can also be studied by ESR spectroscopy. For example, when naphthalene is added to a solution of naphthalene radical anion, an electron exchange reaction occurs which causes a broadening of the hyperfine component of the ESR spectrum. The rate constant for the exchange of electron between naphthalene and naphthalene radical anion can be calculated by making use of this broadening.

(iii) Structure Determination

ESR spectroscopy is not useful for determining molecular structures because the information obtained from superfine structures is mostly about the extent of electron delocalization in the molecule. It does not tell us about the arrangement of atoms in a molecule. Sometimes, ESR gives useful information about the shape of radicals and about the symmetry of molecules. It is also used to obtain structural information about transition metals and their complexes.

(iv) Effect of Ionizing Radiations on Matter

Polymers are generally made more suitable for certain purposes by irradiation with X-rays, γ-rays, stream of electrons, α-particles or neutrons. This is because free radicals are formed during the initial bond breaking by the radiation which subsequently react with one another to form links between the chains. Since free radicals are involved in this process, ESR spectroscopy is useful for the study of the effects of radiations on polymers. For example, polyethylene, on irradiation with γ-rays, exhibits a seven line pattern in its ESR spectrum indicating the formation of the following free radical:

$$-CH_2-\dot{C}-CH_2-$$
$$|$$
$$CH_2$$
$$|$$

(v) Study of Transition Metals and their Complexes

The unpaired electrons of the *d* or *f* sub-shells of transition metals give ESR signals. This makes ESR spectroscopy a powerful tool for determining electron distributions in the immediate vicinity of the atom. There are several cases where evidence for electron delocalization into ligands has been thoroughly studied using ESR spectroscopy.

ESR can be used to determine the valence state of transition metals in cases where it is not obvious on chemical grounds. For example, copper (II) complexes (d^9 system) give a strong ESR signal, whereas copper (I) complexes (d^{10} system) give a much reduced signal. This difference between the ESR signals for two oxidation states of copper has been successfully used to determine the valence state of copper in many complexes and biologically active compounds. For example, ESR studies have shown that copper protein complex has copper in divalent state, whereas it is in monovalent state in some other biologically active copper complexes.

(vi) Analytical Applications

ESR spectroscopy is used for the determination of various transition metal ions like Mn^{2+}, V^{4+}, Cu^{2+}, Cr^{3+}, Gd^{3+}, Fe^{3+}, Ti^{3+} etc. For example, the ESR spectrum of a solution of Mn^{2+} ions exhibits six lines. The multiplicity is given by $2I + 1$, i.e. $2 \times \frac{5}{2} + 1 = 6$, where I for Mn^{2+} ions is $\frac{5}{2}$. This accounts for the six lines in the spectrum. Thus, Mn^{2+} ions can be measured and detected readily even when present in trace quantity. The method provides a very rapid measurement in aqueous solutions over the range from 10^{-8} to 0.1 M.

ESR technique is also used for the estimation of polynuclear hydrocarbons. These are first converted into radical cations and then absorbed in the surface of an activated silica-alumina catalyst for the estimation.

(vii) Biological Applications

The following are some examples of successful applications of ESR spectroscopy in biological system:

1. ESR spectroscopy has demonstrated the presence of free radicals in healthy and diseased tissues.

2. ESR studies of a variety of biological systems like leaves, seeds and tissue preparations have revealed that there is correlation between the concentration of free radicals and the metabolic activity of the material.

3. ESR studies have shown that most of the oxidative enzymes function via one-electron redox reactions involving the production of either enzyme-bound free radicals or by change in the valence state of a transition metal ion.

4. ESR studies on photosynthesis have been generally performed with photosynthetic bacteria. These bacteria on irradiation with near infrared radiation

(700-900 nm), show a single line with $g = 2.0025$ and $\Delta H = \sim 10$ G in their ESR spectra. The g factor (2.0025) is characteristic of an organic free radical arising from a large conjugated system. The oxidation of bacteriochlorophyl *in vitro* exhibits an ESR signal which closely resembles that of the irradiated photosynthetic bacteria.

5. Biological or other compounds having no unpaired electron can also be studied by ESR when they are chemically bonded to a stable free radical. This free radical (spin label) produces a sharp and simple ESR spectrum which gives detailed information about the molecular environment of the *spin label.*

(viii) Miscellaneous Applications

ESR spectroscopy confirms that electrons in impurities (donors) are responsible for conducting properties of semiconductors. It has also been used to detect conduction electrons in solutions of alkali metals in liquid ammonia, alkaline earth metals, alloys etc.

7.12 Comparison Between NMR and ESR Spectroscopy

Following is the comparison between NMR and ESR spectroscopy:

NMR spectroscopy	ESR Spectroscopy
1. Different energy states are produced due to the alignment of the nuclear magnetic moments relative to the applied magnetic field and a transition between these energy states occurs on the application of an appropriate frequency in the radio frequency region.	1. Different energy states are produced due to the alignment of the electronic magnetic moments relative to the applied magnetic field and a transition between these energy states occurs on the application of an appropriate frequency in the microwave region.
2. NMR absorption positions are expressed in terms of chemical shifts.	2. ESR absorption positions are expressed in terms of g values.
3. Nuclear spin-spin coupling causes the splitting of NMR signals.	3. Coupling of the electronic spin with nuclear spins (hyperfine coupling) causes the splitting of ESR signals.

7.13 Some Solved Problems

Problem 1. Which of the following will show electron spin resonance (ESR) spectrum? Give reason for your choice.

(a) H (b) H_2 (c) Cl (d) Na^+ (e) Cu^+

Solution. (a) Hydrogen atom (H) has electronic configuration $1s^1$, i.e. it has one unpaired electron. Thus, it will show ESR spectrum.

(b) Hydrogen molecule (H_2) has electronic configuration $(\sigma_{1s})^2$, i.e. it has no unpaired electron. Thus, it will not show ESR spectrum.

(c) Chlorine atom (Cl) has electronic configuration $1s^2 2s^2 2p^6 3s^2 3p_x^2\ 3p_y^2\ 2p_x^1$, i.e. it has one unpaired electron. Thus, it will show ESR spectrum.

(d) The electronic configuration of Na^+ is $1s^2 2s^2 2p_x^2\ 2p_y^2\ 2p_z^2$. Since it has no unpaired electron, it will not show ESR spectrum.

(e) The electronic configuration of cuprous ions (Cu^+) is $3d^{10}$. Since it has no unpaired electron, it will not show ESR spectrum.

Problem 2. Calculate the ESR frequency of an unpaired electron in a magnetic field of 3000 G (0.30 T).

Solution. From Eq. (7.2), we have

$$\Delta E = h\nu = g\,\beta H_0$$

or

$$\nu = \frac{g\beta H_0}{h}\ \text{Hz}$$

where $g = 2.00$, $\beta = 9.273 \times 10^{-24}\ \text{JT}^{-1}$, $H_0 = 0.30$ T and $h = 6.626 \times 10^{-34}$ Js

Therefore

$$\nu = \frac{2.00 \times 9.273 \times 10^{-24}\ \text{JT}^{-1} \times 0.30\text{T}}{6.626 \times 10^{-34}\ \text{Js}}$$

$$= 8.397 \times 10^9\ \text{Hz} = 8397\ \text{MHz} = 8.397\ \text{kMHz}$$

Problem 3. Calculate the g value if the methyl radical shows ESR signal at 3290 G (0.3290 T) in a spectrometer operating at 9230 MHz.

Solution From Eq. (7.2)

$$h\nu = g\beta H_0$$

Hence,

$$g = \frac{h\nu}{\beta H_0} = \frac{(6.626 \times 10^{-34}\ \text{Js})\,(9.230 \times 10^9\ \text{Hz})}{(9.273 \times 10^{-24}\ \text{JT}^{-1})(0.3290\,\text{T})} = 2.004641$$

Problem 4. Predict the number of lines in the ESR spectrum of each of the following radicals:

(a) $[CF_2H]^{\cdot}$ (b) $[^{13}CH_3]^{\cdot}$ (c) $[CF_2D]$ (d) $[CClH_2]^{\cdot}$ (e) $[^{13}CF_2H]^{\cdot}$

Solution

The number of lines in an ESR signal = $(2nI_i + 1)\,(2mI_j + 1)\ldots$ (7.7)

where n and m are the number of different kinds of coupled nuclei having spin I_i and I_j, respectively.

(a) The spin of F = $\frac{1}{2}$, i.e. $I_i = \frac{1}{2}$ and H = $\frac{1}{2}$, i.e. $I_j = \frac{1}{2}$; n and m for the radical $[CF_2H]^{\cdot}$ are 2 and 1, respectively.
According to Eq. (7.7)

$(2 \times 2 \times \frac{1}{2} + 1)\,(2 \times 1 \times \frac{1}{2} + 1) = 6$ lines

(b) $[^{13}CH_3]^{\cdot}$
Here $n = 1$, $I_i = \frac{1}{2}$; $m = 3$, $I_j = \frac{1}{2}$
According to Eq. (7.7)

$(2 \times 1 \times \frac{1}{2} + 1)\,(2 \times 3 \times \frac{1}{2} + 1) = 8$ lines

(c) $[CF_2D]^{\cdot}$

Here $n = 2$, $I_i = \frac{1}{2}$; $m = 1$, $I_j = 1$ (spin of $D = 1$)

According to Eq. (7.7)

$$(2 \times 2 \times \tfrac{1}{2} + 1)\,(2 \times 1 \times 1 + 1) = 9 \text{ lines}$$

(d) $[CClH_2]^{\cdot}$

Here $n = 1$, $I_i = \frac{3}{2}$ (spin of Cl $= \frac{3}{2}$); $m = 2$, $I_j = \frac{1}{2}$

According to Eq. (7.7)

$$(2 \times 1 \times \tfrac{3}{2} + 1)\,(2 \times 2 \times \tfrac{1}{2} + 1) = 12 \text{ lines}$$

(e) $[^{13}CF_2H]^{\cdot}$

Here $n = 1$, $I_i = \frac{1}{2}$; $m = 2$, $I_j = \frac{1}{2}$; $o = 1$, $I_j = \frac{1}{2}$

According to Eq. (7.7)

$$(2 \times 1 \times \tfrac{1}{2} + 1)\,(2 \times 2 \times \tfrac{1}{2} + 1)\,(2 \times 1 \times \tfrac{1}{2} + 1) = 12 \text{ lines}$$

Problem 5. Which of the following will show ESR spectra and why?

(a) Benzene, C_6H_6 (b) Benzene anion, $C_6H_6^-$

(c) Cyclopentadienyl cation, $C_5H_5^+$ (d) Cyclopentadienyl anion, $C_5H_5^-$

Solution. (a) Benzene, C_6H_6

From Hückel molecular-orbital (HMO) calculations on benzene, the π-electron distribution may be represented as

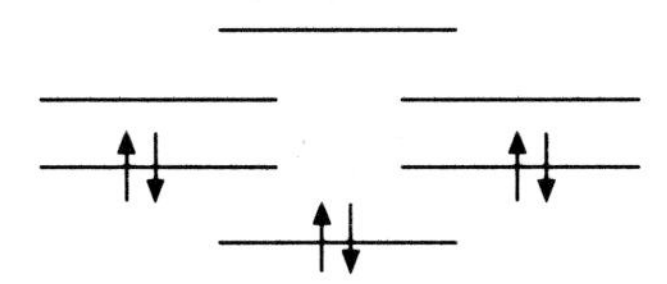

Since there is no unpaired electron, benzene will not show ESR spectrum.

(b) Benzene anion, $C_6H_6^-$

From HMO calculations on $C_6H_6^-$, the π-electron distribution can be represented as

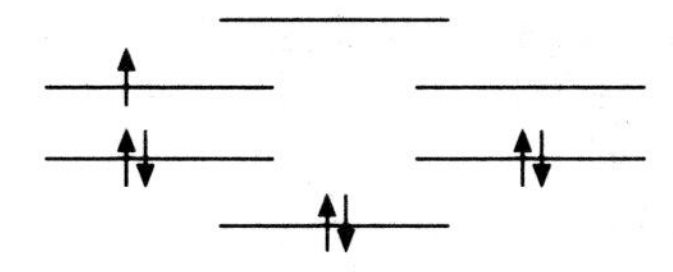

Since there is one unpaired electron, $C_6H_6^-$ will show ESR spectrum.

(c) Cyclopentadienyl cation, $C_5H_5^+$

The π-electron distribution in the molecular orbitals of $C_5H_5^+$ is as follows:

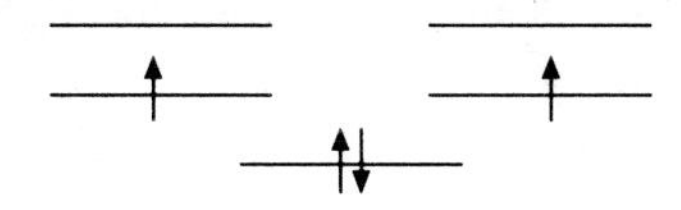

Since there are two unpaired electrons, $C_5H_5^+$ will show ESR spectrum.

(d) Cyclopentadienyl anion, $C_5H_5^-$
The π-electron distribution in the molecular orbitals of $C_5H_5^-$ is

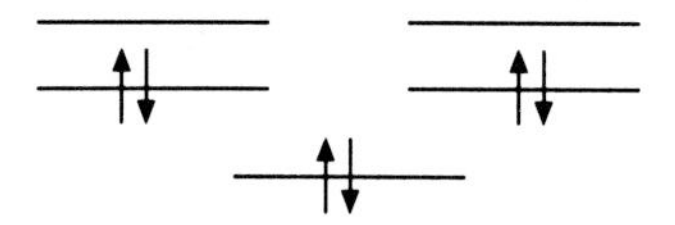

Since there is no unpaired electron, $C_5H_5^-$ will not show ESR spectrum.

Problem 6. ESR spectra of free radicals formed from three substrates during enzyme oxidation by peroxide H_2O_2 are given in Fig. 7.15. Match each of the spectra with the corresponding free radical.

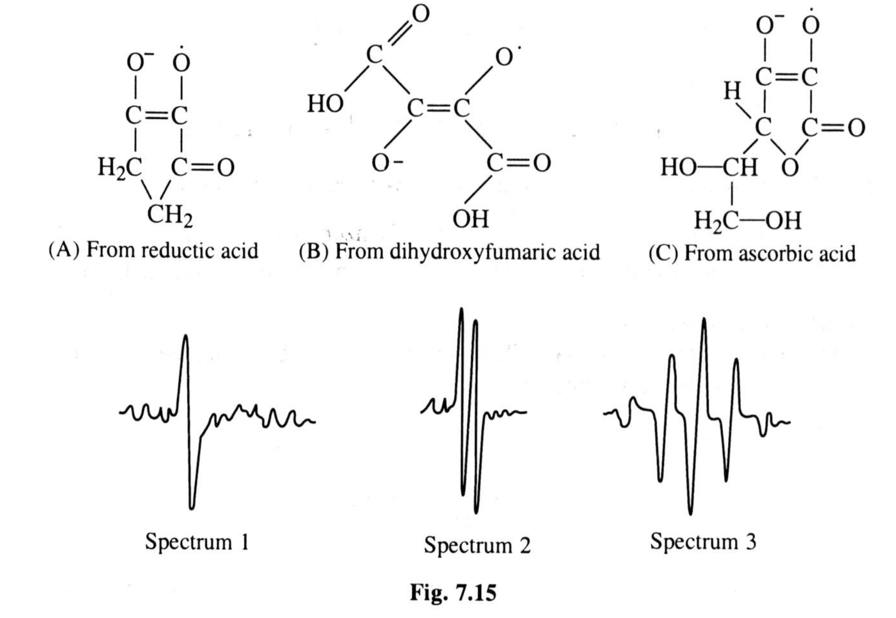

Fig. 7.15

Solution. Spectrum 1 contains a single line, thus it is of the free radical (B) obtained from dihydroxyfumaric acid. In this free radical, there is no possibility of hyperfine coupling. Hence, a single line is observed. It should be noted that due to rapid proton exchange, the OH proton contributes no detectable hyperfine splitting.

Spectrum 2 shows two lines due to the free radical (C) obtained from ascorbic acid because in this free radical the unpaired electron can couple with a single proton to split the ESR signal into two lines.

Spectrum 3 shows five lines due to the free radical (A) obtained from reductic acid because in this case the unpaired electron can couple with four equivalent protons resulting in the splitting of the ESR signal into $2nI + 1$, i.e. $2 \times 4 \times \frac{1}{2} + 1 = 5$ lines.

Problem 7. Predict the number of lines in the ESR spectrum of naphthalene negative ion (naphthalene radical anion) produced by the addition of sodium to naphthalene.

Solution. There are two different sets, each of four equivalent protons ($\overset{a}{H}$ and $\overset{b}{H}$), in naphthalene. The unpaired electron is delocalized over the entire naphthalene ring.

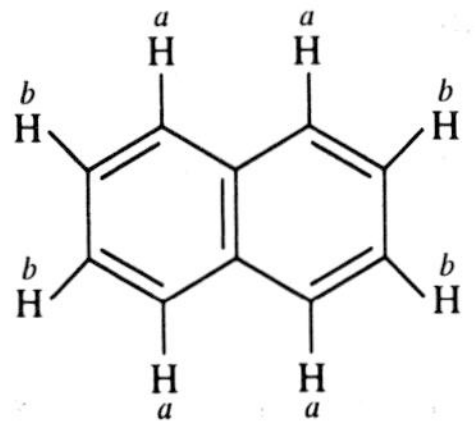

If the protons responsible for hyperfine splitting are not equivalent, then the number of lines in the hyperfine structure will be equal to $(n + 1)(n' + 1)$, and so on. Thus, in case of naphthalene radical anion the number of lines will be $(4 + 1)(4 + 1) = 25$. This has also been found experimentally.

Problem 8. Trichloromethyl radical $[CCl_3]^{\bullet}$ shows ten lines in the hyperfine structure of its ESR spectrum. Calculate the spin of the Cl nucleus.

Solution. The radical $[CCl_3]^{\bullet}$ has one unpaired electron which interacts with the nuclear spin of Cl. We know that the number of lines are equal to $2nI + 1$. Hence, $2nI + 1 = 10$ (n is the number of interacting nuclei and I is their spin).

Or $$2 \times 3 \times I + 1 = 10$$

or $$I = \tfrac{9}{6} = \tfrac{3}{2}$$

Therefore, $$\text{spin of the Cl nucleus} = \tfrac{3}{2}$$

PROBLEMS

1. Which of the following will show ESR spectrum? Give reason for your choice.

   (a) Cu^{2+} (b) CO_2 (c) NO_2 (d) Cl^-

2. Compare NMR and ESR spectroscopy. Comment on the magnitudes of energy separation in NMR and ESR transitions.
3. Predict the number of lines in the ESR spectrum of each of the following radicals:

   (a) $CH_3\dot{C}H_2$ (b) $(CH_3)_2\dot{C}H$ (c) $(CH_3)_3\dot{C}$ (d) $\dot{C}D_3$

4. Write explanatory notes on the following:

   (a) The *g* factor (b) Hyperfine structure (c) ENDOR

5. Predict the number of lines and their relative intensities in the ESR spectra of the following radicals:

   (a) $\dot{D}$ (b) 1,4-benzosemiquinone

   (c) $\dot{C}H_2OH$ (d) Cyclopentadienyl radical

   Hint: Due to rapid proton exchange, the OH proton does not contribute to hyperfine splitting.
6. Which of the following will not show ESR spectrum?

   (a) Cyclopentadienyl anion (b) Naphthalene anion
   (c) Benzene anion (d) Anthracene anion

7. Which of the following will show ESR spectrum?

   (a) Cyclopentadiene (b) Benzene dianion $(C_6H_6)^{2-}$
   (c) Cyclopentadienyl anion (d) Cyclopropene

8. Malonic acid $CH_2(COOH)_2$ gives two products *A* and *B* on irradiation with X-rays. The product *A* exhibits a doublet and *B* a triplet in its ESR spectrum. Identify *A* and *B*.
9. How many lines are expected in the ESR spectrum of the following?

   (a) Radical anion derived from *p*-xylene.
   (b) 1,3,5-cycloheptatrienyl radical.

10. Predict the number of lines and their relative intensities in the ESR spectra of the following:

    (a) Benzene cation $C_6H_6^{\dot{+}}$ (b) $\dot{C}H_3$

    (c) Benzene anion $C_6H_6^{\dot{-}}$ (d) $\dot{C}D(COOH)_2$

11. Draw an energy level diagram showing the transitions responsible for the ESR spectrum of the methyl free radical.
12. Calculate the magnetic field at which resonance occurs if the ESR spectrometer is operating at 9302 MHz and the *g* value for benzene anion is 2.0025. Hint: Apply Eq. (7.2).
13. Which of the following will not exhibit ESR spectrum?

    (a) NO (b) $CO_2^{\dot{-}}$ (c) $C_6H_5^{\dot{+}}$ (d) O_2

14. Deduce the structure of the chemical species which is obtained by the alkali metal reduction of benzene at –70°C and exhibits seven lines in its ESR spectrum.
15. Will $^{13}CO_2^{\dot{-}}$ exhibit ESR spectrum? If yes, predict the number of lines and their relative intensity in its ESR spectrum.
16. Triiodomethyl radical $[CI_3]^{\bullet}$ exhibits sixteen lines in its ESR spectrum. Calculate the spin of the iodine nucleus.

References

1. C.N. Banwell, Fundamentals of Molecular Spectroscopy, McGraw-Hill, 1983.
2. D.J.E. Ingram, Free Radicals as Studied by Electron Spin Resonance, Butterworth, 1958.

3. F. Gerson, High Resolution E.S.R. Spectroscopy, John Wiley, 1971.
4. J.E. Wertz and J.R. Boulton, Electron Spin Resonance: Elementary Theory and Practical Applications, McGraw-Hill, 1972.
5. L. Kevan and R.N. Schwatz (Eds.), Time Domain Electron Spin Resonance, Wiley, 1979.
6. M. Symons, Chemical and Biological Aspects of Electron-Spin Resonance Spectroscopy, Van Nostrand Reinhold Company, New York, 1978.
7. R. Bersohn and J.C. Baird, An Introduction to Electron Paramagnetic Resonance, Benjamin, Inc., New York, 1966.
8. The NMR-EPR Staff of Varian Associates, NMR and EPR Spectroscopy, Pergamon, New York, 1960.

8

Mass Spectroscopy (MS)

8.1 Introduction

The technique of mass spectrometry was first used by J.J. Thompson in 1911 to provide the conclusive proof for the existence of isotopes. However, the extensive application of mass spectroscopy in solving structural problems only began around 1960. Mass spectroscopy is based on a single principle, i.e. it is possible to determine the mass of an ion in the vapour phase. Thus, a mass spectrometer ionizes the sample into a beam of ions in the vapour phase, separates the ions according to their mass to charge ratios (m/e or m/z values) and records the mass spectrum as a plot of m/e of ions against their relative abundances (Fig. 8.1). Actually, m/e is obtained from a mass spectrum, but for all practical purposes it is equal to the mass m of the ion because multicharged ions are very much less abundant than those with a single charge (e or $z = 1$).

Mass spectroscopy is useful for characterization of organic compounds in two ways:

(i) It can give the exact molecular masses and thus, the exact molecular formulae.

(ii) It can show the presence of certain structural units and their points of attachment in the molecule, and thus gives idea about the structure of the molecule.

8.2 Ionization Methods

The most important ionization methods are: (i) electron impact and (ii) chemical ionization.

(i) Electron Impact (EI) Method

In this method, the sample is bombarded in the vapour phase with a beam of high energy electrons (70 eV).

1. On collision of a molecule M with an electron e, the molecule M is highly energized and ejects an electron to give a radical cation $M^{\pm}$ which is known as the *molecular* or *parent ion*. This is the most probable process. The molecular ion (often denoted as M^{+}) has the same mass as the initial molecule M. Thus, its mass gives a direct measure of molecular weight of a compound.

$$M + e \rightarrow \underset{\text{Molecular or parent ion}}{M^{\dot{+}}} + 2e$$

2. The alternative and far less probable process (less probable by a factor of ~10^{-2}) involves the capture of an electron by a molecule to give a radical anion M^-

$$M + e \rightarrow M^{\dot{-}}$$

3. A least probable process gives multiply charged ions

$$M + e \rightarrow M^{n+} + (n + 1)e$$

Mass spectrometers are generally set up to detect only positive ions, but negative-ion mass spectrometry is also possible.

(ii) Chemical Ionization (CI) Method

In this method the sample is introduced into the ionization chamber near the atmospheric pressure with a large excess of an intermediate substance (methane, isobutane or NH_3) also called the *carrier gas*. Methane is generally used as the carrier gas which is first ionized by electron impact to *primary ions*

$$CH_4 \xrightarrow{e} \overset{+}{C}H_4^{\cdot} + 2e$$

$$\downarrow -H^{\cdot}$$

$$^{+}CH_3$$

Primary ions react with excess of methane to give *secondary ions*

$$\overset{+}{C}H_4^{\cdot} + CH_4 \rightarrow CH_5^+ + \dot{C}H_3$$

$$CH_3^+ + CH_4 \rightarrow C_2H_5^+ + H_2$$

The sample (RH) is then ionized largely by collision with these secondary ions rather than electrons

$$CH_5^+ + RH \rightarrow RH_2^+ + CH_4$$

$$C_2H_5^+ + RH \rightarrow RH_2^+ + C_2H_4$$

These $RH_2^+ (M + 1)$ ions are often prominent and are called *quasimolecular ions*. They can lose H_2 to give prominent $M - 1$ ions. This method causes less fragmentation and so the mass spectra can be interpreted more easily. Further, this technique is very useful for locating the molecular ion peak for compounds which give either no molecular ion peak or a very weak parent peak.

Besides the above ionization methods, the following desorption ionization techniques are generally used for involatile or thermally unstable compounds:

(i) FAB (fast atom bombardment)
(ii) Laser desorption
(iii) Thermospray
(iv) ^{252}Cf plasma desorption

8.3 Molecular and Fragment Ions

As discussed in Section 8.2(i), the bombardment with a beam of high energy electrons (70 eV) usually removes one electron from the molecule M to give the molecular or parent ion $\overset{+}{M}{}^{\cdot}$ in the vapour phase. The molecular ion gives the molecular mass of the sample because the mass of the electron lost from the molecule is negligible. The electron with the lowest ionization potential in the molecule is lost to give the molecular ion. The formation of molecular ions from some very common organic compounds are

$$C_6H_6 + e \longrightarrow C_6H_6^{+\cdot} + 2e \quad (M^{+\cdot})$$

$$R\ddot{N}H_2 + e \rightarrow R\overset{+}{\dot{N}}H_2 + 2e \quad (M^{+\cdot})$$

$$R{-}\ddot{\underset{..}{O}}{-}H + e \rightarrow R{-}\overset{+}{\dot{\underset{..}{O}}}{-}H + 2e \quad (M^{+\cdot})$$

$$>C{=}C< + e \longrightarrow >\overset{+}{\dot{C}}{-}C< + 2e \quad (M^{+\cdot})$$

$$>C{=}\ddot{O}{:} + e \longrightarrow \left[>C{=}\dot{O}^{+}{:} \longleftrightarrow >\overset{\cdot}{C}{}^{+}{-}\ddot{O}{:}\right] + 2e \quad (M^{+\cdot})$$

$$-C{\equiv}C- + e \longrightarrow -C{=}\overset{+}{C}{\cdot}- + 2e \quad (M^{+\cdot})$$

$$H_3C{-}CH(CH_3){-}CH_3 + e \longrightarrow H_3C + {\cdot}CH(CH_3){-}CH_3 + 2e \quad (M^{+\cdot})$$

The order of the energy required to remove an electron from a molecule to give the molecular ion is

$$\text{lone pair} < \text{conjugated } \pi < \text{non-conjugated } \pi < \sigma \text{ electron}$$

In alkanes, the removal of an electron from C—C σ bonds is easier than that from C—H bonds.

The energy required for removing one electron from neutral organic molecules is about 10-12 eV. In practice, much higher energy ~70 eV (1 eV = 95 kJ mol^{-1}) is used which causes further fragmentation of the molecular ion resulting in *fragment* or the *daughter* ions. Generally, fragment ions are formed from the molecular ion in the following two ways:

$$\underset{}{M^{\dot{+}}} \rightarrow \underset{\text{Even-electron cation}}{A^{+}} + \underset{\text{Radical}}{B^{\cdot}}$$

$$M^{\dot{+}} \rightarrow \underset{\text{Radical cation}}{C^{\dot{+}}} + \underset{\text{Neutral molecule}}{D}$$

Since only species bearing a positive charge are detected in the mass spectrometer, the mass spectrum will show signals due to $M^{\dot{+}}$, A^{+}, $C^{\dot{+}}$ and also due to fragment ions resulting from subsequent fragmentation of A^{+} and $C^{\dot{+}}$, but there will be no signal due to $B^{\cdot}$ and D. Fragmentation of A^{+} usually gives an even-electron ion E^{+} and a neutral molecule G, whereas $C^{\dot{+}}$ fragments in two ways similar to $M^{\dot{+}}$. It should be noted that the mass spectrometer records signals due to radical cations and cations only. Any species may fragment in a variety of ways; so a mass spectrum consists of many signals. Fragmentation processes of various classes of organic compounds are discussed in Section 8.15.

8.4 Instrumentation

The main components of the schematic diagram of a mass spectrometer given in Fig. 8.1 are discussed as follows.

(i) Inlet or Sample Handing System

This system allows the introduction of a small quantity (ranging from several milligrams to less than a microgram) of vapour of the sample under analysis into the ionization chamber. The pressure in the inlet system is greater than that in the ionization chamber.

(ii) Ionization Chamber

The vapour of the sample from the inlet system enters the ionization chamber (operated under high vacuum, at about 10^{-6} to 10^{-5} Torr) where it is bombarded by a beam of electrons of about 70 eV energy. The various positive ions thus produced are first accelerated by a repeller potential applied between A and B. Then they are accelerated to their final velocities by applying a large (~8 kV) accelerating potential V between B and C (Fig. 8.1).

(iii) Mass Analyzer

This part of the mass spectrometer separates ions according to their masses, (strictly according to their m/e. The accelerated positively charged ions from the ionization chamber enter the mass analyzer where they pass through a uniform perpendicular magnetic field H.

The kinetic energy of an ion of mass m and velocity v is $\frac{1}{2}mv^2$. Let the accelerating potential be V, then the energy given to each singly charged ion is eV. Hence, the kinetic energy possessed by the ion should be equal to this given energy, i.e.

$$\frac{1}{2}mv^2 = eV \tag{8.1}$$

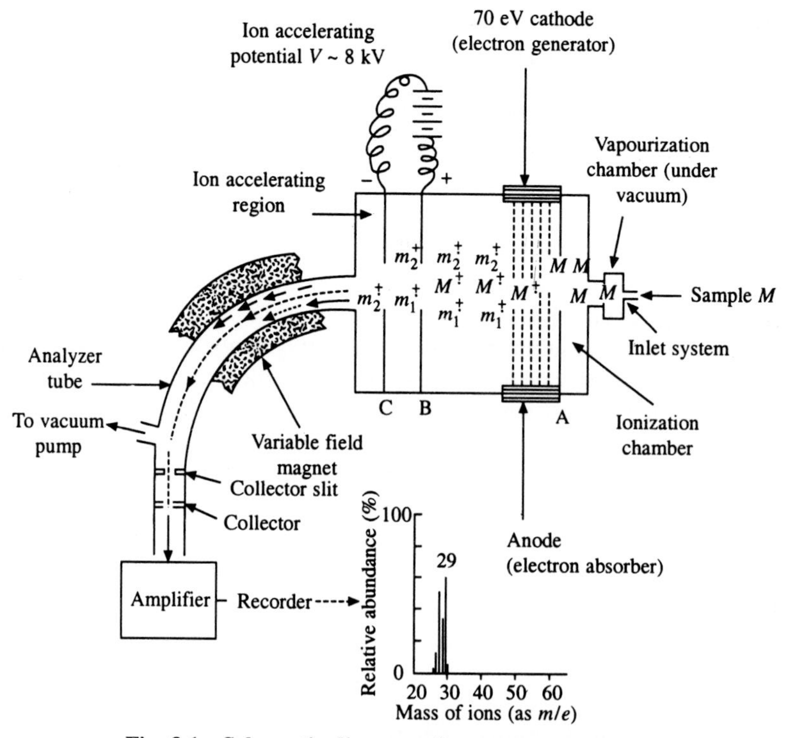

Fig. 8.1 Schematic diagram of a mass spectrometer

The magnetic field H causes the ions to move in a circular path of radius r. The force of attraction (centripetal force) of the magnetic field is given by Hev which is balanced by the centrifugal force $\frac{mv^2}{r}$ at the equilibrium (Newton's second law of motion), i.e.

$$Hev = \frac{mv^2}{r} \tag{8.2}$$

Squaring both sides, we get

$$H^2e^2v^2 = \frac{m^2v^4}{r^2}$$

or

$$H^2e^2 = \frac{m^2v^2}{r^2} \tag{8.3}$$

From Eq. (8.1)

$$\tfrac{1}{2}mv^2 = eV$$

Therefore

$$mv^2 = 2eV$$

Putting the value of mv^2 in Eq. (8.3), we get

$$H^2e^2 = \frac{m \cdot 2eV}{r^2}$$

or

$$H^2e = \frac{2mV}{r^2}$$

or

$$m/e = \frac{H^2r^2}{2V} \tag{8.4}$$

It is clear from Eq. (8.4) that the radius r of the ion path can be changed by varying the magnetic field H at a constant accelerating voltage V or by varying V at a constant H which allows magnetic or voltage scanning of the mass spectrum, respectively. The spectrum is generally obtained by magnetic scanning, i.e. H is increased keeping V constant. On increasing H, the ions of progressively higher m/e attain the necessary radius to pass through the collector slit sequentially.

(iv) Ion Detector and Spectrum Recorder

The streams of ions of different m/e impinge into the ion collector and the amplified current from the ion collector is recorded as relative abundances of ions on y-axis of the recorder. The magnetic or voltage scanning is synchronized with the x-axis of the recorder and calibrated to appear as m/e.

8.5 Double Focusing Mass Spectrometers

The mass spectrometer described in Section 8.4 is a single focusing instrument and it fails to discriminate between small mass differences. This limitation in resolving power is because of small variations in kinetic energies of ions of a given species due to initial spread of their translational energy. Double focusing mass spectrometers with high resolution power are available. In these instruments the stream of ions is passed through a radial electrostatic field prior to entering the mass analyzer. Thus, the ions of the same kinetic energy are selected and allowed to enter the mass analyzer. By use of a double focusing mass spectrometer, mass of an ion can be determined up to the accuracy of ± 0.0001 amu. Thus, it is possible to distinguish between the ions which have the same integral (nominal) mass but differ in their exact masses.

8.6 Mass Spectrum and the Base Peak

The mass spectrum of a compound is graphically represented by plotting m/e values of the various ions (molecular as well as fragment ions) against their relative abundances (i.e. intensities) as percentages of the base peak (Fig. 8.1). The largest (most intense) peak in the spectrum is called the *base peak* which is arbitrarily assigned the value of 100% abundance, while the intensities of the other peaks (including the parent peak) are represented as percentages of the base peak. Sometimes the molecular ion peak or parent peak itself may be the base peak. Generally, mass spectra are represented in the form of a bar graph (Fig. 8.1). A tabular representation of mass spectra is also used in which m/e values of the various ions and their relative abundances are recorded (Table 8.1).

Table 8.1 Tabular representation of the mass spectrum of methane

m/e	Relative abundance (%)
1	3.4
2	0.2
12	2.8
13	8.0
14	16.0
15	86.0
16	100.0 (Base peak)
17	1.1

8.7 Recognition of the Molecular Ion (Parent) Peak and Detection of Isotopes

The recognition of the molecular ion (M^+) peak (parent peak) in the mass spectrum of an organic compound is very important because it gives the molecular mass of the compound from which the molecular formula of the compound can be derived. The molecular ion peak is the peak of highest mass number except for the isotope peaks at mass numbers $M + 1$, $M + 2$ etc. (M = mass of the molecular ion). The isotope (satellite) peaks appear because of the presence of certain number of molecules containing heavier isotopes than the common isotopes. Since the natural abundance of heavier isotopes is generally much more less than that of the lightest isotope, the intensities of isotope peaks are very low relative to the parent peak. Thus, $M + 1$ isotope peak appearing due to ^{13}C is about 1.1% of the molecular ion peak because the natural abundance of ^{13}C isotope is about 1.1% as compared to 98.9% natural abundance of ^{12}C. For example, in the mass spectrum of methane (Fig. 8.1 and Table 8.1), the $M + 1$ isotope peak appears at m/e 17 and its abundance relative to the molecular ion peak is 1.1%. It should be noted that in this particular case the molecular ion peak itself is the base peak. The presence of ^2H(D) will make an additional but very small (0.016%) contribution to the $M + 1$ peak because the natural abundance of deuterium is about 0.016% as compared to 99.984% natural abundance of ^1H.

Because of very low natural abundance of ^{13}C and ^2H the probability of finding two ^{13}C or ^2H in the same molecule is so low that $M + 2$ peaks due to ^{13}C and ^2H are often negligible. Thus, for most of the organic compounds M+2 peaks are too small to be considered. However for compounds containing chlorine, bromine or sulphur, the $M + 2$ isotope peak is important. The $M + 2$ isotope peak due to the presence of a chlorine or a bromine atom is about 33% or 50% of the molecular ion peak, respectively. Thus, in a chloro or bromo compound the ratio of intensity of M^+ and $M + 2$ peaks will be 3 : 1 or 1 : 1, respectively. This is because of the 24.6% natural abundance of ^{37}Cl as compared to 75.4% of ^{35}Cl, and almost equal abundance of ^{79}Br and ^{81}Br (Table 8.2). Similarly, if one sulphur atom is present in a molecule, then according to the natural abundance (4.2%) of ^{34}S, the $M + 2$ peak will be about 4.4% of the parent peak (the natural abundance of ^{32}S is 95.06%). Thus, the presence of Cl, Br or S is easily detectable on the basis of the intensity of $M + 2$ peak relative to the parent peak. Fluorine and iodine have no isotope, i.e. they are isotopically pure. Natural abundances of isotopes of some common elements are given in Table 8.2.

Table 8.2 Natural abundance of isotopes of some common elements

Isotope	Natural abundance (%)	Isotope	Natural abundance (%)
^{1}H	99.984	^{19}F	100
^{2}H	0.016	^{31}P	100
^{12}C	98.9	^{32}S	95.06
^{13}C	1.1	^{33}S	0.74
^{14}N	99.64	^{34}S	4.2
^{15}N	0.36	^{35}Cl	75.4
^{16}O	99.76	^{37}Cl	24.6
^{17}O	0.04	^{79}Br	50.57
^{18}O	0.2	^{81}Br	49.43
		^{127}I	100

The intensity (height) of the molecular ion peak depends on the stability of the molecular ion. The lower the ionization potential of a molecule, and the stabler is the molecular ion, the more intense will be the molecular ion peak. Since the stability of a chemical species is strongly affected by its structure, the intensity of the parent peak shows a great variability. Thus, in some cases the parent peak has the highest intensity, i.e. it is the base peak, whereas in other cases it is not the base peak and may be even so small that a lot of effort is made to locate it. The parent peak will be of low intensity if the molecules contain bonds which are readily cleaved. In general, aromatic compounds, conjugated olefins, saturated cyclic compounds, certain sulphur compounds and short, straight-chain hydrocarbons give an intense molecular ion peak. Straight chain aldehydes, ketones, esters, acids, amides, ethers and halides give a recognizable parent peak. Aliphatic alcohols, amines, nitrites, nitrates, nitro compounds, and highly branched compounds give a parent peak which is frequently not detectable.

There are cases where the identification of molecular ion peak is difficult because it is very weak or does not appear. In such cases the spectrum should be run at maximum sensitivity using larger quantity of the sample to get recognizable molecular ion peak. If the molecular ion peak is present but it is one of several peaks, then the energy of the bombarding electron beam should be decreased. This reduces the intensities of all peaks, but increases the intensity of the molecular ion peak relative to other peaks, including the fragment ion peaks (but not parent peaks) of impurities.

8.8 Confirmation of the Recognized Molecular Ion Peak

The useful points in confirming the recognized molecular ion peak are discussed as follows.

(i) Index of Hydrogen Deficiency (IHD) or Double Bond Equivalents (DBE) or Ring Rule

The molecular ion must be an odd-electron ion. Thus, if the elemental composition of the ion is known, the index of hydrogen deficiency may be used to ascertain whether the ion in question, i.e. recognized as the molecular ion, is an odd-electron ion. It should be noted that there may be other odd-electron fragment ions besides the molecular ion in the spectrum which arise from rearrangement reactions. The index of hydrogen deficiency for a molecular ion (or any other odd-electron fragment) must be a whole number (an integer).

The index of hydrogen deficiency is the number of pairs of hydrogen atoms which must be removed from the saturated formula (e.g. $C_n H_{2n+2}$ for alkanes) to give the molecular formula in question. Thus, the index of hydrogen deficiency is the sum of the number rings, the number of double bonds and twice the number of triple bonds. For a molecule $I_y II_n III_z IV_x$ (e.g. $C_xH_yN_zO_n$),

$$\text{The index of hydrogen deficiency } = x - \frac{y}{2} + \frac{z}{2} + 1 \qquad (8.5)$$

where I is any monovalent atom (e.g., H, F, Cl, Br, etc.), II is the O, S or any other divalent atom, III is the N, P or any other trivalent atom and IV is the C, Si or any other tetravalent atom.

For example, acetone has an index of hydrogen deficiency of $(3 - \frac{6}{2} + 1) = 1$, because it has one C=O. Similarly, the index of hydrogen deficiency for pyrrole would be $(4 - \frac{5}{2} + \frac{1}{2} + 1) = 3$. This is because it has two C=C and one ring. The calculated index of hydrogen deficiency for dimethyl sulphoxide (DMSO) is zero which is in agreement with its following polar structure:

$$H_3C-\overset{\overset{\large O^-}{|}}{S^+}-CH_3$$

Eq. (8.5) can be applied to molecules, molecular ions and fragment ions. The index of hydrogen deficiency is always a whole number for odd-electron ions, whereas it has a non-integral value (odd multiple of 0.5) for even-electron (all electrons paired) ions. Thus, the values of the index of hydrogen deficiency can be applied to molecular ions as well as fragment ions to get useful information. For example, $C_7H_5O^+$ has an index of 5.5 which suggests its following reasonable structure:

$$C_6H_5-\overset{+}{C}=O$$

(ii) Nitrogen Rule

This rule given by Beynon (1960) states that a molecule containing an even number of nitrogen atoms or no nitrogen atom has even-numbered molecular mass; an odd-number of nitrogen atoms require an odd numbered molecular mass. This relationship is applicable to molecular ions as well as to all odd-electron ions. Similarly, an even electron ion with an even number of nitrogen atoms or no nitrogen atom will have an odd mass number.

A corollary of nitrogen rule states that fragmentation of an odd mass-numbered molecular ion at a single bond gives an even mass-numbered ion fragment, and an odd mass-numbered ion fragment from an even mass-numbered molecular ion. This is true only when the ion fragment contains all the nitrogen atoms (if any) of the molecular ion.

Nitrogen rule holds for all compounds containing C, H, N, O, S, halogens, P, B, Si, As and the alkaline earths. The nitrogen rule is based on the fact that for most of the elements present in organic compounds, there is a correspondence between the mass number of the most abundant isotope of the element and its most common valence, i.e. both are even numbered or both are odd numbered. For example, the most abundant isotope of carbon is ^{12}C which has both even numbered, i.e. the mass number 12 and the valence 4. Similarly, for hydrogen, the mass number is 1 and the valence is 1 (both odd numbered); for oxygen, the mass number is 16 and the valence is 2 (both even numbered). Nitrogen is an exception having mass number 14 (even numbered) and the valence 3 (odd numbered). This exception (mass number 14 and valence 3) is the basis of the nitrogen rule (discussed above) according to which the molecular mass to be used is the sum of most abundant isotopes.

For example, CH_4 having no nitrogen has even numbered molecular mass 16 and its molecular ion appears at *m/e* 16. Similarly, $H_2NCH_2CH_2NH_2$ having even number of nitrogen atoms has even numbered molecular mass 60 and its molecular ion appears at *m/e* 60. $C_2H_5NH_2$ having odd number of nitrogen atoms has odd numbered molecular mass 45 and its molecular ion appears at *m/e* 45.

Thus, the nitrogen rule is often helpful in confirming the recognized molecular ion peak because many peaks can be ruled out as possible molecular ion peaks simply on grounds of reasonable structure requirements.

(iii) *M*-15, *M*-18 or *M*-31

The presence of an *M*-15 peak (loss of CH_3) or a *M*-18 peak (loss of H_2O) or a *M*-31 peak (loss of OCH_3 from methyl esters) etc. is used as conformation of a molecular ion peak.

8.9 Multiply Charged Ions

In mass spectrometry, the formation of ions carrying a single positive charge is the most probable process. The formation of multiply charged (M^{n+}) ions is the least probable process. Hence, these are generally of no importance in mass spectrometry. However, sometimes peaks due to doubly or even triply charged ions are observed and these may be useful in the interpretation of mass spectra. A doubly charged ion is recorded at one-half ($m/2$) and a triply charged ion at one-third ($m/3$) value of its actual mass number because the mass spectrometers record mass m to charge e ratios (m/e values) of ions. The *m/e* of singly charged ions ($e = 1$) gives their actual mass number ($m/1$). The formation of multiply charged ions is common in polynuclear hydrocarbons and heteroaromatic compounds. For example, naphthalene shows its molecular ion peak due to the singly charged molecular ion ($M^{+\cdot}$) at *m/e* 128, whereas a peak due to the corresponding doubly charged ion M^{2+} appears at *m/e* 64, i.e. at one-half value of the actual molecular mass of naphthalene. The doubly charged molecular ions undergo fragmentation to give doubly charged fragment ions which are recorded at *m/e* values numerically equal to one-half of their actual mass numbers.

8.10 Metastable Ions or Peaks

If an ion m_1^+ passes the accelerating region of the mass spectrometer without decomposition, then it is recorded as m_1^+. If m_1^+ decomposes into m_2^+ before entering the accelerating region, then it is recorded as m_2^+. However, if m_1^+ decomposes into m_2^+ during acceleration then it is neither recorded as m_1^+ nor m_2^+ but is recorded as a metastable ion m^* causing the metastable peak. Thus, the peaks due to m_1^+ and m_2^+ will be accompanied by a metastable peak at mass m^*. The numerical value of m^* is given by

$$m^* = \frac{(m_2)^2}{m_1}$$

where m_1, m_2 and m^* are the masses (strictly *m/e*) of the ions m_1^+, m_2^+ and m^*, respectively, and $m_1 > m_2$.

Metastable peaks are of low intensity, much broader and appear at non-integral mass values. The presence of a metastable peak gives information that m_2^+ is derived from m_1^+ in one step by loss of a neutral fragment. However, the absence of a metastable peak does not mean that the one-step decomposition of m_1^+ to m_2^+ does not occur. In a mass spectrum, m_1, m_2 and m^* are almost equidistant, i.e. the distances from m_1 to m_2 and that from m_2 to m^* are of similar magnitude on the mass scale.

Let us take an example of the mass spectrum of acetophenone which shows a metastable peak at *m/e* 92.1 and gives evidence for the decomposition of the molecular ion of acetophenone (*m/e* 120) to $C_6H_5C{\equiv}\overset{+}{O}$ (*m/e* 105) in one step because the predicted mass $m^* = \frac{(105)^2}{120} = 91.88$.

$$\underset{m/e\ 120\ (\overset{+}{M})}{\left[C_6H_5COCH_3\right]^{+\cdot}} \rightarrow \underset{m/e\ 105}{C_6H_5C{\equiv}\overset{+}{O}} + \dot{C}H_3$$

It should be noted that the distance from *m/e* 120 to *m/e* 105 is 15 mass unit which is approximately equal to the distance from *m/e* 105 to *m/e* (m^*) about 92, i.e. 13 mass unit.

8.11 Applications of Mass Spectroscopy

Organic chemists use mass spectroscopy for characterization of organic compounds in two ways: (i) determination of molecular mass and molecular formula and (ii) elucidation of structure.

(i) Determination of Molecular Mass and Molecular Formula

The molecular ion ($\overset{+}{M}$) peak gives the mass (strictly *m/e*) of the molecular ion which is the molecular mass of the compound because usually $e = 1$. The determination of a molecular formula (or fragment formula) is possible using high resolution mass spectrometer. Using high resolution mass spectrometer, we can determine the exact molecular mass, from which the exact molecular formula can be determined. The molecular mass thus obtained is not really exact, but it is much more accurate than the usual integral molecular masses (nominal molecular masses). We accurately measure the mass of a molecular ion, then this 'exact' mass is compared with the masses of the different molecular formulae (available by calculation or in tables) and the formula corresponding to the 'exact' mass is the molecular formula of the compound in question. Tables, algorithms and computer programmes are available for this purpose. Different molecular formulae have different exact masses.

We can distinguish among CO, N_2 and C_2H_4 (all with nominal mass 28) because their 'exact' molecular masses are different. The masses observed for their molecular ions are the sum of the exact masses of the most abundant isotopes of the elements present in them. Thus

CO	N_2	C_2H_4
^{12}C 12.0000 ^{16}O 15.9949	2 × 14.0031	^{12}C 24.0000 $^{1}H_4$ 4 × 1.0078
Exact mass of CO = 27.9949	Exact mass of N_2 = 28.0062	Exact mass of C_2H_4 = 28.0312

Most laboratories have limited access to high-resolution mass spectrometers but this is no problem because the molecular formula of a compound can also be determined by using unit mass resolution instrument. In this case the intensities of M^+, $M + 1$ and $M + 2$ ion peaks are used to arrive at the molecular formula. This is because the contribution of various heavier isotopes differs for various elements.

Beynon's table contains masses and isotopic abundance ratios (M + 1 and M+2 relative to M^+) for various combinations of C, H, N and O. When the values of isotopic abundance ratios of M+1 and M+2 relative to M^+ are available from the spectrum of a compound, these are matched with the corresponding values in the Beynon's table in order to arrive at the molecular formula of the compound.

Part of the Beynon's table

Formula	$M + 1$	$M + 2$	FM (Formula mass)
$C_7H_{10}N_4$	9.25	0.38	150
$C_8H_8NO_2$	9.23	0.78	150
$C_8H_{10}N_2O$	9.61	0.61	150
$C_8H_{12}N_3$	9.98	0.45	150
$C_9H_{10}O_2$	9.96	0.84	150
$C_9H_{12}NO$	10.34	0.68	150
$C_9H_{14}N_2$	10.71	0.52	150

For example, let us determine the molecular formula from the following experimental mass data:

m/e	%
150 (M^+)	100
151 ($M + 1$)	10.2
152 ($M + 2$)	0.88

We match the experimental $M + 1$ (10.2) and $M + 2$ (0.88) values with the corresponding $M + 1$ and $M + 2$ values in the Beynon's table (the relevant part for this case is discussed earlier). Formulae $C_8H_8NO_2$, $C_8H_{12}N_3$ and $C_9H_{12}NO$ can be eliminated on the basis of nitrogen rule because these have odd number of nitrogen atom and the observed formula mass (150) is even. Our $M + 1$ peak is 10.2% and $M + 2$ is 0.88% of the parent peak; these best fit $C_9H_{10}O_2$. Thus, the most probable molecular formula of the compound in question is $C_9H_{10}O_2$.

(ii) Elucidation of Structure

Study of fragmentation processes shows the presence of certain structural units and points of their attachment in the molecule, and thus it gives an idea about the structure of the molecule. The fragmentation processes have been discussed in detail in the following sections.

8.12 Representation of Fragmentation Processes

Most of the fragmentation processes involve elimination of radicals from radical cations. Therefore, these may be regarded as homolytic processes (one-electron shift) and are indicated by a single barbed fishhook (⌒). A heterolytic process (a two-electron shift) is indicated by an arrow (⌒). For example, a homolytic fragmentation is represented as follows:

$$H_3C{-}CH_2{-}\overset{\cdot +}{\underset{\cdot\cdot}{O}}{-}R \longrightarrow \dot{C}H_3 + CH_2{=}\overset{+}{\underset{\cdot\cdot}{O}}{-}R$$

(A)

$$H_3C{-}CH_2{-}\overset{\cdot +}{\underset{\cdot\cdot}{O}}{-}R \longrightarrow \dot{C}H_3 + CH_2{=}\overset{+}{\underset{\cdot\cdot}{O}}{-}R$$

(B)

The homolysis of a bond could be indicated by showing two fishhooks shown in (A), but for brevity and clarity only one is drawn as in (B).

8.13 Factors Governing General Fragmentation Processes

The following factors dominate the general modes of fragmentation:

(i) Weak bonds tend to be broken.
(ii) Stable fragments (ions, radicals and molecules) tend to be formed.
(iii) Ability of ions to assume cyclic transition states—rearrangement processes.

Favourable fragmentation processes naturally occur more frequently and ions thus formed appear as strong peaks in the mass spectrum. General concepts of mechanistic organic chemistry are very useful in predicting and understanding favourable mass spectrometric fragmentation processes.

8.14 Examples of General Fragmentation Modes

Following are the examples of general fragmentation modes:

1. Fragmentation at branch point is favoured in aliphatic carbon skeleton because it gives more substituted, i.e. more stable carbocations (relative stabilities of carbocations are tertiary > secondary > primary). For example,

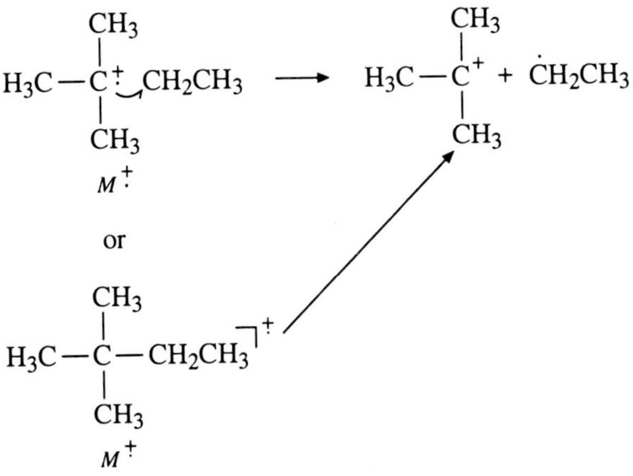

$$H_3C{-}\overset{CH_3}{\underset{CH_3}{C}}{}^{\dot{+}}{\frown}CH_2CH_3 \longrightarrow H_3C{-}\overset{CH_3}{\underset{CH_3}{C}}{}^{+} + \dot{C}H_2CH_3$$

$M^{\dot{+}}$

or

$$\left[H_3C{-}\overset{CH_3}{\underset{CH_3}{C}}{-}CH_2CH_3\right]^{\dot{+}}$$

$M^{\dot{+}}$

Generally, the largest substituent at a branch point is eliminated most readily as a radical because a long-chain radical achieves some stability by delocalization of the lone electron.

2. The cleavage tends to occur β to: (a) double bonds (allylic cleavage), (b) aromatic rings (benzylic cleavage) and (c) hetero atoms singly bonded to a carbon atom because it gives resonance-stabilized carbonations.

(a) Allylic cleavage

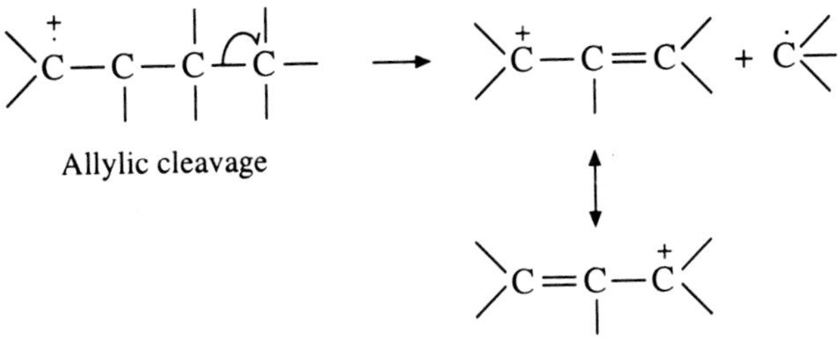

(b) Benzylic cleavage

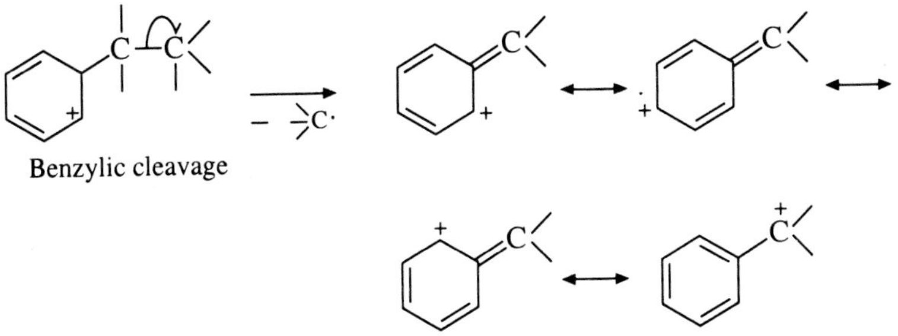

$-\ \geqslant C\cdot$

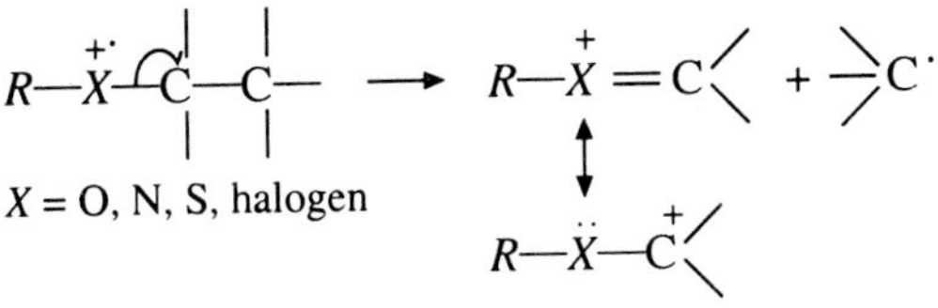

(c) X = O, N, S, halogen

$R{-}\overset{+\cdot}{X}{-}C{-}C{-} \longrightarrow R{-}\overset{+}{X}{=}C\langle\ + \ \geqslant C\cdot$

$\updownarrow$

$R{-}\ddot{X}{-}\overset{+}{C}\langle$

3. Cleavage tends to occur α to a carbonyl group to give resonance-stabilized acylium cations

$$R\text{—}\overset{\overset{+}{\ddot{O}}:}{\overset{\|}{C}}\text{—}C\lessdot \longrightarrow R\text{—}C\equiv\overset{+}{O}: + .C\lessdot \;\longleftrightarrow\; R\text{—}\overset{+}{C}=O$$

R = alkyl, aryl, OH, OR etc.

4. Cleavage may also occur α to heteroatoms singly bonded to a carbon atom, e.g. in case of ethers, sulphides, alcohols, thiols, amines and halogen compounds. This is *heterolytic cleavage.*

$$\text{—}\overset{|}{\underset{|}{C}}\text{—}\overset{+\cdot}{X} \longrightarrow \text{—}\overset{+}{\underset{|}{C}}\text{—} + \dot{X}$$

X = OR, SR, OH, SH, NH_2, NHR, NR_2, halogen

For example

$$RCH_2\text{—}\overset{+\cdot}{\ddot{B}r}: \longrightarrow R\overset{+}{C}H_2 + :\dot{\ddot{B}r}:$$

$$>C\text{—}\overset{+\cdot}{\ddot{O}}\text{—}R \longrightarrow >\overset{+}{C}\text{—} + :\dot{\ddot{O}}\text{—}R$$

5. Cyclohexene and its derivatives undergo characteristic fragmentation through retro-Diels-Alder reaction in which the charge may be with the diene portion or the ethylenic portion.

$$\left[\text{cyclohexene}\right]^{+\cdot} \longrightarrow \left[\text{butadiene}\right]^{+\cdot} + \begin{matrix}CH_2\\ \|\\ CH_2\end{matrix} \quad \text{or} \quad \text{butadiene} + \left[\begin{matrix}CH_2\\ \|\\ CH_2\end{matrix}\right]^{+\cdot}$$

In saturated rings, cleavage tends to occur at the α bond to lose the side chain. This is just a special case of fragmentation at branch point.

$$\left[\text{cyclohexyl—}CH_2CH_3\right]^{+\cdot} \longrightarrow \text{cyclohexyl}^{+} + CH_3\dot{C}H_2$$

6. Ions which have ability to assume cyclic transition states, specially six-membered, tend to undergo *fragmentations accompanied by rearrangements*. Such cases involve bond formation as well as bond breaking. The most important is the McLafferty rearrangement. This involves migration of a γ hydrogen accompanied by cleavage of a β bond through a six-membered cyclic transition state to form a radical cation and a neutral molecule. This rearrangement is characteristic of compounds containing a γ hydrogen with respect to a multiple bond. For example

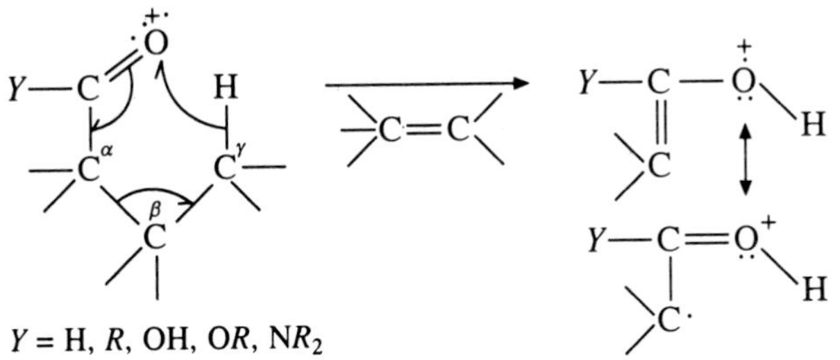

Y = H, R, OH, OR, NR_2

Several other rearrangements occurring during fragmentation processes have been discussed in Section 8.15. Most of the rearrangements are hydrogen transfer rearrangements. The transfer of an aryl, alkyl or alkoxy group occurs in very few cases.

8.15 Fragmentation Modes of Various Classes of Organic Compounds

Various classes of organic compounds exhibit characteristic mass spectral features which are very useful for their identification and structure elucidation.

(i) Hydrocarbons

(a) Alkanes

1. *Straight-chain alkanes*. Their mass spectra are characterized by groups of peaks separated by 14 (CH_2) mass units. The largest peak in each group corresponds to C_nH_{2n+1} fragment, and it is accompanied by much smaller peaks due to C_nH_{2n} and C_nH_{2n-1} fragments formed by loss hydrogens. The most intense peaks are at m/e 43 and 57 due to $C_3H_7^+$ and $C_4H_9^+$ fragment ions, respectively. These ions are highly branched and arise via molecular rearrangements. Any molecular or fragment ion will give a peak at one mass unit higher due to the presence of ^{13}C. Abundance of this peak will be $N \times 1.1\%$ of the abundance of the ^{12}C containing peak, where N is the number of carbon atoms in that ion and 1.1% is the natural abundance of ^{13}C. The molecular ion peak is always present in the mass spectra of straight-chain alkanes but its intensity decreases as the molecular mass increases.

As an example, the mass spectrum of pentane is given in Fig. 8.2 and its mode of fragmentation shown in Scheme 8.1.

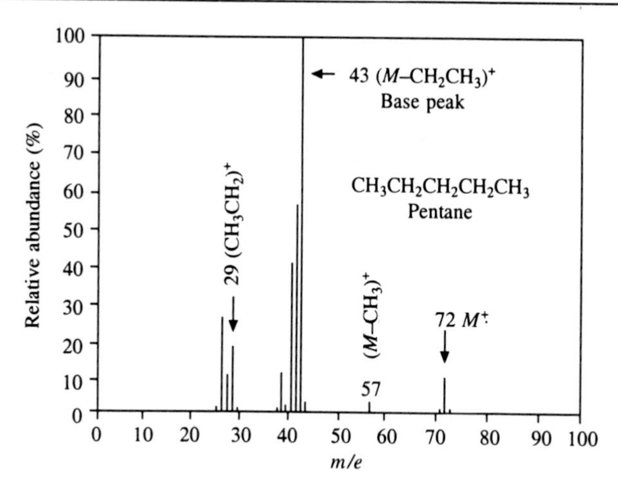

Fig. 8.2 Mass spectrum of pentane

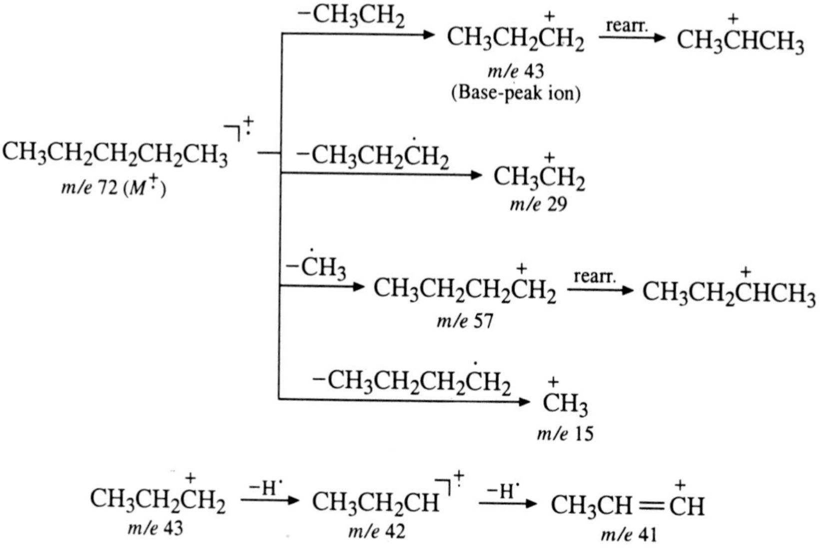

$$CH_3CH_2CH_2CH_2CH_3^{+\cdot} \; (m/e\ 72\ (M^{+\cdot}))$$

$$\xrightarrow{-CH_3\dot{C}H_2} CH_3CH_2\overset{+}{C}H_2 \; (m/e\ 43,\ \text{Base-peak ion}) \xrightarrow{\text{rearr.}} CH_3\overset{+}{C}HCH_3$$

$$\xrightarrow{-CH_3CH_2\dot{C}H_2} CH_3\overset{+}{C}H_2 \; (m/e\ 29)$$

$$\xrightarrow{-\dot{C}H_3} CH_3CH_2CH_2\overset{+}{C}H_2 \; (m/e\ 57) \xrightarrow{\text{rearr.}} CH_3CH_2\overset{+}{C}HCH_3$$

$$\xrightarrow{-CH_3CH_2CH_2\dot{C}H_2} \overset{+}{C}H_3 \; (m/e\ 15)$$

$$CH_3CH_2\overset{+}{C}H_2 \; (m/e\ 43) \xrightarrow{-H^{\cdot}} CH_3CH_2CH^{+\cdot} \; (m/e\ 42) \xrightarrow{-H^{\cdot}} CH_3CH{=}\overset{+}{C}H \; (m/e\ 41)$$

Scheme 8.1 Mode of fragmentaion of pentane

2. *Branched-chain alkanes.* There is large similarity between the mass spectra of straight-chain and branched-chain hydrocarbons. The only difference being in the intensities of peaks. Bond cleavage favourably occur at the branch point because it gives more stable carbocations. Generally, the largest substituent at a branch point is eliminated most readily as a radical because it is stabilized by delocalization of the lone electron. The resulting carbocation gives the base peak (Stevenson's rule). The greater the branching in alkanes, the more rapid and

extensive is the fragmentation. Thus, the molecular ion peak is very less intense and is mostly not observed.

As an example, the fragmentation pattern of 2-methylpentane is given in Scheme 8.2.

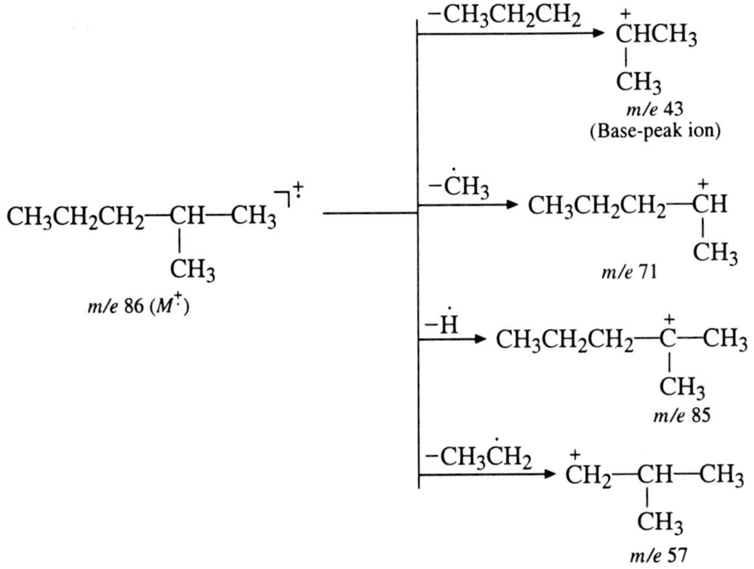

$$CH_3CH_2CH_2-CH(CH_3)-CH_3^{+\cdot} \quad (m/e\ 86\ (M^{+\cdot}))$$

$$\xrightarrow{-CH_3CH_2\dot{C}H_2} \overset{+}{C}HCH_3(CH_3) \quad m/e\ 43\ \text{(Base-peak ion)}$$

$$\xrightarrow{-\dot{C}H_3} CH_3CH_2CH_2-\overset{+}{C}H(CH_3) \quad m/e\ 71$$

$$\xrightarrow{-\dot{H}} CH_3CH_2CH_2-\overset{+}{C}(CH_3)-CH_3 \quad m/e\ 85$$

$$\xrightarrow{-CH_3\dot{C}H_2} \overset{+}{C}H_2-CH(CH_3)-CH_3 \quad m/e\ 57$$

Scheme 8.2 Fragmentation pattern of 2-methylpentane

3. *Cycloalkanes.* In cycloalkanes with side chains, cleavage tends to occur at the α bond, i.e. the fission of bond between the ring and side chain occurs. This is just a special case of fragmentation at branch point. Fragmentation of the ring requires the fission of at least two bonds. Molecular ions in cycloalkanes are more abundant than those of the corresponding alkanes. For example, the mode of fragmentation of ethylcyclohexane and cyclohexane is

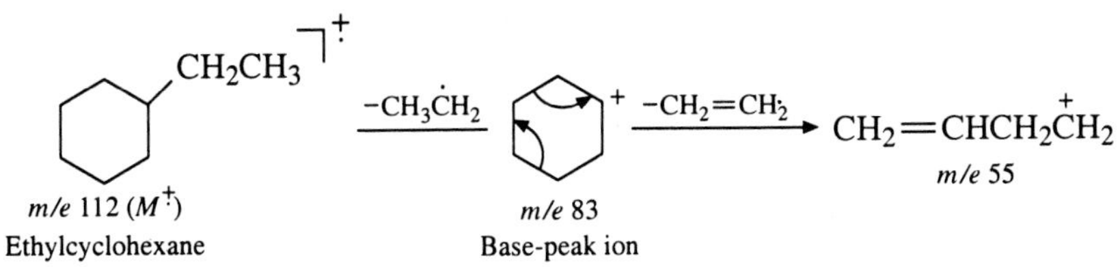

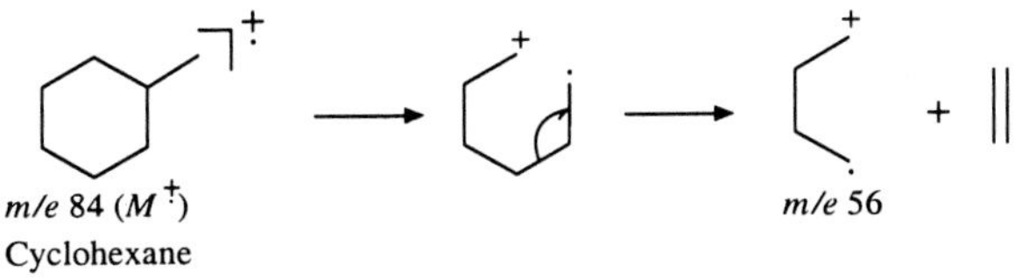

$$C_6H_{11}CH_2CH_3^{+\cdot}\ (m/e\ 112\ (M^{+\cdot})\ \text{Ethylcyclohexane}) \xrightarrow{-CH_3\dot{C}H_2} C_6H_{11}^{+}\ (m/e\ 83\ \text{Base-peak ion}) \xrightarrow{-CH_2=CH_2} CH_2=CHCH_2\overset{+}{C}H_2\ (m/e\ 55)$$

$$C_6H_{12}^{+\cdot}\ (m/e\ 84\ (M^{+\cdot})\ \text{Cyclohexane}) \longrightarrow \longrightarrow m/e\ 56 + CH_2=CH_2$$

(b) Alkenes (Olefins)

Similar to alkanes, acyclic olefins are also characterized by groups of peaks separated by 14 (CH_2) mass units. In these groups, the C_nH_{2n-1} and C_nH_{2n} peaks are more intense than the C_nH_{2n+1} peaks. Molecular ion peaks of unsaturated compounds are more intense than the corresponding saturated analogues. Because of facile migration of the double bond after electron impact ionization, its location is difficult in acyclic olefins. The most common fragmentation of alkenes is the allylic cleavage, i.e. cleavage of the bond β to the double bond resulting in the formation of a resonance stabilized allylic cation. Alkenes having a γ hydrogen with respect to the double bond also undergo the McLafferty rearrangement. For example, the prominent mode of fragmentation of 1-pentene is

$$\underset{m/e\ 70\ (M^{+\cdot})}{[CH_3CH_2CH_2CH{=}CH_2]^{+\cdot}} \xrightarrow[-CH_3\dot{C}H_2]{\beta\ \text{cleavage}} \underset{m/e\ 41}{\overset{+}{C}H_2CH{=}CH_2}$$

$$[CH_2{=}CH{-}CH_2{-}CH_2{-}CH_2{-}H]^{+\cdot} \xrightarrow[\text{rearr.}]{\text{McLafferty}} \underset{m/e\ 42}{[CH_3{-}CH{=}CH_2]^{+\cdot}} + CH_2{=}CH_2$$

The formation of an even mass ion from an even mass molecular ion, and an odd mass ion from an odd mass molecular ion indicates that the fragmentation has occurred through a rearrangement involving the cleavage of two bonds. This is true only when the fragment ion contains all the nitrogen atoms (if any) of the molecular ion.

1. *Cycloalkenes.* Cyclohexene and its derivatives undergo characteristic fragmentation through retro-Diels-Alder reaction in which a diene and a monoene fragments are formed. The positive charge may be on the diene or the monoene fragment but the abundance of the former is greater because of its higher resonance stabilization. For example

$$\text{butadiene} + \underset{m/e\ 28}{[CH_2{=}CH_2]^{+\cdot}} \longleftarrow \underset{m/e\ 82\ (M^{+\cdot})}{[\text{cyclohexene}]^{+\cdot}} \longrightarrow \underset{\substack{m/e\ 54 \\ \text{More intense}}}{[\text{butadiene}]^{+\cdot}} + CH_2{=}CH_2$$

(c) Alkynes

In alkynes, the fragment ions (with composition C_nH_{2n-3}) are generally formed by loss of alkyl radicals. Thus, M-15, M-29, M-43 etc. peaks are commonly present in the mass spectra of alkynes. Alkynes having a γ hydrogen with respect to the triple bond also undergo the McLafferty rearrangement. The molecular ion peaks in alkynes are usually distinct. In case of 1-butyne and 2-butyne the molecular ion peak is the base peak. Taking the example of 1-pentyne, the fragmentation mode of alkynes is

$$\overset{+}{C}H_2C{\equiv}CH \xleftarrow{-CH_3\dot{C}H_2} [CH_3CH_2CH_2C{\equiv}CH]^{+\cdot} \xrightarrow{-\dot{C}H_3} \overset{+}{C}H_2CH_2C{\equiv}CH$$

m/e 39 — *m/e* 68 ($M^{+\cdot}$) — *m/e* 53

$$\text{1-pentyne } (m/e\ 68\ (M^{+\cdot})) \xrightarrow[\text{rearr.}]{\text{McLafferty}} [CH_2{=}C{=}CH_2]^{+\cdot}\ (m/e\ 40) + CH_2{=}CH_2$$

(d) Aromatic Hydrocarbons

Molecular ion peaks in aromatic compounds are fairly large because an aromatic ring stabilizes the molecular ion. These are accompanied by $M + 1$ and $M + 2$ peaks due to ^{13}C and/or D. In alkylbenzenes, the dominant fragmentation involves the cleavage of the bond β to the aromatic ring (benzylic cleavage) because it gives resonance stabilized carbocations.

$$C_6H_6^{+\cdot}\ (m/e\ 78\ (M^{+\cdot})) \xrightarrow{-H^{\cdot}} C_6H_5^{+}\ (m/e\ 77\ \text{(low intensity)}) \xrightarrow{-CH{\equiv}CH} C_4H_3^{+}\ (m/e\ 51)\ \text{or}\ ^{+}C_4H_3$$

$$C_6H_5^{+} \xrightarrow{-C_3H_2} C_3H_3^{+}\ \text{or}\ ^{+}C_3H_3\ (m/e\ 39)$$

Fragmentation pattern of benzene

The low intensity of the peak of phenyl cation is understandable because the positive charge of this cation is in an empty sp^2 orbital and consequently cannot interact with π-electrons of the ring. Hence, the phenyl cation is not stabilized by resonance. Due to its less stability, intensity of its peak is low.

In alkylbenzenes, β-cleavage is most favourable. It gives a benzyl cation which changes to the more stable tropylium ion. When the side chain has at least three carbon atoms, McLafferty rearrangement also takes place. For example, the fragmentation of *n*-propylbenzene is as follows:

$C_6H_5\overset{+}{C}HCH_2CH_3$ (m/e 119 (weak)) ← $-H^{\cdot}$ — $[C_6H_5CH_2CH_2CH_3]^{+\cdot}$ — $-CH_3\dot{C}H_2$ → $[C_6H_5CH_2]^+$ (m/e 91) → Tropylium ion (m/e 91) — $-C_2H_2$ → m/e 65

$[C_6H_5CH_2CH_2CH_3]^{+\cdot}$ — $-CH_3CH_2\dot{C}H_2$ → $C_6H_5^+$ (m/e 77); m/e 77 — $-C_2H_2$ → m/e 51; m/e 77 — $-C_3H_2$ → $\overset{+}{C_3H_3}$ (m/e 39)

m/e 120 ($M^{+\cdot}$) — McLafferty rearr. → m/e 92 + $CH_2{=}CH_2$

Fragmentation pattern of *n*-propylbenzene

(ii) Halogen Compounds

Because of the presence of molecular ions containing the ^{37}Cl isotope, a compound containing one chlorine atom shows a $M + 2$ peak of approximately one-third intensity of the molecular ion peak. A compound containing one bromine atom shows a $M + 2$ peak which is almost equal in intensity to the molecular ion peak because of the presence of molecular ions containing the ^{81}Br isotope. Fluorine and iodine have no isotope, hence do not give such $M + 2$ peaks.

A compound containing two chlorines, or two bromines, or one chlorine and one bromine will show $M + 2$ and $M + 4$ peaks. Thus, the number of chlorine and/or bromine atoms in a molecule can be ascertained by the number of such isotope peaks beyond the parent peak. For example, a compound containing

Table 8.3 Intensities of isotope peaks (relative to the parent peak) for combinations of bromine and chlorine

Halogen present	%					
	$M + 2$	$M + 4$	$M + 6$	$M + 8$	$M + 10$	$M + 12$
Br	97.7					
Br_2	195.0	95.5				
Br_3	293.0	286.0	93.4			
Cl	32.6					
Cl_2	65.3	10.6				
Cl_3	97.8	31.9	3.47			
Cl_4	131.0	63.9	14.0	1.15		
Cl_5	163.0	106.0	34.7	5.66	0.37	
Cl_6	196.0	161.0	69.4	17.0	2.23	0.11
BrCl	130.0	31.9				
Br_2Cl	228.0	159.0	31.2			
Cl_2Br_2	163.0	74.4	10.4			

three chlorines, or three bromines, or one chlorine and two bromines, or one bromine and two chlorines will give peaks at $M + 2$, $M + 4$ and $M + 6$. Intensities of isotope peaks (relative to the parent peak) for combinations of bromine and chlorine have been calculated by Beynon; atoms other than Cl and Br were ignored (for a portion of these results, see Table 8.3).

The halogen-containing fragments can be detected by the ratio of the fragment + 2 peaks + 2 peaks + fragment peaks in monobromides or monochlorides.

(a) Alkyl Halides

The intensities of molecular ion peaks of alkyl halides follow the order:

$$RI > RBr > RCl > RF$$

The intensity of the molecular ion decreases with increase in chain length and increase in branching.

Cleavage of a C—C bond β to the halogen atom is the favourable mode of fragmentation. Alkyl fluorides and chlorides also fragment by loss of HF and HCl to give peaks at M-20 in case of fluorides, and at M-36 and M-38 in case of chlorides. However, in alkyl bromides and iodides, loss of $Br^{\cdot}$ and $I^{\cdot}$ is preferred. For example,

$$R\text{CH}_2\overset{+}{\text{C}}\text{H}_2 \xleftarrow[\text{(if } X = \text{Br or I)}]{-X,\ \alpha\text{-cleavage}} R\text{CH}_2\text{CH}_2\text{—}\overset{+}{\ddot{X}} \xrightarrow[\text{(if } X = \text{F or Cl)}]{-\text{HX, 1,2-elimination}} [R\text{—CH=CH}_2]^{+}$$

$$R\text{CH}_2\text{CH}_2\text{—}\overset{+}{\ddot{X}} \xrightarrow[\beta\text{-cleavage}]{-R\dot{\text{C}}\text{H}_2} \text{CH}_2\text{=}\overset{+}{X}$$

Straight-chain chlorides and bromides longer than C_6 give $C_3H_6X^+$, $C_4H_8X^+$ and $C_5H_{10}X^+$ ions. Of these, $C_4H_8X^+$ gives the most intense (sometimes the base) peak. (m/e 91 and 93 in chlorides, and m/e 135 and 137 in bromides). The intensity of these ions is attributed to the stability of the five-membered cyclic structure.

$$R\text{—CH}_2\text{CH}_2\text{CH}_2\text{CH}_2\text{—}\overset{+}{X} \xrightarrow{-\dot{R}} \text{cyclic } (\text{CH}_2)_4\overset{+}{X}$$

X = Cl or Br

X = Cl, m/e 91 and 93
X = Br, m/e 135 and 137

Small peaks due to HCl^+, HBr^+, Cl^+, Br^+ etc. may also be detected.

(b) Aryl Halides

The molecular ion peaks are strong in aryl halides. M-X peak is intense for all compounds containing X directly attached to the ring. For example, the mode of fragmentation of a halobenzene is

$$C_6H_5X^{+\cdot} \xrightarrow[\text{(if } X = \text{Cl, Br, I)}]{-X^\cdot} C_6H_5^+ \ (m/e\ 77) \xrightarrow{-C_2H_2} C_4H_3^+ \ (m/e\ 51)$$

$$C_6H_5^+ \ (m/e\ 77) \xrightarrow{-C_3H_2} C_3H_3^+ \ (m/e\ 39)$$

$$C_6H_5F^{+\cdot} \xrightarrow[\text{(if } X = \text{F)}]{-C_2H_2} C_4H_3F^{+\cdot} \ (m/e\ 70)$$

Benzyl halides lose halogen to form benzyl cation which changes to the more stable tropylium ion. For example,

$$C_6H_5CH_2Cl^{+\cdot} \xrightarrow{-Cl^\cdot} C_6H_5\overset{+}{C}H_2 \ (m/e\ 91) \longrightarrow C_7H_7^+ \ (m/e\ 91) \xrightarrow{-C_2H_2} C_5H_5^+ \ (m/e\ 65)$$

(iii) Hydroxy Compounds

(a) Aliphatic Alcohols

In all the three classes of alcohols, i.e. primary, secondary and tertiary alcohols, the α-cleavage of C—C bond is characteristic. The largest alkyl group is expelled most readily as a radical because a long-chain radical is stabilized by delocalization of the lone electron. When R and/or R' = H, a M-1 peak is often observed. In long-chain (longer than C_6) alcohols, the fragmentation pattern is dominated by the hydrocarbon pattern.

$$R'—\underset{R''}{\overset{R}{C}}—\overset{+}{\dot{O}}H \xrightarrow{-\dot{R}''} \overset{R}{\underset{R'}{>}}C=\overset{+}{O}—H$$

$R'' > R'$ or R = alkyl

The molecular ion peaks of primary or secondary alcohols are weak and that of tertiary alcohols are either very weak or undetectable. Long-chain primary alcohols show small M-2 and M-3 peaks

$$\underset{M\text{-}2}{R—CH\overset{+}{\dot{O}}} \xleftarrow{-2H} RCH_2\overset{+}{\dot{O}}H \xrightarrow{-3H} \underset{M\text{-}3}{R—C\equiv\overset{+}{O}}$$

The most characteristic fragmentation of higher alcohols is associated with elimination of water to give a distinct peak at *M*-18. This peak is prominent in the mass spectra of primary alcohols

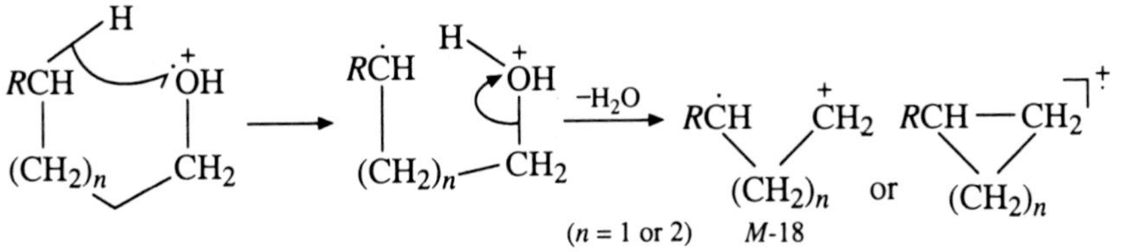

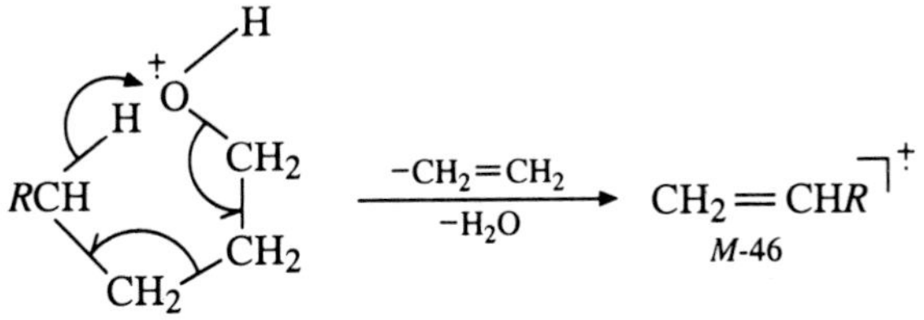

$$RCH(H)-(CH_2)_n-CH_2-\overset{+}{O}H \longrightarrow R\dot{C}H-(CH_2)_n-CH_2-\overset{+}{O}H_2 \xrightarrow{-H_2O} R\dot{C}H\ \overset{+}{C}H_2\ (CH_2)_n \text{ or } [RCH-CH_2(CH_2)_n]^{+\cdot}$$

(n = 1 or 2) *M*-18

$$\xrightarrow[-H_2O]{-CH_2=CH_2} [CH_2=CHR]^{+\cdot}$$

M-46

(b) Aromatic Alcohols

Molecular ion peak is generally strong in benzyl alcohols. Benzylic cleavage occurs and the charge is retained by the aryl group. Prominent *M*-1, *M*-2 and *M*-3 peaks are noticed. As an example, the fragmentation mode of benzyl alcohol is given below:

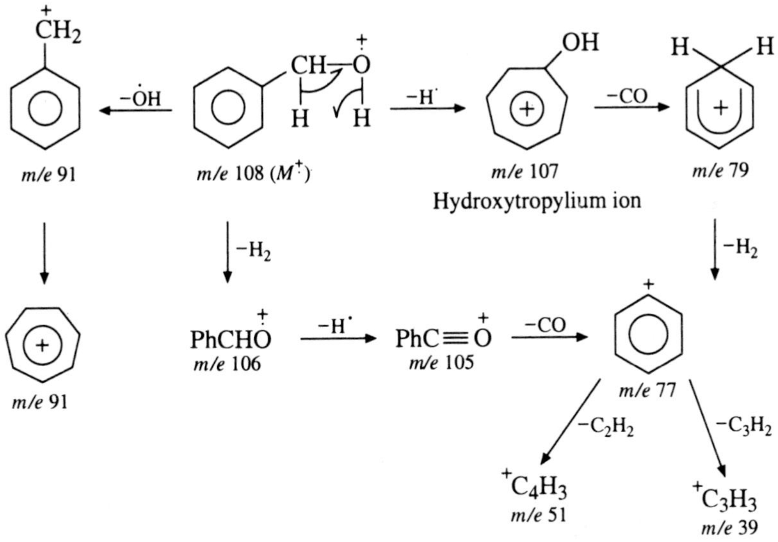

$C_6H_5\overset{+}{C}H_2$ (*m/e* 91) $\xleftarrow{-\dot{O}H}$ $C_6H_5CH_2\overset{+}{O}H$ (*m/e* 108 ($M^{+\cdot}$)) $\xrightarrow{-H^{\cdot}}$ Hydroxytropylium ion (*m/e* 107) $\xrightarrow{-CO}$ *m/e* 79

m/e 91 ⟶ tropylium ion, *m/e* 91

m/e 108 $\xrightarrow{-H_2}$ $PhCH\overset{+}{O}$ (*m/e* 106) $\xrightarrow{-H^{\cdot}}$ $PhC\equiv\overset{+}{O}$ (*m/e* 105) $\xrightarrow{-CO}$ $C_6H_5^+$ (*m/e* 77)

m/e 79 $\xrightarrow{-H_2}$ *m/e* 77

m/e 77 $\xrightarrow{-C_2H_2}$ $^+C_4H_3$ (*m/e* 51)

m/e 77 $\xrightarrow{-C_3H_2}$ $^+C_3H_3$ (*m/e* 39)

Loss of H_2O to give distinct *M*-18 peak is common in some ortho substituted benzyl alcohols, e.g.

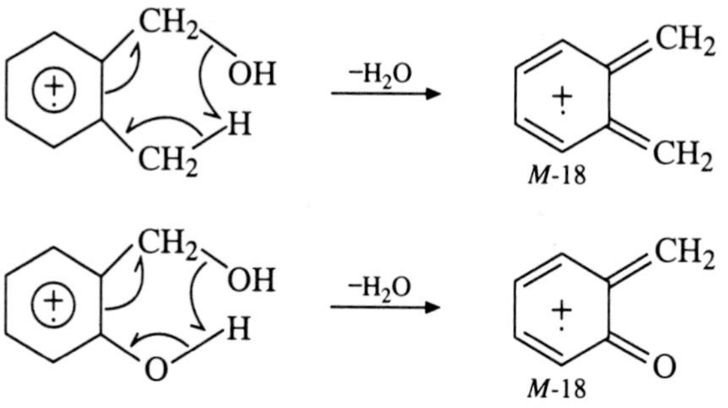

$-H_2O$ → M-18

$-H_2O$ → M-18

(c) Phenols

Phenols exhibit prominent molecular ion peaks. In phenol itself, the molecular ion peak is the base peak. The most common fragmentations in phenols involve the loss of CO (M-28) and CHO (M-29). In addition, a small peak at M-1 is also observed due to loss of hydrogen radical.

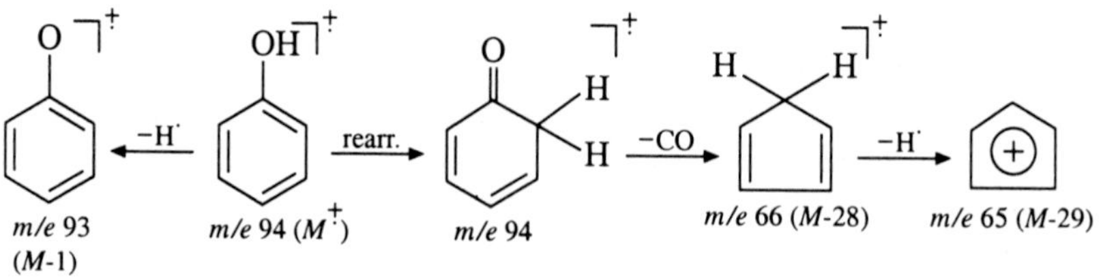

m/e 93 (M-1) ← $-H^{\cdot}$ — m/e 94 ($M^{+\cdot}$) —rearr.→ m/e 94 —$-CO$→ m/e 66 (M-28) —$-H^{\cdot}$→ m/e 65 (M-29)

Phenols with alkyl side chains, undergo the most favourable benzylic cleavage (loss of $H^{\cdot}$ or $R^{\cdot}$) to give hydroxytropylium ions which further fragment to give other daughter ions as

1. o-, m- or p-cresol m/e 108 ($M^{+\cdot}$) —$-H^{\cdot}$→ m/e 107 (M-1) → m/e 107 (M-1) (100%) —$-CO$→ m/e 79 —$-H_2$→ m/e 77 —$-C_2H_2$→ $\overset{+}{C_4H_3}$ m/e 51; —$-C_3H_2$→ $\overset{+}{C_3H_3}$ m/e 39

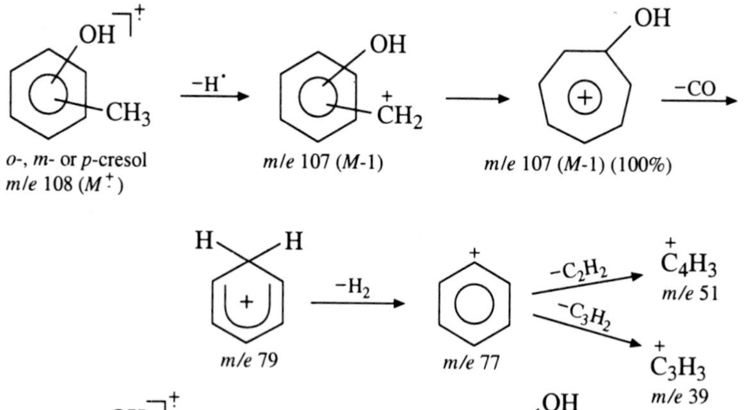

2. $-\dot{R}$ → (100%) when X = H, m/e 107

$-H^{\cdot}$ ↓ $\overset{+}{C}HR$ (3.5%)

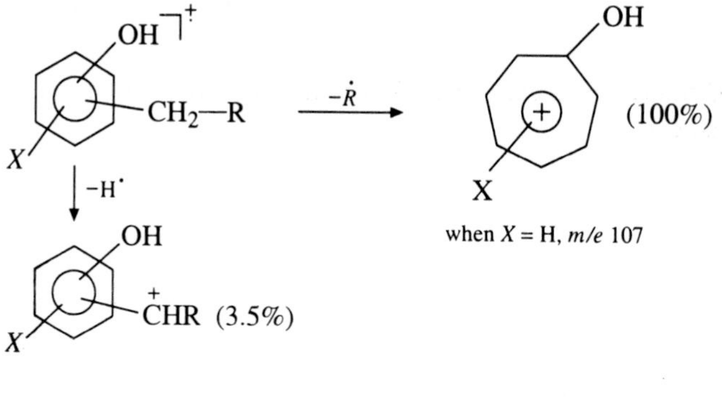

(iv) Ethers, Acetals and Ketals

(a) Aliphatic Ethers

Aliphatic ethers show weak molecular ion peaks. The use of larger sample size makes the molecular ion peak or $M + 1$ peak obvious. The $M + 1$ ion is formed by the bimolecular collision of a parent ion and a neutral molecule; the net effect being the transfer of $H^{\cdot}$ to the molecular ion.

$$\underset{M^{+\cdot}}{RCH_2-\overset{+}{O}-CH_2R} + RCH_2-O-CH_2R \rightarrow \underset{(M+1)^+}{RCH_2-\overset{+}{O}(H)-CH_2R} + RCH_2-O-\dot{C}HR$$

The presence of strong peaks at m/e 31, 45, 59, 73 etc. due to RO^+ and $ROCH_2^+$ fragments shows the presence of an oxygen atom. The following two fragmentation modes are characteristic of ethers:

1. *Cleavage of the C—C bond next to the oxygen atom.* The loss of the larger fragment as a radical is preferred. A weak M-1 peak is also formed by loss of an α-H

$$\underset{m/e\ 73}{CH_3CH_2CH_2-\overset{+}{O}=CH_2} \xleftarrow{-\dot{C}H_3} \underset{m/e\ 88\ (M^{+\cdot})}{CH_3CH_2CH_2-\overset{+}{O}-CH_2CH_3} \xrightarrow{-CH_3\dot{C}H_2} \underset{m/e\ 59}{CH_2=\overset{+}{O}-CH_2CH_3}$$

$$\downarrow -H^{\cdot}$$

$$\underset{m/e\ 87}{CH_3CH_2CH=\overset{+}{O}-CH_2CH_3}$$

The ions thus produced may fragment further to eliminate an alkene if β hydrogen is available. This fragmentation takes place via hydrogen rearrangement, e.g.

$$\underset{m/e\ 59}{CH_2=\overset{+}{O}-CH_2-CH_2-H} \longrightarrow \underset{m/e\ 31}{CH_2=\overset{+}{O}-H} + CH_2=CH_2$$

2. *Cleavage of the C—C bond*

$$R-\overset{+}{O}-R \longrightarrow R^+ + RO^{\cdot}\ \text{(preferred)}$$

$$R-\overset{+}{O}-R \longrightarrow RO^+ + R^{\cdot}$$

As expected, long-chain ethers dominate the fragmentation pattern of hydrocarbons.

(b) Aromatic Ethers

These show prominent molecular ion peaks. Primary fragmentation occurs at the bond β to the ring, and the ion thus formed fragments further, e.g. anisole gives ions of m/e 93 and 65

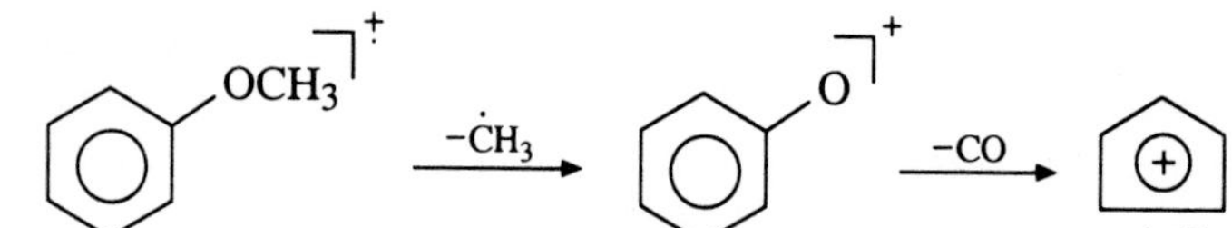

m/e 108 ($M^{+\cdot}$) $\xrightarrow{-\dot{C}H_3}$ m/e 93 $\xrightarrow{-CO}$ m/e 65

The characteristic aromatic peaks at m/e 78 and 77 may arise from anisole via hydrogen rearrangement as

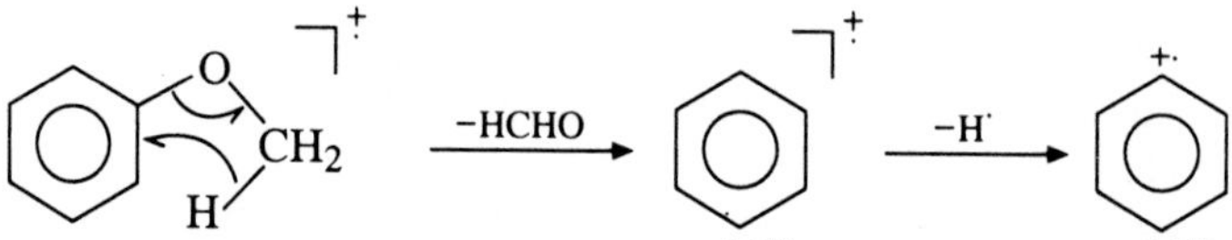

m/e 108 ($M^{+\cdot}$) $\xrightarrow{-HCHO}$ m/e 78 $\xrightarrow{-H^{\cdot}}$ m/e 77

In alkyl aryl ethers, when the alkyl group has two or more carbon atoms, cleavage β to the ring is accompanied by hydrogen rearrangement. Here, C—C cleavage next to the oxygen atom is insignificant.

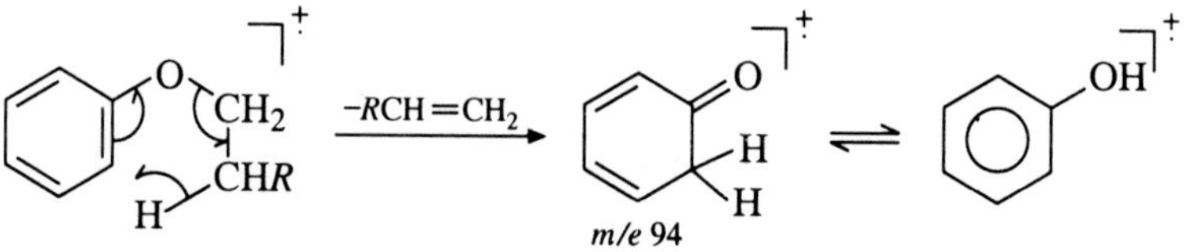

$\xrightarrow{-RCH=CH_2}$ m/e 94 $\rightleftharpoons$ $[C_6H_5OH]^{+\cdot}$

(c) Acetals and Ketals

These are special class of ethers. Their mass spectra show very weak molecular ion peaks. Following is the characteristic fragmentation mode of acetals which is mediated by an oxygen atom and this facile.

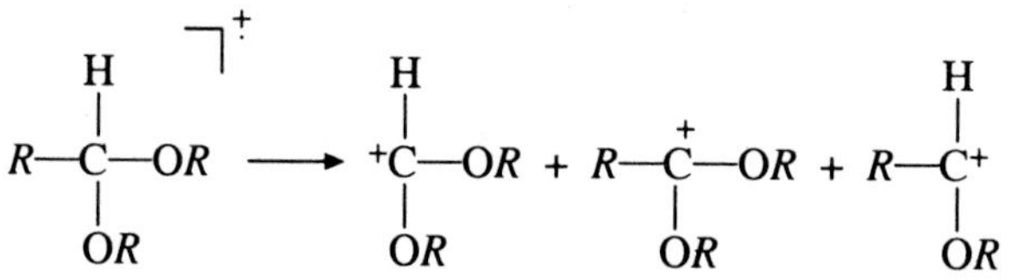

$$[R-CH(OR)-OR]^{+\cdot} \longrightarrow {}^{+}CH(OR)-OR + R-\overset{+}{C}H-OR + R-\overset{+}{C}H-OR$$

Ketals follow similar fragmentation mode.

(v) Aldehydes and Ketones

Their molecular ion peaks are prominent and they undergo following two important types of cleavage:

1. *α-Cleavage.* This involves the cleavage of the C—C bond next to the oxygen atom to give resonance-stabilized acylium ions as

$$H{-}C{\equiv}\overset{+}{O} \xleftarrow{-R^{\cdot}} R{-}\overset{\overset{+\cdot}{O}}{\overset{\|}{C}}{-}H \xrightarrow{-H^{\cdot}} R{-}C{\equiv}\overset{+}{O}$$

m/e 29 (left); M-1 (right)

$$R'{-}C{\equiv}\overset{+}{O} \xleftarrow{-R^{\cdot}} R{-}\overset{\overset{+\cdot}{O}}{\overset{\|}{C}}{-}R' \xrightarrow{-R'^{\cdot}} R{-}C{\equiv}\overset{+}{O}$$

The M-1 peak is characteristic of aldehydes. In aliphatic aldehydes, $H{-}C{\equiv}\overset{+}{O}$ ($M - R$ ion) is more abundant than the M-1 (M – H) ion. In aliphatic ketones, the larger alkyl group is preferably lost as a radical. The α-cleavage in ketones may also occur with charge retention by the alkyl group

$$\overset{+}{R} \xleftarrow{-R'\dot{C}=O} R{-}\overset{\overset{+\cdot}{O}}{\overset{\|}{C}}{-}R' \xrightarrow{-R\dot{C}=O} \overset{+}{R'}$$

2. *β-Cleavage (McLafferty rearrangement)*. Aldehydes and ketones containing a γ-hydrogen atom undergo β-cleavage via McLafferty rearrangement as

$$Y{-}C(=\overset{+\cdot}{O})(\cdots H){-}CHR^1{-}CHR^2{-}CHR^3 \xrightarrow{-R^2CH=CHR^3} Y{-}\dot{C}(=CHR^1){-}\overset{+}{O}{-}H \longleftrightarrow Y{-}C(=\overset{+}{O}H){-}\dot{C}HR^1$$

An aldehyde Y = H; a ketone Y = alkyl or aryl group

Aliphatic ketones undergo double McLafferty rearangement if each of the alkyl groups attached to C=O group contains a three carbon or longer chain. For example, 4-heptanone:

$$CH_3CH_2CH_2{-}C(=\overset{+\cdot}{O}){-}CH_2{-}CH_2{-}CH_2{-}H \xrightarrow{-CH_2=CH_2} CH_2{-}CH_2{-}CH_2{-}C(=CH_2){-}\overset{+\cdot}{O}{-}H$$

First McLafferty rearrangement — Second McLafferty rearrangement

$$\xrightarrow{-CH_2=CH_2} CH_2{=}C(CH_3){-}\overset{+\cdot}{O}{-}H$$

Aromatic aldehydes and ketones undergo α-cleavage to give characteristic $ArC{\equiv}\overset{+}{O}$ ion which usually accounts for the base peak. Aromatic aldehydes show prominent *M*-1 peak, i.e. the loss of the aldehydic hydrogen from the molecular ion is favoured due to the stability of $ArC{\equiv}\overset{+}{O}$. Both benzaldehyde (a) and acetophenone (b) show strong peaks at *m/e* 105 due to $PhC{\equiv}\overset{+}{O}$ and this further fragments as follows:

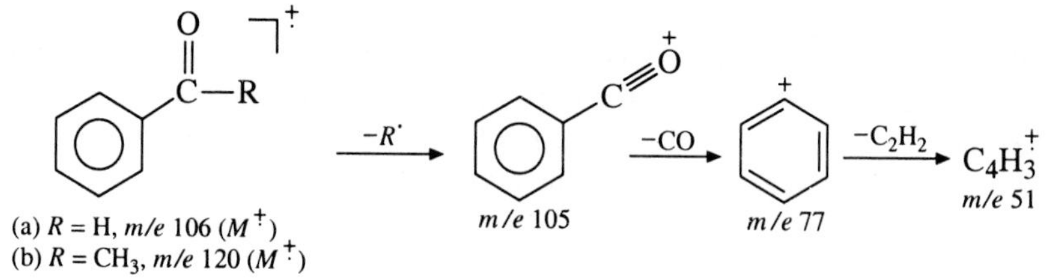

(a) $R = H$, *m/e* 106 ($M^{+\cdot}$)
(b) $R = CH_3$, *m/e* 120 ($M^{+\cdot}$)

$\xrightarrow{-R^{\cdot}}$ *m/e* 105 $\xrightarrow{-CO}$ *m/e* 77 $\xrightarrow{-C_2H_2}$ $C_4H_3^+$ *m/e* 51

Cyclic Ketones. These show prominent molecular ion peak. Primary cleavage of the C—C bond next to the oxygen atom produces an ion which undergoes hydrogen rearrangement from a primary radical to a more stable conjugated secondary radical followed by fragmentation. For example, the fragmentation modes of cyclopentanone and cyclohexanone are

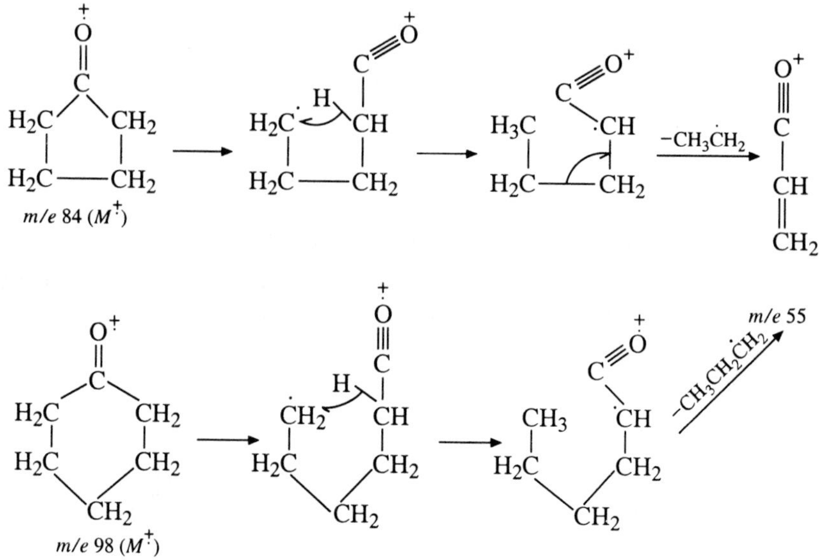

m/e 84 ($M^{+\cdot}$) → → $\xrightarrow{-CH_3\dot{C}H_2}$ $\overset{+}{O}{\equiv}C{-}CH{=}CH_2$ *m/e* 55

m/e 98 ($M^{+\cdot}$) → → $\xrightarrow{-CH_3CH_2\dot{C}H_2}$ *m/e* 55

(vi) Carboxylic Acids, Esters and Amides

(a) Aliphatic

Usually, these give detectable molecular ion peaks. Aliphatic acids, esters and amides undergo α-cleavage, i.e. cleavage of bonds next to C=O, and the positive charge may remain with *R* or *Y*. R^+ and Y^+ may also appear. In short-chain acids, peaks at *M*-OH and *M*-COOH are prominent.

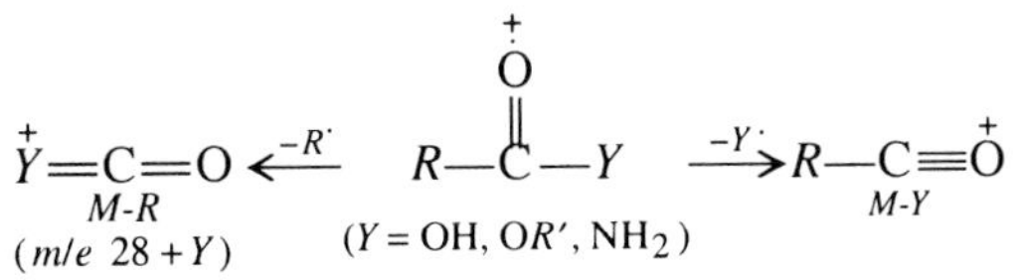

$$\overset{+}{Y}{=}C{=}O \xleftarrow{-R^{\cdot}} R{-}\overset{\overset{+\cdot}{O}}{\overset{\|}{C}}{-}Y \xrightarrow{-Y^{\cdot}} R{-}C{\equiv}\overset{+}{O}$$

M-R (*m/e* 28 + *Y*)   (*Y* = OH, OR′, NH_2)   *M-Y*

Long-chain acids exhibit two series of peaks resulting from cleavage at each C—C bond with retention of charge either on the oxygen-containing fragment (*m/e* 45, 59, 73, 87, . . .) or on the alkyl fragment (*m/e* 29, 43, 57, 71, 85 . . .). For example, this fragmentation pattern occurs in hexanoic acid (caproic acid) as

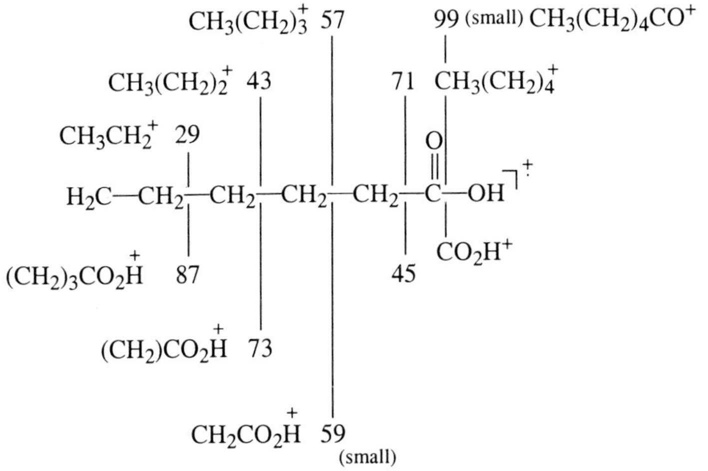

$H_2C{-}CH_2{-}CH_2{-}CH_2{-}CH_2{-}C(=O){-}OH^{+\cdot}$

$CH_3CH_2^+$ 29; $CH_3(CH_2)_2^+$ 43; $CH_3(CH_2)_3^+$ 57; 71 $CH_3(CH_2)_4^+$; 99 (small) $CH_3(CH_2)_4CO^+$

$(CH_2)_3CO_2H^+$ 87; $(CH_2)CO_2H^+$ 73; $CH_2CO_2H^+$ 59 (small); 45 CO_2H^+

Methyl esters of long-chain acids also follow the same fragmentation pattern.

Carboxylic acids, esters and amides undergo McLafferty rearrangement if they contain γ hydrogen.

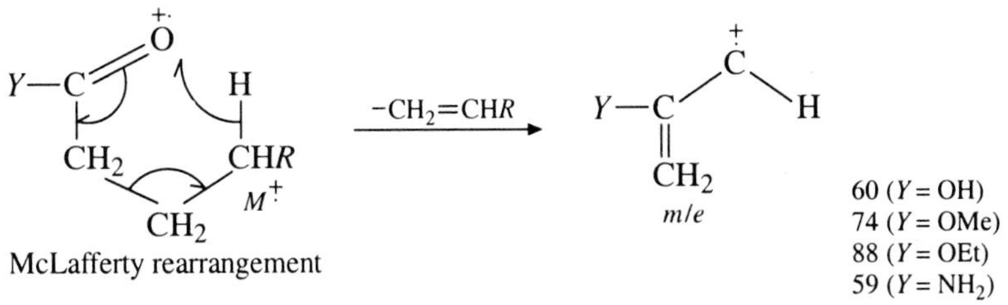

$$Y{-}C(=\overset{+\cdot}{O})\cdots H{-}CHR{-}CH_2{-}CH_2 \;(M^{+\cdot}) \xrightarrow{-CH_2=CHR} Y{-}C(=CH_2){-}\overset{+\cdot}{O}{-}H$$

McLafferty rearrangement

m/e
60 (*Y* = OH)
74 (*Y* = OMe)
88 (*Y* = OEt)
59 (*Y* = NH_2)

Esters containing ethoxy or longer alkoxy group also undergo another type of McLafferty rearrangement as

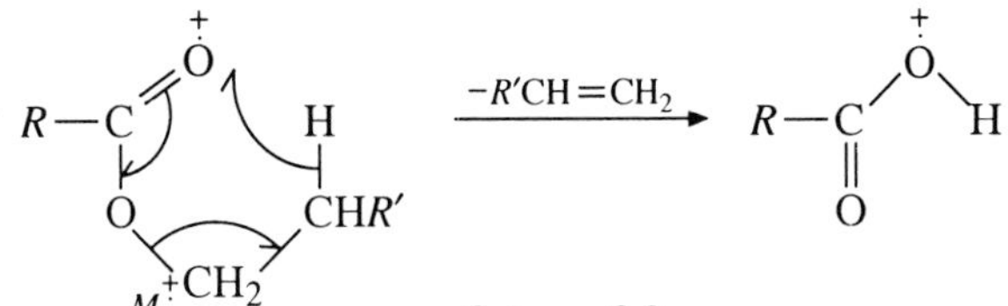

$$R{-}C(=\overset{+}{O})\cdots H{-}CHR'{-}CH_2{-}O \;(M^{+\cdot}) \xrightarrow{-R'CH=CH_2} R{-}C(=O){-}\overset{+\cdot}{O}{-}H$$

Scheme 8.3

Besides the above two types of McLafferty rearrangements, esters of long-chain alcohols show diagnostic peak at *m/e* 61, 75 or 89 from elimination of alkyl moiety and transfer of two hydrogen atoms (double hydrogen rearrangement) as

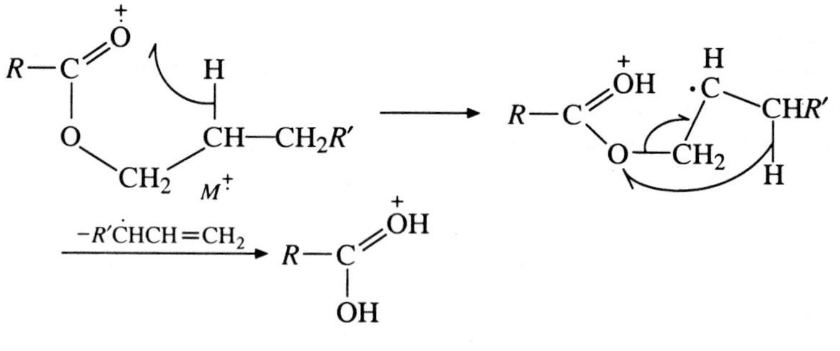

$$R{-}C(=\overset{+}{\dot{O}}){-}O{-}CH_2{-}CH(H){-}CH_2R' \quad M^{+\cdot} \longrightarrow R{-}C(=\overset{+}{O}H){-}O{-}CH_2{-}\dot{C}H{-}CH(H)R' \xrightarrow{-R'\dot{C}HCH=CH_2} R{-}C(=\overset{+}{O}H){-}OH$$

Scheme 8.4

Acetates eliminate CH_3COOH and unsaturated esters eliminate CO_2 as

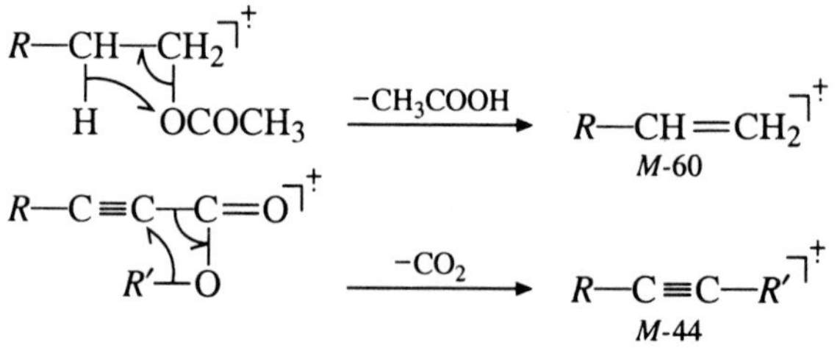

$$R{-}CH(H){-}CH_2{-}OCOCH_3 \;]^{+\cdot} \xrightarrow{-CH_3COOH} R{-}CH{=}CH_2 \;]^{+\cdot} \quad M\text{-}60$$

$$R{-}C{\equiv}C{-}C(=O){-}O{-}R' \;]^{+\cdot} \xrightarrow{-CO_2} R{-}C{\equiv}C{-}R' \;]^{+\cdot} \quad M\text{-}44$$

In addition to α-cleavage and McLafferty rearrangement (β-cleavage), suitable amides also undergo $\gamma\delta$ C—C bond cleavage, e.g.

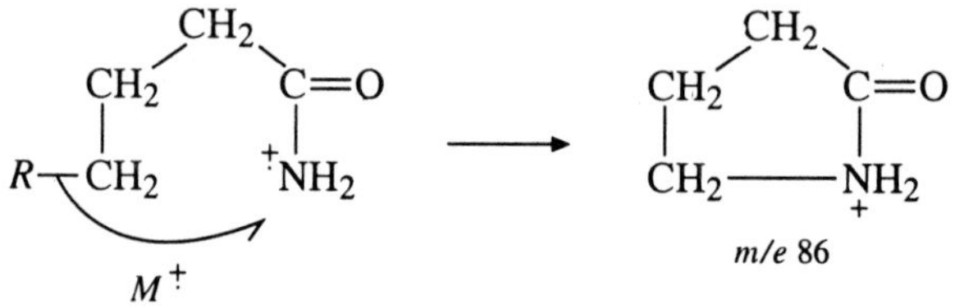

$$R{-}CH_2{-}CH_2{-}CH_2{-}C(=O){-}\overset{+}{N}H_2 \quad M^{+\cdot} \longrightarrow \underset{m/e\ 86}{\text{cyclic } CH_2{-}CH_2{-}CH_2{-}C(=O){-}\overset{+}{N}H_2{-}(CH_2)}$$

When in secondary and tertiary amides, N-alkyl groups are ethyl or longer and the alkyl group attached to the C=O group is shorter than propyl, the following fragmentation mode predominates. This involves cleavage of the N-alkyl group β to the nitrogen atom accompanied by hydrogen rearrangement.

$$O{=}C(CH_2R){-}\overset{+\cdot}{N}H{-}CH_2{-}R' \quad M^{+\cdot} \xrightarrow{-\dot{R}'} O{=}C(R{-}CH(H){-}H){-}\overset{+}{N}H{=}CH_2 \xrightarrow{-RHC=C=O} \overset{+}{N}H_2{=}CH_2 \quad m/e\ 30$$

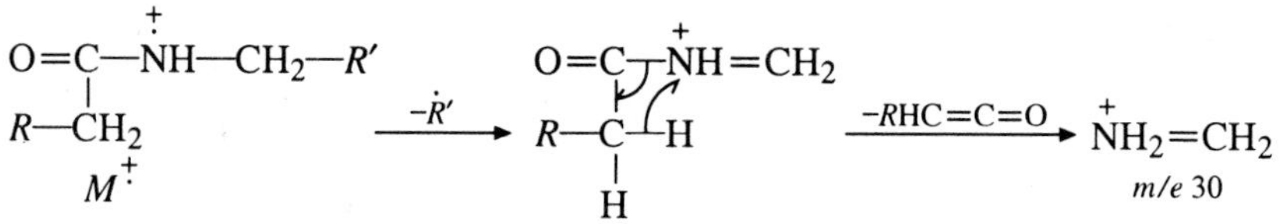

(b) Aromatic

These show intense molecular ion peaks. These compounds undergo fragmentation mode similar to that of aromatic aldehydes and ketones as given in Scheme 8.2, R = OH for acids, R = OMe for esters and R = NH_2 for amides.

Ortho effect. When a substituent and a hydrogen are in close proximity to form a six-membered transition state, the loss of neutral molecules of H_2O, ROH or NH_3 occurs. This is called *ortho* effect. For example, the *ortho* effect is observed in aromatic carboxylic acids, esters and amides if an *ortho* substituent containing a hydrogen atom is present as shown below:

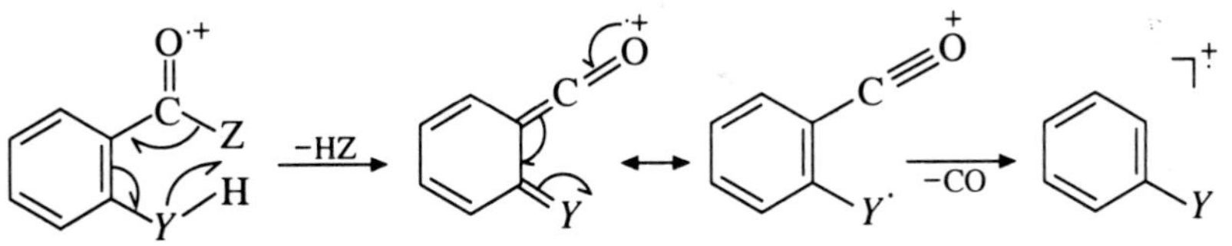

Z = OH, OR, NH_2; Y = CH_2, O, NH

Aromatic esters with ethyl or higher alkyl group eliminate alkene through hydrogen rearrangement similar to ethers.

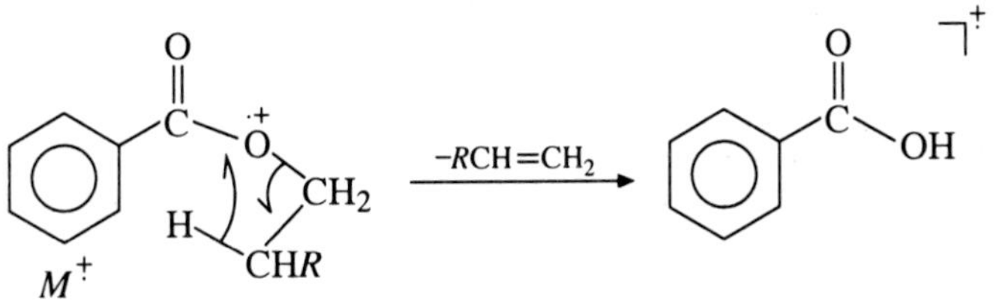

Benzyl and phenyl acetates eliminate ketenes to form the base peak. In case of benzyl acetate, peaks due to $CH_3C{\equiv}\overset{+}{O}$ (m/e 43) and tropylium ion (m/e 91) are also prominent.

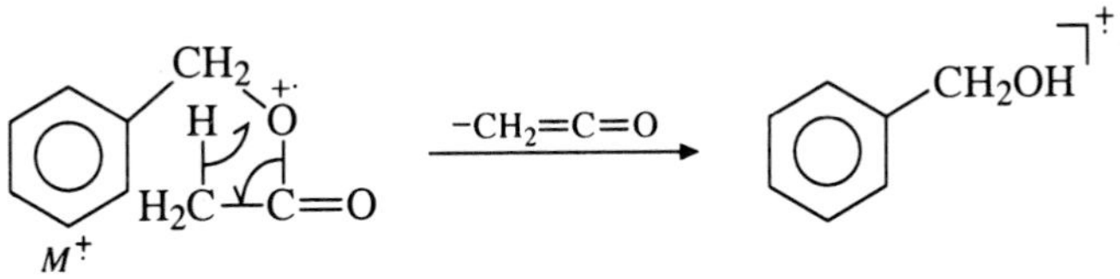

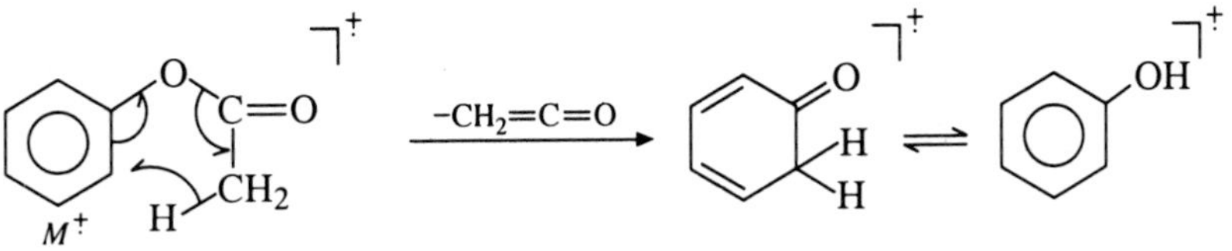

In esters of aromatic acids, as the alkyl moiety increases in length, three modes of cleavage become important:

(i) McLafferty rearrangement (Scheme 8.3), R = Ar
(ii) Double hydrogen rearrangement (Scheme 8.4), R = Ar
(iii) Retention of the positive charge with the alkyl group

$$\mathrm{Ar{-}\overset{O}{\overset{\|}{C}}{-}\overset{+\cdot}{O}{-}R} \xrightarrow{\mathrm{R{-}\overset{O}{\overset{\|}{C}}{-}\dot{\ddot{O}}:}} R^+$$

(vii) Amines

(a) Aliphatic

The cleavage of the C—C bond next to the nitrogen atom is the most favourable fragmentation pattern of amines. The cleavage is so favourable that the molecular ion peak is often not observed. This mode of fragmentation is similar to that of alcohols and ethers. In case of all primary, secondary and tertiary amines that are not branched at the α-carbon, the ion I accounts for the base peak. The largest alkyl group is lost as a radical in this cleavage. In amines containing an α hydrogen, usually M-1 peak is visible

$$R{-}\overset{|}{\underset{|}{C}}{-}\overset{+}{N}\langle \xrightarrow{-R^{\cdot}} \rangle C{=}\overset{+}{N}\langle \quad \text{(I)}$$

The primary fragment (I) from a secondary or tertiary amine undergoes fragmentation via hydrogen rearrangement similar to that described for aliphatic ethers to give a peak at m/e 30, 44, 58 or 72, . . . , e.g.

$$\underset{M^{+\cdot}}{R{-}CH_2{-}\underset{H}{\overset{+}{N}}{-}CH_2CH_2R'} \xrightarrow{-R'^{\cdot}} CH_2{=}\underset{H}{\overset{+}{N}}{-}CH_2{-}\underset{H}{}CHR' \longrightarrow \underset{m/e\ 30}{CH_2{=}\overset{+}{N}H_2} + R'CH{=}CH_2$$

Longer-chain amines give cyclic fragments:

$$R{-}CH_2{-}(CH_2)_n{-}\overset{+}{N}H_2 \xrightarrow{-R^{\cdot}} \text{cyclic } CH_2{-}(CH_2)_n{-}\overset{+}{N}H_2$$

n = 3, 4; m/e 72, 86

Primary straight-chain amines show homologous series of peaks at m/e 30, 44, 58, . . . resulting from cleavage at C—C bonds successively removed from the nitrogen atom with retention of the charge on the N-containing fragment.

(b) Cyclic

Unlike acyclic amines, cyclic amines usually show intense molecular ion peaks. Their primary cleavage occurs at the bond next to the nitrogen atom leading to loss of an α-hydrogen atom to give a strong M-1 peak, or to opening of the ring. The latter process is followed by elimination of ethylene to give $\dot{C}H_2—\overset{+}{N}H{=}CH_2$ (*m*/*e* 43, base peak). As an example, the fragmentation pattern of pyrrolidine is as follows:

$$\text{pyrrolidinium (}m/e\ 70) \xleftarrow{-H^\cdot} \text{pyrrolidine}^{+\cdot}\ (m/e\ 71\ (M^{+\cdot})) \longrightarrow \text{ring-opened } \overset{+}{N}H{=}CH_2\ (m/e\ 71) \xrightarrow{-C_2H_4} \dot{C}H_2—\overset{+}{N}H{=}CH_2\ (m/e\ 43\ \text{(base peak)}) \xrightarrow{-H^\cdot} CH_2{=}\overset{+}{N}{=}CH_2\ (m/e\ 42)$$

(c) Aromatic

They show intense molecular ion peak. Loss of one of the amino hydrogens of aniline gives an intense *M*-1 peak. Loss of HCN followed by loss of a H atom also gives prominent peaks. As an example, the fragmentation pattern of aniline is given as follows:

$$C_6H_5NH^{+}\ (m/e\ 92) \xleftarrow{-H^\cdot} C_6H_5NH_2^{+\cdot}\ (m/e\ 93\ (M^{+\cdot})) \xrightarrow{-HCN} C_5H_6^{+\cdot}\ (m/e\ 66) \xrightarrow{-H^\cdot} C_5H_5^{+}\ (m/e\ 65)$$

The case of N-alkylanilines (alkyl aryl amines), cleavage of C—C bond next to the nitrogen atom is dominant, e.g.

$$C_6H_5\overset{+\cdot}{N}H—CH_2—R\ (M^{+\cdot}) \xrightarrow{-R^\cdot} C_6H_5\overset{+}{N}H{=}CH_2\ (m/e\ 106)$$

(iii) Aliphatic Nitriles

The molecular ion peaks of aliphatic nitriles are weak and sometimes undetectable. *M*-1 peak formed by loss of α-hydrogen is weak but very useful for their detection

$$RCH(H)—C{\equiv}\overset{+\cdot}{N}\ (M^{+\cdot}) \xrightarrow{-H\cdot} RCH{=}C{=}\overset{+}{N}\ (M\text{-}1)$$

Straight-chain nitriles (C_4—C_9) undergo McLafferty rearrangement and the resulting ion (*m*/*e* 41) is responsible for the base peak.

$$N{\equiv}C—CH_2—CH_2—CHR(H)\ (M^{+\cdot}) \xrightarrow{-\overset{+}{R}CH{=}CH_2} CH_2{=}C{=}\overset{+\cdot}{N}H^-\ (m/e\ 41\ \text{(base peak)})$$

Straight-chain nitriles (C_8 and higher) udergo fragmentation via-hydrogen rearrangement to give a characteristic and intense peak at *m/e* 97 as

$$\overset{+}{N}{\equiv}C{-}CH_2{-}CH_2{-}CH_2{-}CH_2{-}CH_2{-}CH_2{-}CHR \xrightarrow{-RCH=CH_2} \text{cyclic } H\overset{+}{N}{=}C(CH_2)_5 \; (m/e\ 97)$$

These peaks are accompanied by the usual peaks of the hydrocarbon pattern.

(ix) Nitro Compounds

(a) Aliphatic

Their molecular ion peaks are weak or absent. Presence of a nitro group is indicated by peaks at *m/e* 30 ($\overset{+}{N}O$) and 46 ($\overset{+}{N}O_2$). The main peaks are due to the hydrocarbon fragments up to *M*-46.

(b) Aromatic

Their molecular ion peak are strong. Elimination of an NO_2 radical gives *M*-46 peak (the base peak in nitrobenzene) and neutral NO molecule is lost to give strong *M*-30 peak due to the phenoxy cation (*m/e* 93). For example, the fragmentation pattern of nitrobenzene is given as

$$C_6H_5NO_2^{+\cdot}\ (m/e\ 123,\ M^+) \xrightarrow{\text{Rearrangement}} C_6H_5{-}O{-}N{=}O^{+\cdot}\ (m/e\ 123) \xrightarrow{-\dot{N}O} C_6H_5O^+\ (m/e\ 93) \xrightarrow{-CO} C_5H_5^+\ (m/e\ 65)$$

$$C_6H_5NO_2^{+\cdot}\ (m/e\ 123) \xrightarrow{-\dot{N}O_2} C_6H_5^+\ (m/e\ 77) \xrightarrow{-C_2H_2} C_4H_3^+\ (m/e\ 51)$$

Peaks due to $\overset{+}{N}O$ and $\overset{+}{N}O_2$ are also observed at *m/e* 30 and 46, respectively, in the mass spectrum of nitrobenzene.

When a substituent is present in the *m*- or *p*-positions, e.g. in *m*- and *p*-nitroanilines, the fragmentation pattern similar to that of nitrobenzene is observed. However, when a hydrogen-containing substituent is present *ortho* to the *nitro* group, *M*-OH peak is also observed, e.g.

m/e 138 ($M^{+\cdot}$) $\xrightarrow{-\dot{O}H}$ m/e 121

(x) Sulphur Compounds

If one sulphur atom is present in a molecule, then according to the natural abundance (4.2%) of ^{34}S, the $M + 2$ peak will be about 4.4% of the parent peak (the natural abundance of ^{32}S is 95.06%). Thus, the contribution of the ^{34}S to the $M + 2$ peak usually shows the presence and the number of sulphur atoms. For example, if a compound shows $M + 2$ peak whose intensity is 8.8% of the parent peak, then this compound contains (8.8/4.4 = 2) two sulphur atoms. The molecular ion peaks of sulphur compounds is generally much more intense than the corresponding oxygen-containing compounds. This is because the ionization energy of the non-bonding sulphur electron is lower than that of oxygen.

(a) Thioalcohols (Thiols or Mercaptans)

These show fragmentation modes very similar to that of alcohols, i.e. α-cleavage, the cleavage with loss of hydrogen sulphide (M-34) and the cleavage with loss of H_2S together with elimination of an olefin are characteristic

$$RR'CH-\overset{+\cdot}{S}H \xrightarrow{-\dot{R}'} RCH{=}\overset{+}{S}H$$

m/e 47, 61, 75, ...

In secondary and tertiary mercaptans, the largest alkyl group from the α-C is lost to give a prominent peak M-CH_3, M-C_2H_5, M-C_3H_7, ... A peak at M-33 (loss of $\dot{S}H$) is generally present in secondary mercaptans. In long-chain mercaptans, the hydrocarbon fragmentation pattern is also observed.

(b) Thioethers (Sulphides)

These show fragmentation modes very similar to that of ethers. Cleavage of α, β C—C bonds occur with favourable loss of the largest group. The first-formed ions thus form fragment further via hydrogen rearrangement with loss an olefin to give $RCH{=}\overset{+}{S}H$. Sulphides undergo cleavage of C—S bond with retention of charge on sulphur to show characteristic peak due to RS^+ ion, e.g.

$$CH_3CH_2-\overset{+\cdot}{S}-CH_2CH_3 \xrightarrow{-\dot{C}H_3} CH_2{=}\overset{+}{S}-CH_2-CH_2-H$$

m/e 90 ($M^{+\cdot}$) → m/e 75

$$CH_3CH_2-\overset{+\cdot}{S}-CH_2CH_3 \xrightarrow{-CH_3\dot{C}H_2} CH_3CH_2-S^+ \quad (m/e\ 61)$$

$$m/e\ 75 \xrightarrow{-CH_2{=}CH_2} CH_2{=}\overset{+}{S}H \quad (m/e\ 47)$$

(xi) Heterocyclic Compounds

Aromatic heterocyclic compounds show intense molecular ion peak. They undergo fragmentation similar to benzene, e.g. benzene eliminates C_2H_2 from its molecular ion, whereas pyrrole and pyridine loose HCN. Similarly, thiophene eliminates $CH\dot{S}$ and furan $CH\dot{O}$ from their parent ion. Pyrrole, thiophene and furan also eliminate $C_3\dot{H}_3$ from their molecular ions to give $HC{\equiv}\overset{+}{N}$, $HC{\equiv}\overset{+}{S}$ and $HC{\equiv}\overset{+}{O}$ ions, respectively.

m/e 79 ($M^{+\cdot}$) —(−HCN)→ $C_4H_4^{+\cdot}$ *m/e* 52

m/e 67 ($M^{+\cdot}$) —(−HCN)→ $C_3H_4^{+\cdot}$ *m/e* 40

X^+ (three-membered ring) ←($-C_2H_2$; X = S, NH)— ring $X^{+\cdot}$ —($-C_3H_3^{\cdot}$; X = O, S, NH)→ $HC{\equiv}\overset{+}{X}$

In alkyl substituted heteroaromatics, cleavage of the bond β to the ring is favoured similar to the alkylbenzenes. The fragment ions thus formed undergo ring expansion as benzyl cation changes to tropylium ion. This process is followed by loss of HCN in nitrogen heterocycles. Some examples are

m/e 81 ($M^{+\cdot}$) —($-H^{\cdot}$)→ *m/e* 80 (base peak) → —(−HCN)→ $C_4H_5^+$ *m/e* 53

m/e 112 ($M^{+\cdot}$) —($-\dot{C}H_3$)→ *m/e* 97

m/e 82 ($M^{+\cdot}$) —($-H^{\cdot}$)→ *m/e* 81 → —(−CO)→ $C_4H_5^+$ *m/e* 53

m/e 121 ($M^{+\cdot}$) —($-\dot{C}H_3$)→ *m/e* 106 → *m/e* 106 —(−HCN)→

m/e 79 → *m/e* 79 —($-H_2$)→ *m/e* 77 —($-C_2H_2$)→ $C_4H_3^+$ *m/e* 51

8.16 Some Solved Problems

Problem 1. Explain the formation of prominent ion peaks at *m/e* 72, 71, 57 and 43 in the mass spectrum of 2-methylbutane. Indicate the ion responsible for the base peak.

Solution. The mode of formation of the prominent ions responsible for peaks at *m/e* 72, 71, 57, 55, 43 and 41 is

$$CH_3CH_2CH(CH_3)-CH_3 \xrightarrow[-2e]{e} [CH_3CH_2CH(CH_3)-CH_3]^{+\cdot} \quad m/e\ 72\ (M^+)$$

$$[CH_3CH_2CH(CH_3)CH_3]^{+\cdot} \xrightarrow{-H^\cdot} CH_3CH_2-\overset{+}{C}(CH_3)-CH_3 \quad m/e\ 71$$

$$[CH_3CH_2CH(CH_3)CH_3]^{+\cdot} \xrightarrow{-\dot{C}H_3} CH_3CH_2\overset{+}{C}HCH_3 \quad m/e\ 57$$

$$[CH_3CH_2CH(CH_3)CH_3]^{+\cdot} \xrightarrow{-CH_3\dot{C}H_2} CH_3\overset{+}{C}HCH_3 \quad m/e\ 43 \text{ Base peak ion}$$

$$\underset{m/e\ 57}{CH_3CH_2\overset{+}{C}HCH_3} \xrightarrow{-H_2} \underset{m/e\ 55}{CH_3CH{=}\overset{+}{C}-CH_3}$$

$$\underset{m/e\ 43}{CH_3\overset{+}{C}HCH_3} \xrightarrow{-H_2} \underset{m/e\ 41}{CH_3\overset{+}{C}{=}CH_2}$$

Problem 2. The mass spectrum of 2-pentene exhibits prominent peaks at *m/e* 70, 55, 41, 39, 29 and 27. Explain the formation of the ions corresponding to these peaks.

Solution. The formation of the ions responsible for the observed peaks is explained by the following fragmentation pattern:

$$CH_3CH_2CH{=}CHCH_3 \xrightarrow[-2e]{e} [CH_3CH_2CH{=}CHCH_3]^{+\cdot} \quad m/e\ 70\ (M^+)$$

$$[M]^{+\cdot} \xrightarrow{-CH_3\dot{C}H_2} \underset{m/e\ 41}{\overset{+}{C}H{=}CH-CH_3} \xrightarrow{-H_2} \underset{m/e\ 39}{\overset{+}{C}{\equiv}C-CH_3}$$

$$[M]^{+\cdot} \xrightarrow{-CH_3CH{=}\dot{C}H} \underset{m/e\ 29}{CH_3\overset{+}{C}H_2} \xrightarrow{-H_2} \underset{m/e\ 27}{CH_3{=}\overset{+}{C}H}$$

$$[M]^{+\cdot} \xrightarrow{-\dot{C}H_3} \underset{m/e\ 55}{\overset{+}{C}H_2CH{=}CHCH_3}$$

Problem 3. An organic compound gave a peak at *m/z* 122 (M^+) and another peak of nearly equal intensity at *m/z* 124 in its mass spectrum. What is the likely molecular formula of the compound?

Solution. The compound in question shows the molecular ion (M^+) peak at *m/z* 122 and $M + 2$ peak at *m/z* 124 in the intensity ratio 1 : 1. This is characteristic of a monobromo compound. M^+ and $M + 2$ peaks are due to the contribution of ^{79}Br and ^{81}Br isotopes, respectively.

On deduction of the mass of ^{79}Br from the molecular mass M^+ of the compound, i.e. $122 - 79 = 43$. The alkyl group corresponding to this mass is C_3H_7. Thus, the compound is C_3H_7Br.

Problem 4. How will you distinguish 3-methylcyclohexene and 4-methylcyclohexene using mass spectroscopy?

Solution. Cyclohexene derivatives undergo fragmentation through retro-Diels-Alder reaction. Thus, 3-methylcyclohexene gives a much abundant diene fragment at *m/e* 54, whereas 4-methylcyclohexene gives the diene fragment at *m/e* 68 as

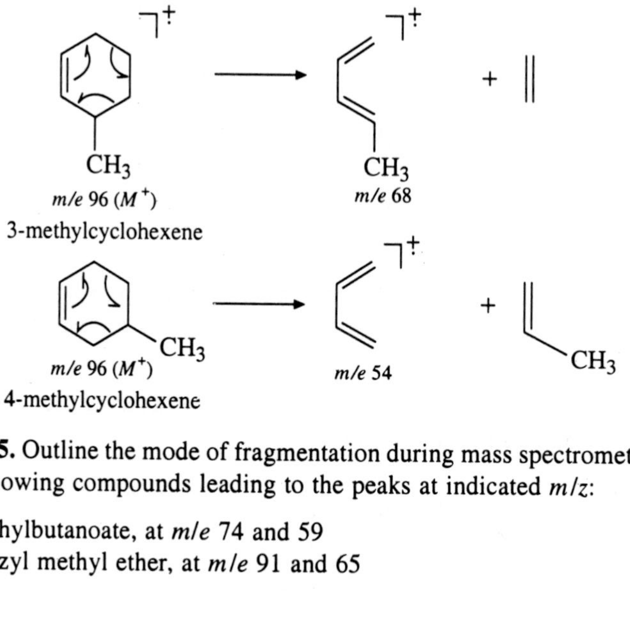

CH3
m/e 96 (M^+)
3-methylcyclohexene
→
CH3
m/e 68
+

CH3
m/e 96 (M^+)
4-methylcyclohexene
→
m/e 54
+ CH3

Problem 5. Outline the mode of fragmentation during mass spectrometric study of the following compounds leading to the peaks at indicated *m/z*:

(i) Methylbutanoate, at *m/e* 74 and 59
(ii) Benzyl methyl ether, at *m/e* 91 and 65

Solution

(i)

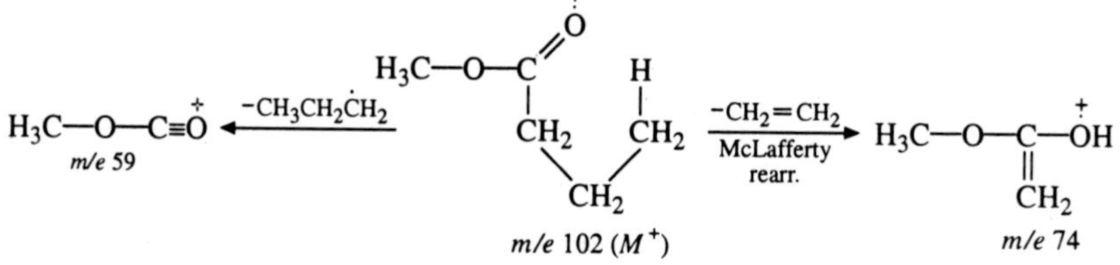

$H_3C-O-C\equiv\overset{+}{O}$ (*m/e* 59) ← $-CH_3CH_2\dot{C}H_2$ — $H_3C-O-C(=\overset{+\cdot}{O})-CH_2-CH_2-CH_2-H$ (*m/e* 102 (M^+)) — $\xrightarrow[\text{McLafferty rearr.}]{-CH_2=CH_2}$ $H_3C-O-C(=CH_2)-\overset{+}{O}H$ (*m/e* 74)

(ii)

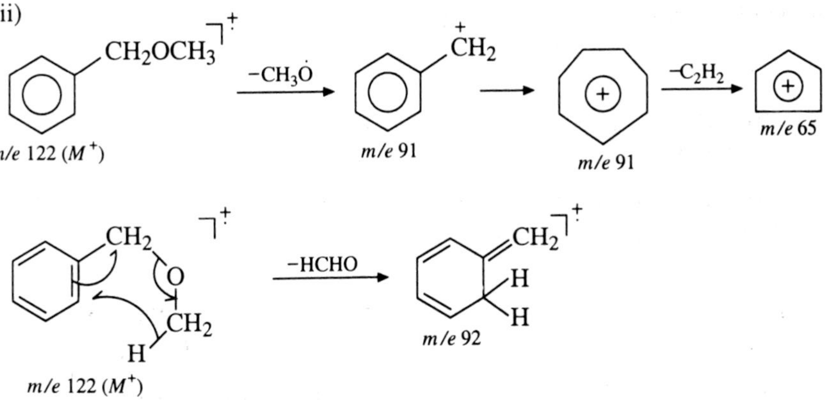

$$C_6H_5CH_2OCH_3^{+\cdot} \ (m/e\ 122\ (M^+)) \xrightarrow{-CH_3\dot{O}} C_6H_5\overset{+}{C}H_2 \ (m/e\ 91) \longrightarrow C_7H_7^+ \ (m/e\ 91) \xrightarrow{-C_2H_2} C_5H_5^+ \ (m/e\ 65)$$

$$C_6H_5CH_2OCH_2H^{+\cdot} \ (m/e\ 122\ (M^+)) \xrightarrow{-HCHO} C_7H_8^{+\cdot} \ (m/e\ 92)$$

Problem 6. Using mass spectrometry how will you distinguish 2-pentanone and 3-pentanone?

Solution. 2-Pentanone will show peaks at *m/e* 71 and 43 due to acylium ions formed by the α-cleavage as shown below:

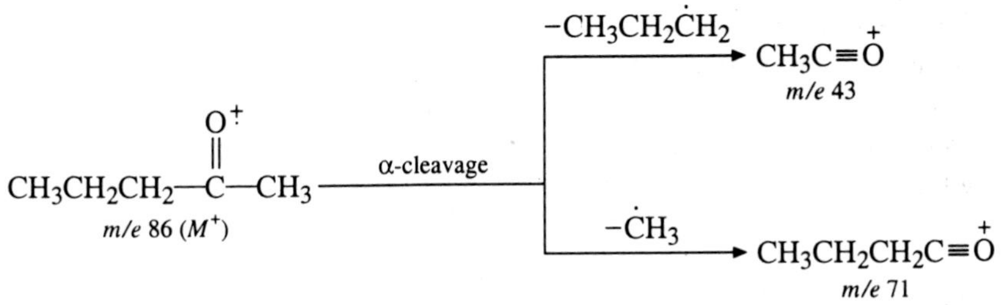

$$CH_3CH_2CH_2-\overset{O^{+\cdot}}{\overset{\|}{C}}-CH_3 \ (m/e\ 86\ (M^+)) \xrightarrow{\alpha\text{-cleavage}} \begin{cases} \xrightarrow{-CH_3CH_2\dot{C}H_2} CH_3C\equiv\overset{+}{O} \ (m/e\ 43) \\ \xrightarrow{-\dot{C}H_3} CH_3CH_2CH_2C\equiv\overset{+}{O} \ (m/e\ 71) \end{cases}$$

In addition, 2-pentanone will also undergo McLafferty rearrangement to give a peak at *m/e* 58

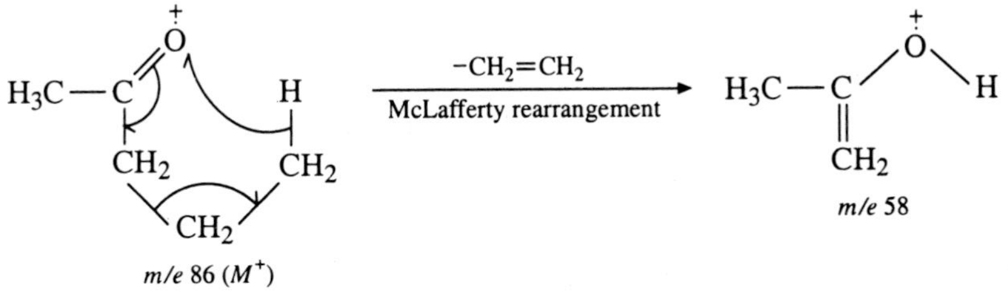

$$H_3C-C(=\overset{+}{O})-CH_2CH_2CH_2H \ (m/e\ 86\ (M^+)) \xrightarrow[\text{McLafferty rearrangement}]{-CH_2=CH_2} H_3C-C(=CH_2)-\overset{+}{O}-H \ (m/e\ 58)$$

On the other hand, 3-pentanone will undergo α-cleavage to give only one acylium ion showing a peak at *m*/e 57. Unlike 2-pentanone, 3-pentanone will not undergo McLafferty rearrangement.

$$CH_3CH_2-\overset{\overset{+}{O}}{\overset{\|}{C}}-CH_2CH_3 \ (m/e\ 86\ (M^+)) \xrightarrow[-\dot{C}H_3CH_2]{\alpha\text{-cleavage}} CH_3CH_2C\equiv\overset{+}{O} \ (m/e\ 57)$$

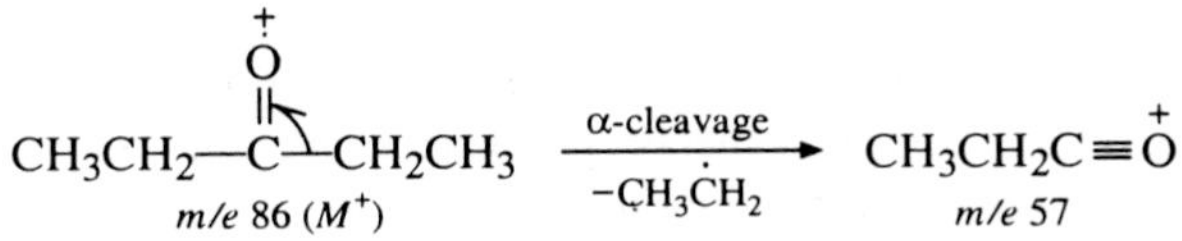

In addition, 3-pentanone will show a peak at *m*/*e* 29 as follows:

$$CH_3CH_2-\overset{\dot{O}^{+}}{\overset{\|}{C}}-CH_2CH_3 \longleftrightarrow CH_3CH_2-\overset{\overset{+}{\ddot{O}}:}{\dot{C}}-CH_2CH_3 \longrightarrow CH_3\overset{+}{C}H_2 + CH_3CH_2C{=}\ddot{O}$$

m/*e* 86 (M^+) *m*/*e* 29

Problem 7. Rationalize the formation of peaks at *m*/*z* 122 (35%, M^+), 92 (65%), 91 (100%), 65 (15%), and metastable peaks at *m*/*z* 46.4 and 69.4 in the mass spectrum of 2-phenylethanol.

Solution. The presence of a metastable peak at mass 69.4 indicates that the ion with *m*/*z* 122 has fragmented into the ion *m*/*z* 92 ($m_1^+ \rightarrow m_2^+$) with loss of a neutral fragment (HCHO) of mass 30 in one step because $69.4 = \frac{(92)^2}{122}$, i.e. $m^* = \frac{(m_2)^2}{m_1}$.

Similarly, the presence of another metastable peak at *m*/*z* 46.4 indicates that *m*/*z* 91 ion has decomposed into *m*/*z* 65 ion in one step with loss of a neutral fragment (C_2H_2) of mass 26.

The peak at an even mass *m*/*z* 92 arising from an even mass (*m*/*z* 122) molecular ion suggest that the fragmentation has occurred through a rearrangement involving the cleavage of two bonds. The peak at *m*/*z* 91 has resulted through *β*-cleavage of C—C bond with loss of $\dot{C}H_2OH$. The fragmentation mode leading to all the ions responsible for the observed peaks is outlined below:

$C_6H_5CH_2CH_2OH \xrightarrow[-2e]{e} [C_6H_5CH_2-CH_2OH]^{+\cdot}$ *m*/*z* 122 ($M^{+\cdot}$) $\xrightarrow[\beta\text{-cleavage}]{-CH_2OH} C_6H_5\overset{+}{C}H_2$ *m*/*z* 91 $\longrightarrow$

Tropylium ion *m*/*z* 91 (100% intensity, base peak) $\xrightarrow{-C_2H_2}$ cyclopentadienyl cation *m*/*z* 65

McLafferty rearr. *m*/*z* 122 (M^+) $\xrightarrow{-HCHO}$ *m*/*z* 92 ($=CH_2$, H, H)

Problem 8. The mass spectrum of 1-pentanol shows prominent peaks at *m*/*z* 88 (M^+), 70, 55, 42, 31, 29 and a metastable peak at *m*/*z* 43.3. Give the mode of formation of the ions corresponding to these peaks.

Solution. The mode of formation of the ions responsible for observed peaks is

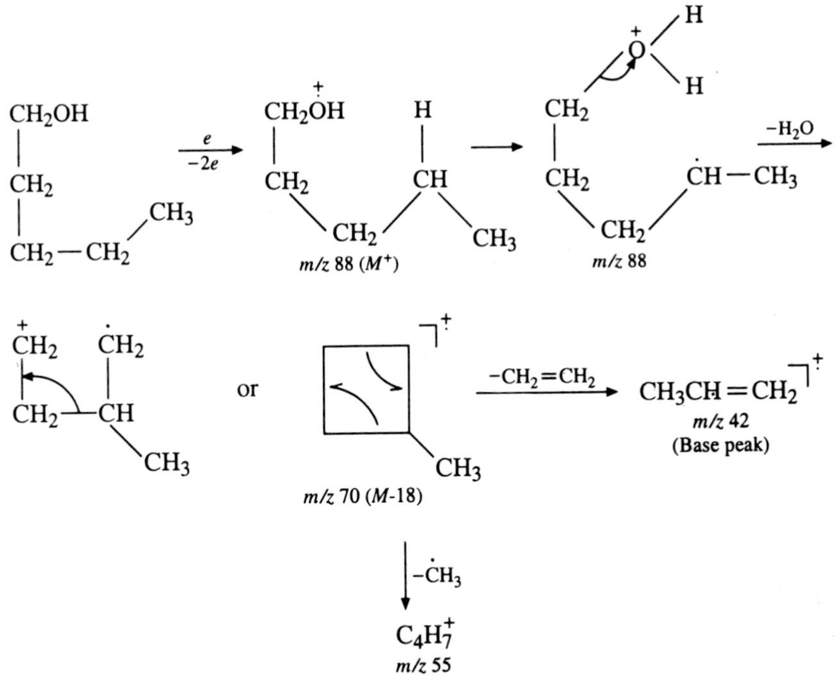

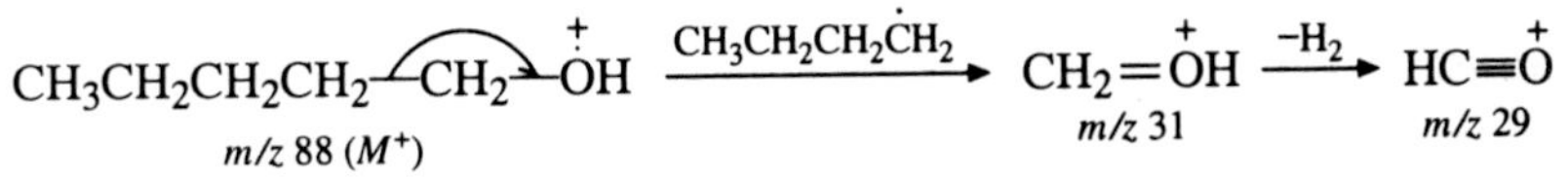

$$CH_2OH-CH_2-CH_2-CH_2-CH_3 \xrightarrow[-2e]{e} \overset{+}{C}H_2\dot{O}H \ldots \; m/z\ 88\ (M^+) \longrightarrow m/z\ 88 \xrightarrow{-H_2O}$$

$$m/z\ 70\ (M\text{-}18) \xrightarrow{-CH_2=CH_2} CH_3CH=CH_2^{\cdot+}\quad m/z\ 42\ \text{(Base peak)}$$

$$m/z\ 70 \xrightarrow{-\dot{C}H_3} C_4H_7^+\quad m/z\ 55$$

$$CH_3CH_2CH_2CH_2-CH_2-\overset{+}{\dot{O}}H\ (m/z\ 88\ (M^+)) \xrightarrow{CH_3CH_2CH_2\dot{C}H_2} CH_2=\overset{+}{O}H\ (m/z\ 31) \xrightarrow{-H_2} HC\equiv\overset{+}{O}\ (m/z\ 29)$$

The metastable peak at m/z 43.3 is accompanied by the peaks at m/z 70 and 55 because

$$m^* = \frac{m_2^2}{m_1} = \frac{(55)^2}{70} = 43.2$$

This shows that the ion of m/z 70 has decomposed to the ion of m/z 55 in one step with the loss of a neutral fragment ($\dot{C}H_3$) of mass 15.

Problem 9. Outline the mode of fragmentation leading to the ions causing peaks at indicated m/e in the mass spectra of the compounds:

(a) Pentanoic acid, at m/e 60
(b) Phenetole, at m/e 94
(c) Diethylamine, at m/e 30

Solution

(a)

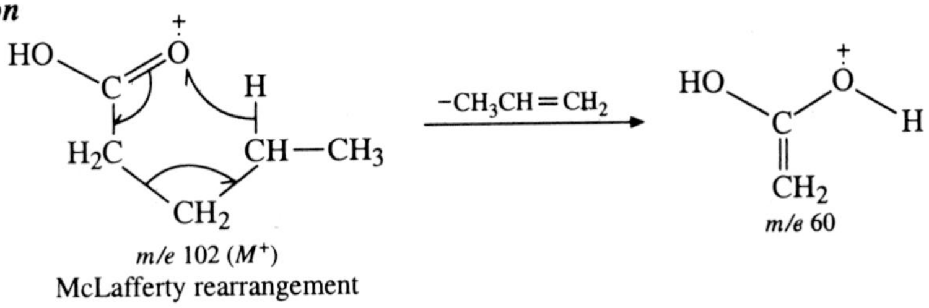

$$\text{HO}-C(=\overset{+}{\dot{O}})-CH_2-CH_2-CH(H)-CH_3\ \ (m/e\ 102\ (M^+)) \xrightarrow{-CH_3CH=CH_2} \text{HO}-C(=CH_2)-\overset{+}{\dot{O}}-H\ \ (m/e\ 60)$$

McLafferty rearrangement

(b) $C_6H_5\overset{+\cdot}{O}-CH_2-CH_2-H$ $\xrightarrow{-CH_2=CH_2}$ $C_6H_5OH^{+\cdot}$

m/e 122 (M^+)
Hydrogen rearrangement

m/e 94

(c) $CH_3CH_2-\overset{+\cdot}{N}H-CH_2-CH_3$ $\xrightarrow{-\dot{C}H_3}$ $\dot{C}H_2-\overset{+}{N}H=CH_2$ (CH_2-H)

m/e 73 (M^+)

m/e 58

Hydrogen rearrangement

$\downarrow -CH_2=CH_2$

$\overset{+}{N}H_2=CH_2$

m/e 30

Problem 10. In the mass spectrum of an unsaturated hydrocarbon, the molecular ion M^+ peak has relative intensity 70.0, the $M + 1$ peak 4.7, and the base peak a relative intensity of 100. How many carbon atoms are there in the hydrocarbon per molecule.

Solution

$$\text{Number of C atoms} = \frac{\text{Intensity of } M{+}1 \text{ peak as the percentage of } M^+ \text{peak}}{1.1}$$

$$= \frac{4.7 \times 100}{70 \times 1.1} = 6.1 \equiv 6$$

The hydrocarbon contains 6 carbon atoms.

PROBLEMS

1. The mass spectrum of 3,3-dimethylheptane shows prominent peaks at m/e 113, 99, 71, 57, 55, 43, 41, 29 and 27. Give the fragmentation pattern leading to the ions corresponding to these peaks.
2. Draw the conclusions and identify the structure of (A) and (B) on the basis of the following MS data:

   (i) Compound (A) shows M^+ and $M + 2$ ion peaks at m/e 60 and 62 in the intensity ratio 3 : 1.

   (ii) An organic ester (B) shows m/e 60 (M^+), 31 and 29.
3. Rationalize the formation of peaks at m/e 122 (M^+), 107 (base peak), 79, 77 and 51 in the mass spectrum of 1-phenylethanol.
4. Outline the mode of fragmentation of 4-n-butyltoluene involving benzylic cleavage and McLafferty rearrangement.

5. (a) Give a brief account of McLafferty rearrangement.
(b) Discuss nitrogen rule as applied to mass spectrometry.

6. Define the following terms used in mass spectroscopy:

(i) Base peak (ii) Parent peak
(iii) $M + 1$ and $M + 2$ peaks (iv) Metastable peak

7. The mass spectrum of ethyl butanoate shows two characteristic peaks due to odd-electron ions at *m/e* 88 and 60, and another peak at *m/e* 71. Explain the fragmentation pattern.

8. How will you account for the formation of ions at *m/z* 102, 87, 59, 45 and 43 during the mass spectrometric study of di-isopropyl ether?

9. What are the most probable species responsible for peaks at *m/z* 46, 45, 31, 29, 17 and 15 in the mass spectrum of ethanol?

10. How will you distinguish the following pairs of compounds using the mass spectrometry:

(a) Butanoic acid and 2-methylpropanoic acid
(b) Pentanal and 2-methylbutanal
(c) Diethylamine and *N,N*-dimethylethylamine

11. Outline the mode of fragmentation during mass spectrometric study of the following compounds leading to the peak at indicated *m/z*:

(i) 1-butanol, at *m/z* 56
(ii) 1-hexanol, at *m/z* 56
(iii) *cis*-methyl crotonate, at *m/z* 68.

12. What is probable molecular formula of molecule having parent peak at *m/e* 142, and a $M + 1$ peak whose intensity is 1.1% of the parent peak?
Hint:

$$\text{Number of C atoms} = \frac{\text{Intensity of } M + 1 \text{ peak as the percentage of parent peak}}{1.1}$$

13. Give the fragmentation mode of 2-methylpentanal showing peaks at *m/e* 100, 71, 58, 43, 41 and 29 in its mass spectrum.

14. A hydrocarbon exhibits peaks at *m/e* 100 (M^+), 85, 71, 57, 43 (100%), 41 and 29 in its mass spectrum. Deduce the structure of the hydrocarbon and give its fragmentation modes.

References

1. A. Frigerio, Essential Aspects of Mass Spectrometry, Spectrum Publishers, New York, 1947.
2. B.S. Middeditch, Ed., Practical Mass Spectrometry, Plenum, New York, 1979.
3. D.H. Williams and I. Howe, Principles of Mass Spectrometry, McGraw-Hill, London, 1972.
4. J.H. Beynon and A.E. Willams, Mass and Abundance Tables for Use in Mass Spectrometry, Elsevier, Amsterdam, 1963.
5. J.T. Watson, Introduction to Mass Spectrometry, 2nd Ed., Raven Press, New York, 1985.

6. M. Hamming and N. Foster, Interpretation of Mass Spectra of Organic Compounds, Academic, New York, 1972.
7. M.E. Rose and R.A.W. Johnstone, Mass Spectrometry for Organic Chemists, 2nd Ed., University Press, Cambridge, 1982.
8. R. Davis and M. Frearson, Mass Spectrometry, Wiley, Chichester, 1987.
9. R.M. Silverstein, G.C. Bassler and T.C. Morril, Spectrometric Identification of Organic Compounds, 3rd Ed., Wiley, New York, 1974.
10. S.R. Shrader, Introduction to Mass Spectrometry, Allyn and Bacon, Boston, 1971.

9

Spectroscopic Solutions of Structural Problems

9.1 Introduction

Various spectroscopic methods described in this book provide sufficient information for the structure determination of organic compounds. A general approach to solving structural problems by a combination of spectroscopic methods is given as follows:

1. Molecular formula and molecular weight of a compound is known from its elemental analysis and mass spectrum. The molecular formula gives an idea about the number and kinds of possible functional groups.
2. Index of hydrogen deficiency is determined from the molecular formula that gives the sum of multiple bonds and rings in the compound (Section 8.8(i)). It helps in limiting the possibilities of structures for further consideration.
3. The UV spectrum gives indication about the presence (or absence) of a conjugated system, an aromatic ring, a carbonyl group (aldehydic or ketonic) etc.
4. The IR spectrum shows the presence (or absence) of certain functional groups, e.g. carbonyl groups, hydroxyl groups, –NH– etc., and $C{\equiv}C$ and $C{=}C$ bonds.
5. The PMR spectrum reveals the environment of hydrogen atoms in the molecule. It gives the number and kinds of protons present in the molecule. Spin-spin splitting tells about the neighboring protons. In brief, we should examine the PMR spectrum for the presence of $-CH_3$, CH_3CH_2-, $-CH_2-CH_2$, $(CH_3)_2CH-$, $-CH{=}CH-$, $-C{\equiv}CH$ etc. groups; aromatic protons, and protons on heteroatoms, i.e. exchangeable protons. Thus, the PMR spectrum leads to some extent to the molecular skeleton.
6. The CMR spectrum gives the number of kinds of carbons, and the number of methine carbons, methylene carbons, methyl carbons, and carbons having no hydrogen. Thus, the CMR suggests the carbon skeleton of the molecule.
7. The mass spectrum, in addition to molecular weight, shows the presence of certain structural units, and the fragmentation pattern indicates their points of attachment in the molecule.

The following solved problems demonstrate the use of various spectral data in structure determination.

9.2 Some Solved Problems

Problem 1. An organic compound with molecular formula $C_6H_{12}O_2$ gave the following spectral data:

UV: λ_{max} 283 nm, ε_{max} 27 (hexane).
IR: Significant absorption bands at 1705, 2900 and 3450 cm^{-1}.
^1H NMR: δ 1.3 (six proton singlet), 2.2 (three proton singlet) and 3.8 (one proton singlet, exchangeable with D_2O).
^{13}C NMR (Off-resonance decoupled) : Two singlets, one triplet and two quartets, one of the singlets is at δ 210 and the other at δ 70.
Mass: Prominent peaks at *m/z* 116, 58 and 43.

Deduce the structure of the compound and explain the spectral data.

Solution. The index of hydrogen deficiency (Section 8.8(i)) calculated for the given molecular formula $C_6H_{12}O_2 = x - \frac{y}{2} + \frac{z}{2} - 1 = 6 - \frac{12}{2} + 1 = 1$.

UV: The compound shows λ_{max} 283 nm with ε_{max} 27 which indicates the presence of an aldehydic or a ketonic $>C{=}O$ group. This band is due to forbidden transition $n \rightarrow \pi^*$ of the $>C{=}O$ group.

IR: Absorption bands at 1705, 2900 and 3450 cm^{-1} may be assigned to ketonic $>C{=}O$, alkyl and hydroxyl groups, respectively.

^1H NMR: The compound shows four signals indicating the presence of four kinds of protons. The six proton singlet at δ 1.3 indicates the presence of two equivalent —CH_3 groups with no neighboring proton. The three proton singlet at δ 2.2 shows the presence of another methyl group with no adjacent proton. The one proton singlet at 3.8 (exchangeable with D_2O) shows that the proton is attached to a heteroatom, in this case to oxygen, i.e. the compound contains an —OH group.

^{13}C NMR (Off-resonance decoupled): The presence of two singlets shows the presence of two nonequivalent carbons having no hydrogen, one of these may be the carbonyl carbon (δ 210) and the other the carbon containing the —OH groups. One triplet shows the presence of a —CH_2— group, and two quartets indicate the presence of two nonequivalent —CH_3 groups.

Mass: The peak at *m/z* 116 is due to molecular ion as indicated by the given molecular formula ($C_6H_{12}O_2$). The presence of a peak at *m/z* 43 is due to $CH_3C{\equiv}\overset{+}{O}$. The formation of a fragment ion peak of even mass number from the even mass numbered molecular ion shows that the fragmentation has occurred through a rearrangement involving the cleave of two bonds.

From the above explained spectral data, the deduced structure of the compound is

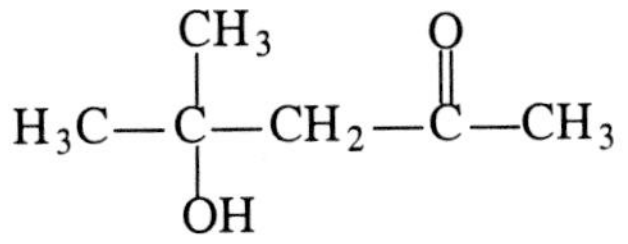

$$H_3C-\underset{OH}{\overset{CH_3}{C}}-CH_2-\overset{O}{\overset{\|}{C}}-CH_3$$

The calculated index of hydrogen deficiency also corresponds to this structure. The presence of peaks at *m/z* 116, 58 and 43 in the mass spectrum of the compound is explained by the formation of corresponding ions as follows:

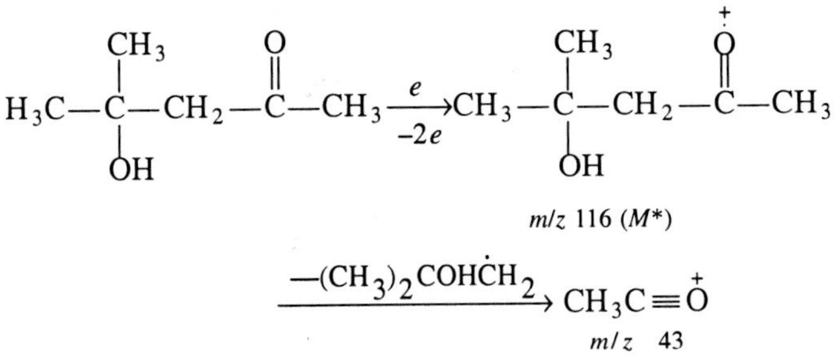

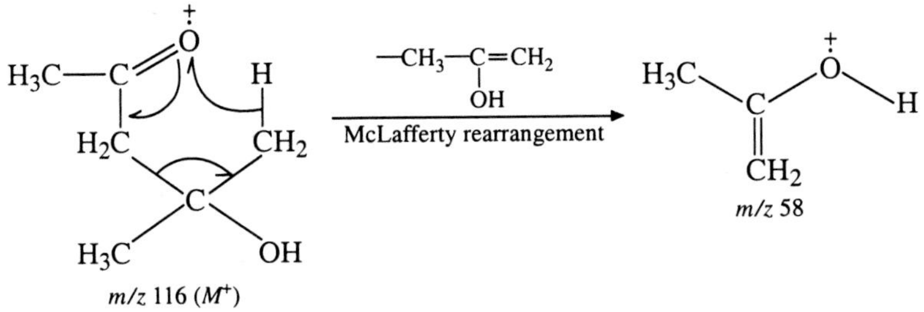

$$H_3C-\underset{OH}{\overset{CH_3}{C}}-CH_2-\overset{O}{\overset{\|}{C}}-CH_3 \xrightarrow[-2e]{e} CH_3-\underset{OH}{\overset{CH_3}{C}}-CH_2-\overset{\overset{+}{\ddot{O}}}{\overset{\|}{C}}-CH_3$$

m/z 116 (*M**)

$$\xrightarrow{-(CH_3)_2COH\dot{C}H_2} CH_3C\equiv\overset{+}{O}$$

m/z 43

$$\underset{m/z\ 116\ (M^+)}{H_3C-C(=\overset{+}{\ddot{O}})-CH_2-C(CH_3)(OH)-CH_2-H} \xrightarrow[\text{McLafferty rearrangement}]{-CH_3-C(OH)=CH_2} \underset{m/z\ 58}{H_3C-C(=CH_2)-\overset{+}{\ddot{O}}-H}$$

Problem 2. An organic compound $C_8H_{14}O_4$ gave the following spectral data:

UV: No significant absorption above 210 nm.

IR: Significant absorption bands at 1735 and 2950 cm^{-1}.

1H NMR: δ 1.2 (six proton triplet, $J = 7$ Hz), 2.6 (four proton singlet) and 4.2 (four proton quartet, $J = 7$ Hz).

^{13}C NMR (Off-resonance): One singlet, two triplets, and one quartet. The singlet is at δ 175.

Mass: Prominent peaks at *m/z* 174 (M^+), 129 and 101.

Deduce the structure of the compound and explain the spectral data.

Solution

UV: The absence of absorption band above 210 nm indicates the absence of conjugation, aldehydic and ketonic carbonyl groups and an aromatic ring.

IR: Absorption bands at 1735 and 2950 cm^{-1} indicate the presence of an ester and alkyl groups, respectively.

^1H NMR: The presence of a six proton triplet at δ 1.2 and a four proton quartet at δ 4.2 (both having $J = 7$ Hz) indicates the presence of two equivalent

CH_3CH_2— groups. The four proton singlet at δ 2.6 may be due to a —CH_2—CH_2— group with equivalent protons. Thus, the compound contains two equivalent CH_3CH_2— groups and one —CH_2—CH_2— group.

^{13}C NMR (Off-resonance): One singlet at δ 175 may be due to the carbonyl carbon of the ester group. Two triplets and one quartet indicate that there are two kinds of —CH_2— groups and one kind of CH_3— group.

Mass: The molecular ion (M^+) peak at m/z 174 corresponds to the molecular formula $C_8H_{14}O_4$ of the compound.

On the basis of the spectral data, as explained above, and the given molecular formula, the deduced structure of the compound is

$$\begin{array}{l} CH_2COOCH_2CH_3 \\ | \\ CH_2COOCH_2CH_3 \end{array}$$

The index of hydrogen deficiency calculated for $C_8H_{14}O_4 = 8 - \frac{14}{2} + 1 = 2$ is also in agreement with this structure because it has two double bonds.

The formation of ions corresponding to the peaks at m/z 174 (M^+), 129 and 101 is as follows:

$$CH_2(COOCH_2CH_3)—CH_2(COOCH_2CH_3) \xrightarrow[-2e]{e} CH_2(\overset{+\cdot}{O}{=}C—OCH_2CH_3)—CH_2(COOCH_2CH_3)$$

m/z 174 (M^+)

$-CH_3CH_2O$: $CH_2—C{\equiv}\overset{+}{O}$ / $CH_2—C(=O)—CH_2CH_3$ — m/z 129

$-CH_3CH_2OC{=}\dot{O}$: $\overset{+}{C}H_2$ / $CH_2—C(=O)—O—CH_2CH_3$ — m/z 101

Problem 3. A compound containing C, H, O and a halogen shows molecular ion peak at m/e 108/110 in the intensity ratio 3 : 1. The IR spectrum shows a very broad band in the range of 2500-3300 cm^{-1} and centering around 2900 cm^{-1}, and a strong intensity band at 1705 cm^{-1}. In the ^{1}H NMR spectrum of the compound, two triplets at δ 2.8 and 3.8 and a singlet at δ 12 are found in the intensity ratio 2 : 2 : 1. One of the ^{13}C NMR peaks (proton-noise decoupled) is a low intensity peak at δ 170. Deduce the structure of the compound and explain the spectral data.

Solution. The compound shows molecular ion peak at m/e 108/110, i.e. the M^+ peak at m/e 108 and $M + 2$ peak at m/e 110 in the intensity ratio 3 : 1. This clearly indicates that the compound contains a chlorine atom. The presence of a broad IR band in the range 2500-3300 cm^{-1} indicates the presence of an —OH group, and the strong band at 1705 cm^{-1} may be due to carboxylic carbonyl group, i.e. the compound contains a —COOH group. The presence of two PMR triplets at δ 2.8 and 3.8 show the presence of two adjacent and nonequivalent —CH_2— groups, i.e. —CH_2—CH_2— group, and the presence of a very low field singlet at δ 12 indicates the presence of a carboxylic proton. These are found in the intensity ratio 2 : 2 : 1. The presence of a low intensity peak at δ 170 in the ^{13}C NMR (proton-noise decoupled) spectrum supports the presence of the carbonyl group of the carboxylic acid.

Thus, the compound under investigation contains —Cl, —CH_2—CH_2— and —COOH groups, and its structure is

$$Cl{-}CH_2{-}CH_2{-}COOH$$

This structure is strictly in accordance with the given spectral data as explained above.

Problem 4. A compound with molecular weight 130 gave a negative iodoform test. It absorbs at 292 nm, ε_{max} 16 in the UV spectrum. It shows significant bands at 3042, 2941, 2862, 2740, 1722, 1605, 1575 and 1462 cm^{-1} in its IR spectrum. In the PMR spectrum, three signals are present at τ 2.73 (multiplet), 7.2 (doublet) and 0.22 (triplet) in the intensity ratio 5 : 2 : 1, respectively. Determine the structural formula of the compound.

Solution. The UV absorption at 292 nm, ε_{max} 16 is characteristic of an aldehydic or ketonic carbonyl group arising from $n \rightarrow \pi^*$ forbidden transition. The IR absorption band at 1722 cm^{-1} also supports the presence of the carbonyl group. This is an aldehydic carbonyl group because the compound shows characteristic C—H stretching absorption of —CHO group at 2862 and 2740 cm^{-1}. The absorptions at 3042 and 2941 cm^{-1} are due to =C—H saturated (or aromatic C—H) and C—H stretching, respectively. The absorptions at 1605, 1575 and 1462 cm^{-1} are due to aromatic ring C—C stretching.

The PMR spectrum shows a multiplet at τ 2.73 which indicates the presence of an aromatic ring; a doublet and a low field triplet are present at τ 7.2 and 0.22, respectively, showing the presence of —CH_2CHO group. The signals are in the intensity ratio 5 : 2 : 1, respectively.

The above discussion and the molecular weight (130) of the compound suggest that its structure is

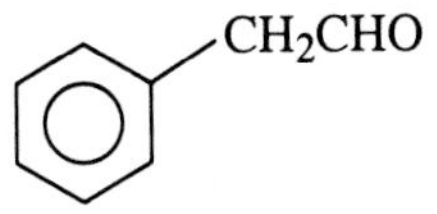

CH_2CHO

Problem 5. An organic compound with molecular formula $C_6H_{10}O_2$ furnished the following spectral data:

UV: No significant absorption above 210 nm.
IR: Significant absorption bands at 1760 and 2950 cm^{-1}.
PMR: δ 1.5 (3H, *d*), 2.2 (6H, *s*) and 6.8(1H, *q*)
^{13}C NMR (Off-resonance decoupled): One singlet at δ 165, one doublet and two quartets.
MS: Prominent peaks at *m/e* 146, 87 and 43.

Deduce the structure of the compound and explain the spectral data.

Solution

UV: No absorption above 210 nm indicates the absence of conjugation, aldehydic and ketonic >C=O group, and aromatic ring.

IR: Absorption bands at 1760 and 2950 cm^{-1} are due to stretching of an ester >C=O and an alkyl C—H band, respectively.

PMR: The presence of a 3H doublet at δ 1.5 and 1H quartet at δ 6.8 indicate the presence of a CH_3CH< group. The 6H singlet at δ 2.2 indicates the presence of two equivalent CH_3— groups with no neighboring protons. The remaining two carbon and four oxygen atoms constitute two —COO— groups.

^{13}C NMR (Off-resonance decoupled): The singlet at δ 165 is due to ester carbonyl carbon. One doublet is due to the methine (>CH—) carbon and two quartets are due to two kinds of —CH_3 carbons.

MS: The peak at *m/e* 146 is due to molecular ion as shown by the molecular formula. The peak at *m/e* 43 is due to $CH_3C{\equiv}\overset{+}{O}$.

From the above discussion it is clear that the compound contains one CH_3CH<, two CH_3— and two —COO— groups. Thus, the structure of the compound is

$$CH_3CH\begin{cases} O-\overset{\overset{\displaystyle O}{\|}}{C}-CH_3 \\ O-\underset{\underset{\displaystyle O}{\|}}{C}-CH_3 \end{cases}$$

This structure of the compound is strictly in accordance with the spectral data as explained above and the molecular formula.

The peaks at *m/e* 146, 87 and 43 in the MS are

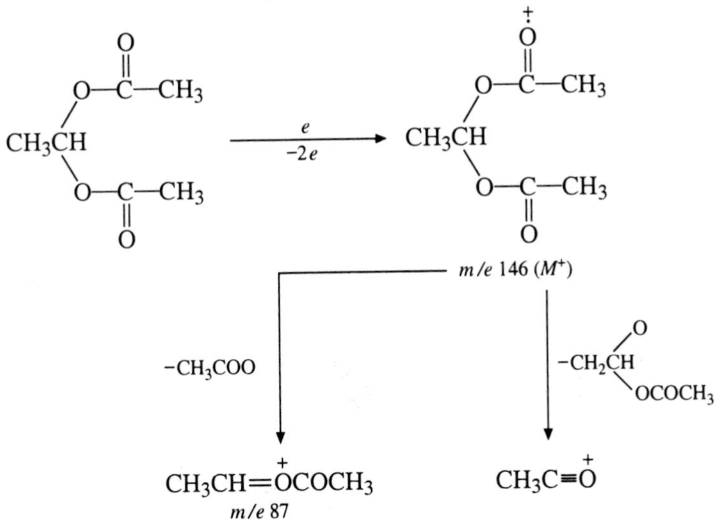

$$CH_3CH(OCOCH_3)_2 \xrightarrow[-2e]{e} CH_3CH(O\!-\!\overset{+\cdot}{C}(=O)CH_3)(OCOCH_3)$$

m/e 146 (M^+)

$-CH_3COO$ → $CH_3CH{=}\overset{+}{O}COCH_3$ *m/e* 87

$-CH_2CH(O)(OCOCH_3)$ → $CH_3C{\equiv}\overset{+}{O}$

Problem 6. An organic compound with molecular mass gave the following spectral data:

UV: No significant absorption band above 200 nm.
IR: Significant absorption bands at 2940, 2270 and 1460 cm^{-1}.
PMR: 2.72 (septet, J = 6.7 cps) and 1.33 (doublet, J = 6.7 cps) in the intensity ratio 1 : 6.

Deduce the structure of the compound and explain the spectral data.

Solution. The IR absorption bands at 2270, 2940 and 1460 are due to C≡N, C—H and C—C stretching, respectively.

Two PMR signals, a septet and a doublet with the same values of coupling constant suggest the presence of $(CH_3)_2CH$— groups. This corresponds to 43 mass units. The remaining 69 – 43 = 26 mass units are due to C≡N group as also indicated by the IR spectrum. Thus, the compound contains $(CH_3)_2CH$— and C≡N structural units, and its structure is

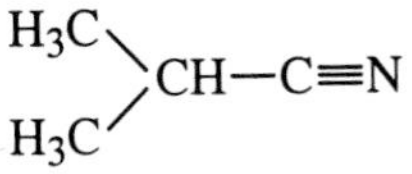

$$(H_3C)_2CH{-}C{\equiv}N$$

Problem 7. An organic compound containing C, H and O gave the following spectral data:

Mass: Molecular ion (M^+) peak at *m/z* 158.
UV: λ_{max} 225 nm, ε_{max} 50 (hexane).
IR: Significant absorption bands at 1757, 1828 and 2857-3077 cm^{-1}.
^1H NMR: δ 2.70 (septet, J = 6.7 Hz) and 1.20 (doublet, J = 6.7 Hz) in the intensity ratio 1 : 6, respectively.
^{13}C NMR (Off-resonance decoupled): One singlet, one doublet and one quartet.

The singlet is at about δ 170.

Deduce the structure of the compound and explain the spectral data.

Solution

Mass: The molecular ion (M^+) peak at m/z 158 shows that the molecular mass of the compound is 158.

UV: $\lambda_{max}{}^{225}$ nm with δ_{max} 50 indicates a carboxylic, ester or anhydride $\rangle C{=}O$ group.

IR: Bands at 1757 and 1828 cm^{-1} are the characteristic of anhydride (—CO—O—CO—) grouping. The absorption at 2857-3077 cm^{-1} is due to alkyl C—H stretching.

^1H NMR: A septet and a doublet with the same coupling constant ($J = 6.7$ Hz) show the presence of $(CH_3)_2$ CH— group.

^{13}C NMR (Off-resonance decoupled): One singlet at δ 170 may be due to carbonyl group of the an hydride function: the doublet and quartet indicate the presence of $\rangle CH$— and CH_3— groups.

Thus, the spectral data as explained above, indicate the presence of $(CH_3)_2CH$— and —CO—O—CO— groups whhich amounts to 115 mass units. The remaining 158 – 115 = 43 mass units may be another $(CH_3)_2$ CH— group. Hence, the structure of the compounds is

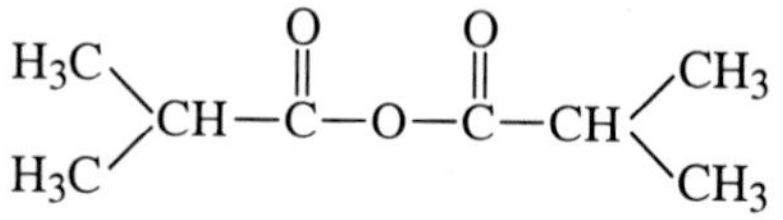

$$(H_3C)_2CH-\overset{O}{\overset{\|}{C}}-O-\overset{O}{\overset{\|}{C}}-CH(CH_3)_2$$

Problem 8. Explain the following spectral data systematically and deduce the structure of an organic molecule containing C, H and O:

UV: λ_{max} 278 and 319 nm.

IR: Significant absorption bands at 3070-3010, 2970-2860, 1685, 1605, 1580 and 1450 cm^{-1}.

PMR: δ 2.1 (3H, *s*) and 7.5 (5H, *m*).

^{13}C NMR: δ 198 and 137 (two singlets), 134, 129 and 128 (three doublets) and 26 (one quartet).

MS: m/e 120 (M^+), 105, 77, 51 and 43.

Solution

UV: λ_{max} 278 and 319 indicate the presence of a carbonyl (an aldehyde or a ketone) group and an aromatic ring.

IR: Absorption bands at 3070-3010 and 2970-2860 cm^{-1} are due to the aromatic and alkyl C—H stretch, respectively. The band at 1685 cm^{-1} may be due to the aromatic ketonic $\rangle C{=}O$ group. The absorption bands at 1605, 1580 and 1450 cm^{-1} are characteristic of an aromatic ring.

PMR: The presence of a five proton multiplet at δ 7.5 indicates the presence of a phenyl group and the singlet at δ 2.1 shows the presence of a CH_3— group.

^{13}C NMR: The presence of two singlets, three doublets and one quartet indicates the presence of six kinds of carbons including two kinds of >CH— and one CH_3— groups. The singlet appearing at δ 198 (low field) is due to the ketonic groups.

MS: M^+ at *m/e* 120 shows that the molecular mass of the compound is 120.

The above discussion clearly shows that the compound contains C_6H_5—, >C=O (*keto*) and CH_3— structural units which amount to 120 mass units, i.e. the molecular mass of the compound. Thus, the structure of the compound is

$$C_6H_5-\overset{\overset{\displaystyle O}{\|}}{C}-CH_3$$

The peaks at *m/e* 105, 77, 51 and 43 are explained by the following fragmentation modes:

$$\underset{m/e\ 120\ (M^+)}{C_6H_5-\overset{\overset{+}{\ddot{O}}}{\|}C-CH_3} \xrightarrow{-\dot{C}H_3} \underset{m/e\ 105}{C_6H_5-C\equiv\overset{+}{O}} \xrightarrow{-CO} \underset{m/e\ 77}{C_6H_5^+} \xrightarrow{-C_2H_2} \underset{m/e\ 51}{\boxed{\oplus}}$$

$$m/e\ 120\ (M^+) \xrightarrow{-C_6H_5^{\cdot}} CH_3C\equiv\overset{+}{O}\quad m/e\ 43$$

Problem 9. Explain the following spectral data given by an organic compound $C_9H_{10}O_2$ and deduce the structure of the compound.

UV: λ_{max} 257 nm, ε_{max} 194.

IR: Significant absorption bands at 3040, 2950, 1740, 1480, 1440 and 1220, 700 and 750 cm^{-1}.

PMR: δ 1.96 (3H, *s*), 5.00 (2H, *s*) and 7.22 (5H, *s*).

^{13}C NMR (Off-resonance): Two singlets, one triplet, one quartet and three doublets. One of the singlets is at δ 171 and the other is at δ 136.

MS: Prominent peaks at *m/e* 150 (M^+), 108, 91, 79, 78 and 77.

Solution

UV: λ_{max} 257 nm, ε_{max} 194 indicate the presence of an aromatic ring.

IR: Absorption bands at 3040, 1480, 1440, 700 and 750 indicate the presence of a monosubstituted benzene ring. The presence of a band at 1740 together with another band at 1220 cm^{-1} is characteristic of an acetate. The band at 2950 cm^{-1} is due to alkyl C—H stretching.

PMR: The presence of five proton singlet at δ7.22 is due to aromatic protons. The singlet nature of this peak shows that the compound is a monosubstituted benzene derivative and the substituent has nearly the same electronegativity as the ring carbon. A three proton singlet at δ 1.96 indicates the presence of a CH_3— group on a carbonyl carbon. The other singlet (2H, δ 5.00) shows the presence of a CH_2 group and its chemical shift is in accordance with its attachment to a benzene ring on one side and oxygen atom on the other.

^{13}C NMR (Off-resonance): The quartet and triplet indicate the presence of a CH_3— and a —CH_2— group, respectively. The three doublets are due to three kinds of =CH— groups of the benzene ring. One singlet at δ 136 is the ring carbon containing a substituent and the other singlet at δ 171 is due to the ester carbonyl carbon.

MS: The peak at *m/e* 150 is due to the molecular ion and at *m/e* 91 is characteristic of benzyl (or tropylium) ion. The peaks at *m/e* 79, 78 and 77 are additional proof for the benzene ring while the peak at *m/e* 43 is due to $CH_3C{\equiv}\overset{+}{O}$.

Thus, following is the structure of compound consistent with its given molecular formula and spectral data

The following fragmentation modes explain the formation of fragment ions

$C_6H_5CH_2OC(=O)—CH_3$

corresponding to the observed peaks in the mass spectrum of the compound:

$C_6H_5\overset{+}{C}H_2$ (*m/e* 91) ← $-CH_3COO^{\cdot}$, β-cleavage — $C_6H_5CH_2\overset{+}{O}C(=O)CH_3$ (*m/e* 150 (M^+)) — $-CH_2{=}C{=}O$ → $[C_6H_5CH_2OH]^{+\cdot}$ (*m/e* 108)

m/e 91 → tropylium ion (*m/e* 91)

m/e 108 → $-H^{\cdot}$ → hydroxytropylium ion (*m/e* 107) → $-CO$ → $[C_6H_6]^{+\cdot}$ (*m/e* 78)

m/e 78 → $-H^{\cdot}$ → $C_6H_7^+$ (*m/e* 79) → $-H_2$ → $C_6H_5^+$ (*m/e* 77)

Problem 10. A hydrocarbon containing 85.7% C and 14.3% H is transparent above 210 nm in UV spectrum. It shows IR absorption bands at 3020, 1675 and 965 cm^{-1}. Its PMR spectrum reveals a doublet at τ 8.40 and a quartet at τ 4.45 in the integral area ratio 3 : 1, respectively. Determine the structural formula of the compound.

Solution. From the given C and H percentage, the empirical formula of the compound comes to be C_4H_8.

The UV spectrum is transparent above 210 nm. This indicates the absence of conjugation, carbonyl group and aromatic ring.

Since aromatic ring is absent, IR absorption at 3020 is due to =C—H stretching and that at 1675 due to C=C stretching.

In the PMR spectrum, the doublet and quartet may be due to CH_3—CH= group.

Since the empirical formula (C_4H_8) is just double of it, the probable structure of the compound is CH_3—CH=CH—CH_3. The presence of an IR band at 965 cm^{-1} is

$$\begin{array}{ccc} H_3C & & H \\ & \diagdown C=C \diagup & \\ & \diagup \quad \diagdown & \\ H & & CH_3 \end{array}$$

Problem 11. Rationalizing the following spectral data, deduce the structure of a compound with molecular formula C_3H_7NO:

UV: λ_{max} 238 nm, ε_{max} 10500.
IR: Significant absorptions at 3428, 2941-2857, 1681 and 1452 cm^{-1}.
^1H NMR: δ 1.9 (3H, *s*), 2.7 (3H, *s*) and 8.13 (1H, *s*).
^{13}C NMR (Off-resonance decoupled): Two quartets and one singlet. The singlet is at δ 176.

Solution

UV: The absorption at 238 nm, ε_{max} 10500 is characteristic of conjugation.

IR: The bands at 2941-2857 and 3428 cm^{-1} are due to C—H and N—H stretching, respectively. The band at 1681 cm^{-1} is caused by stretching of C=O group which is in conjugation with the lone pair of electrons of nitrogen atom.

^1H NMR: Two singlets, each due to three protons, indicate the presence of two methyl groups, one of which (at δ 2.7) may be due to a methyl group attached to nitrogen atom. The low field one proton singlet is due to —NH—.

^{13}C NMR (Off-resonance decoupled): Two quartets confirm the presence of two methyl groups, and the singlet at δ 176 indicates an amino C=O group.

Thus, the structure of the compound is

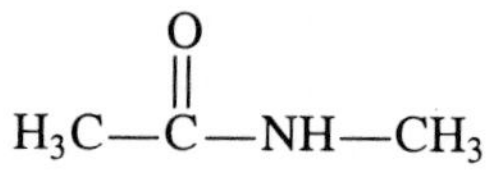

$$H_3C—\overset{\overset{\displaystyle O}{\|}}{C}—NH—CH_3$$

Problem 12. An organic compound C_9H_{12} gave the following spectral data:

UV: λ_{max}268 nm, ε_{max} 480.
IR: Singificant absorption bands at 3065-2910, 1608 and 1473 cm^{-1}.
PMR: δ 2.26 (9H, *s*) and 6.79 (3H, *s*).
^{13}C NMR (Off-resonance decoupled): One quartet, one doublet and one singlet.
Explaining the spectral data, derive the structure of the compound.

Solution

UV: λ_{max} 268, ε_{max} 480 indicate the presence of an aromatic ring.

IR: Absorption band at 3065-2910 cm^{-1} may be due to aromatic and saturated C—H stretching. The bands at 1608 and 1473 cm^{-1} are due to C=C and C—C stretching.

PMR: The singlet at δ 6.79 indicates the presence of three aromatic protons, and the singlet at δ 2.26 is due to nine alkyl protons. Thus, there is a benzene ring containing three methyl groups.

^{13}C NMR (Off-resonance decoupled): One quartet is due to methyl group, one doublet due to =CH of aromatic ring and the singlet is due to the ring carbons to which the methyl group is attached.

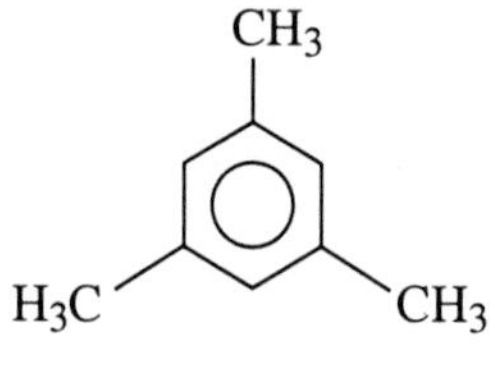

CH3

H3C CH3

Thus, the structure of the compound which fits the given spectral data is

Problem 13. An organic compound with molecular mass 100 gave the following spectral data:

UV: λ_{max} 274 nm, ε_{max} 2050.
IR: Significant absorption bands at 3030, 2940, 1725, 1605, 1505, 1060 and 830 cm^{-1}.
PMR: δ2.35 (3H, *s*), 3.82 (3H, *s*) and 7.20-7.85 (unsymmetrical pattern, 4H).

Rationalizing the spectral data, deduce the structure of the compound.

Solution

UV: λ_{max} 274 nm, ε_{max} 2050 indicates the presence of conjugation or aromatic ring.

IR: Band at 3030 and 2940 cm^{-1} are due to =CH and alkyl C—H stretching, respectively. The band at 1725 cm^{-1} may be due to C=O stretching of an ester group. This is supported by the presence of another band at 1060 cm^{-1} due to C—O stretching. The presence of absorption bands at 1605 and 1505 cm^{-1} indicate the presence of an aromatic ring.

PMR: The presence of a four proton signal with unsymmetrical pattern at δ 7.20-7.85 indicates the presence of a disubstituted benzene ring. Singlets at δ 2.35 and 3.82 (3H each) show the presence of two methyl groups.

Thus, there are two methyl groups, one disubstituted benzene ring and an ester group in the molecule. The IR band at 830 cm^{-1} is characteristic of *para*-disubstituted benzene ring. This suggests that the following is the structure of the compound:

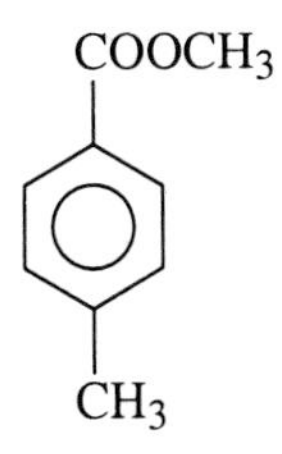

Problems 14. Explain the spectral data and deduce the structure of the compound from its UV, IR, PMR, CMR and mass spectra given in Fig. 9.1 (a-e).

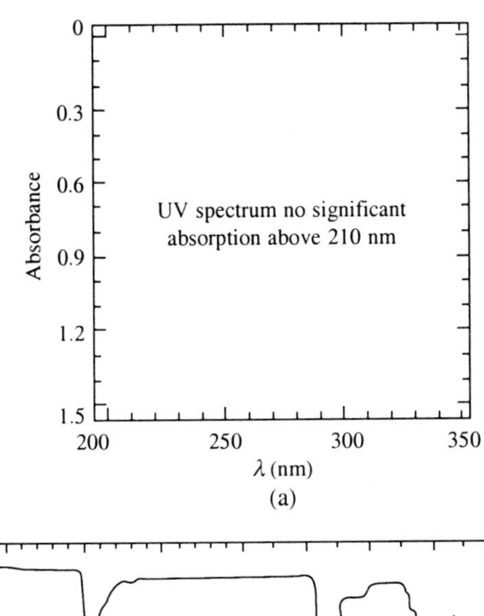

(a)

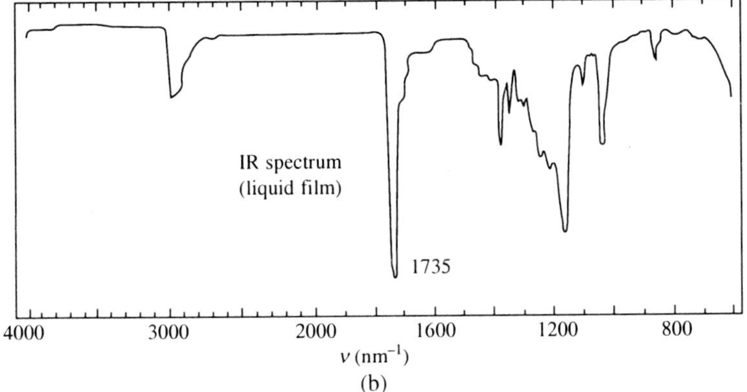

(b)

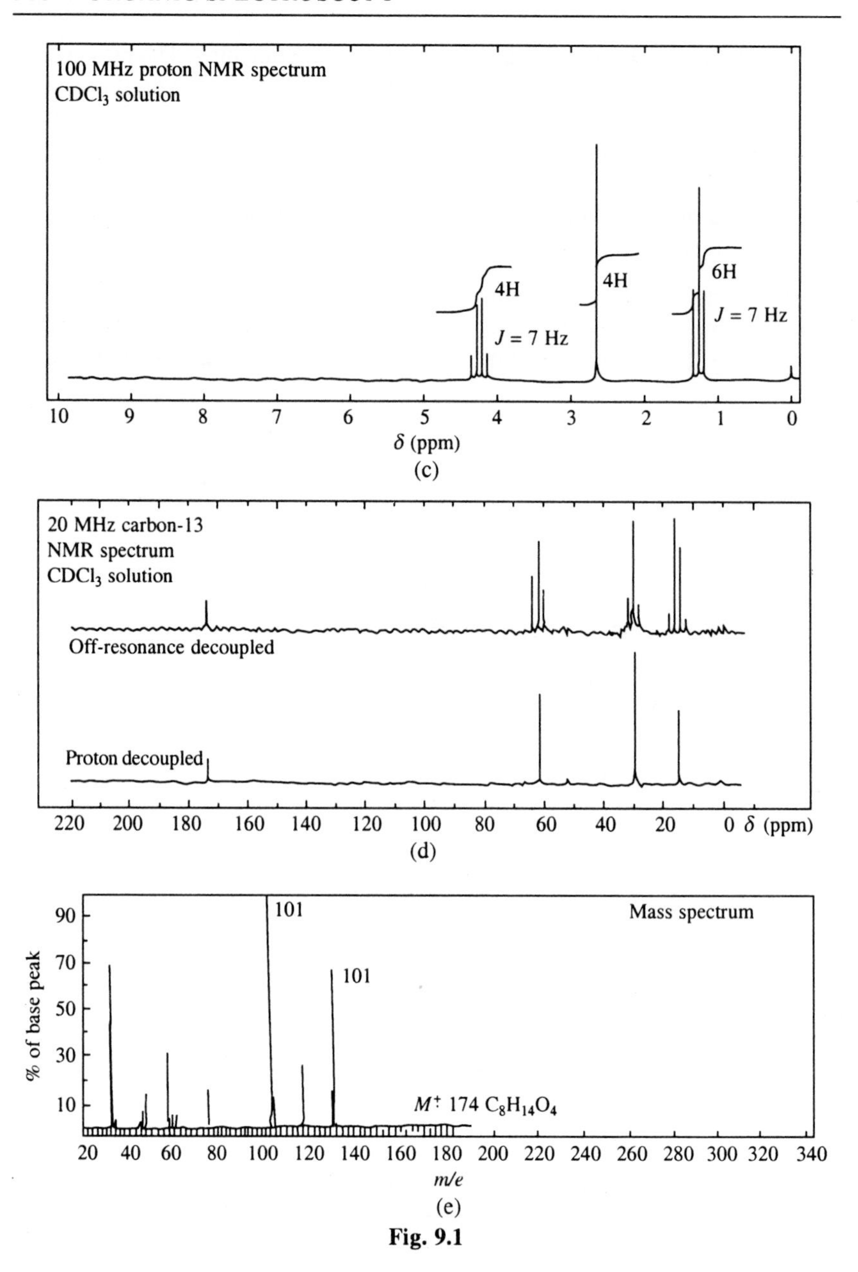

Fig. 9.1

Solution

UV: No significant absorption above 210 nm indicates the absence of conjugation, aldehydic or ketonic carbonyl group and an aromatic ring.

IR: The presence of a significant band at 1735 cm^{-1} indicates the presence of an ester group.

^1H NMR: The presence of a triplet (6H) at δ 1.2 and a quartet (4H) at δ 4.2 both having the same value of the coupling constant ($J = 7$ Hz) indicates the presence of two equivalent CH_3CH_2— groups. The four proton singlet at δ 2.6 may be due to —CH_2CH_2— group with equivalent protons. Thus, the compound contains two equivalent CH_3CH_2— groups and one —CH_2—CH_2— group.

^{13}C NMR (Proton decoupled): The presence of four signals shows the presence of four kinds of carbon atoms in the compound.

^{13}C NMR (Off-resonance decoupled): One singlet at δ 175 may be due to the carbonyl carbon of the ester group. Two triplets and one quartet indicate the presence of two kinds of —CH_2— and one kind of —CH_3 groups.

Mass: The molecular ion ($M^{+\cdot}$) peak at m/z 174 corresponds to the molecular formula $C_8H_{14}O_4$ of the compound.

On the basis of the spectral data explained above and the given molecular formula, the deduced structure of the compound is

$$\begin{array}{l} CH_2COOCH_2CH_3 \\ | \\ CH_2COOCH_2CH_3 \end{array}$$

The formation of ions corresponding to the indicated peaks in the MS at m/z 174 (M^+), 129 and 101 is

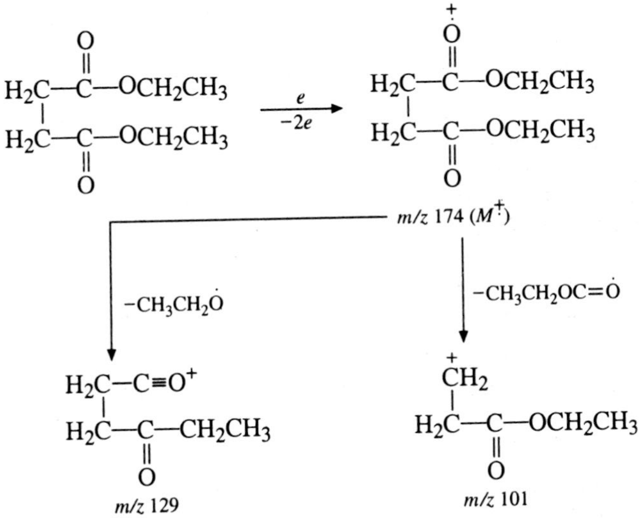

$$\begin{array}{l} H_2C—C(=O)—OCH_2CH_3 \\ | \\ H_2C—C(=O)—OCH_2CH_3 \end{array} \xrightarrow[-2e]{e} \begin{array}{l} H_2C—C(=\overset{+}{\dot{O}})—OCH_2CH_3 \\ | \\ H_2C—C(=O)—OCH_2CH_3 \end{array}$$

m/z 174 ($M^{+\cdot}$)

$\xrightarrow{-CH_3CH_2\dot{O}}$ $\begin{array}{l} H_2C—C{\equiv}O^+ \\ | \\ H_2C—C(=O)—CH_2CH_3 \end{array}$ m/z 129

$\xrightarrow{-CH_3CH_2OC=\dot{O}}$ $\begin{array}{l} \overset{+}{C}H_2 \\ | \\ H_2C—C(=O)—OCH_2CH_3 \end{array}$ m/z 101

PROBLEMS

1. An organic compound with molecular formula $C_6H_{12}O_3$ gave the following spectral data:

   UV: λ_{max} 283 nm, ε_{max} 27 (hexane).
   IR: Significant absorption bands at 1715 and 2900 cm^{-1}.

PMR: δ 2.2 (three proton singlet), 2.6 (two proton doublet, $J = 7$ Hz), 3.4 (six proton singlet) and 4.8 (one proton triplet, $J = 7$ Hz).
^{13}C NMR (Off-resonance decoupled): One singlet, one doublet, one triplet and two quartets. The singlet is at δ 205.
Mass: Prominent peaks at *m/z* 132, 101, 75 and 43.

Deduce the structure of the compound and explain the spectral data.

2. A compound with the molecular weight 116 gave the following spectral information:

UV: λ_{max} 283, ε_{max} 22.
IR: A very broad band in the region 2500-3000 cm^{-1} and a strong band at 1715 cm^{-1}.
PMR: δ 2.12 (3H, *s*), 2.60 (2H, *t*), 2.25(2H, *t*) and 11.1 (1H, *s*).

Deduce the structure of the compound and explain the spectral data.

3. A compound with molecular formula $C_6H_{12}O$ gave the following spectral data:

UV: λ_{max} 280 nm, ε_{max} 25.
IR: Significant bands at 1715 and 2900 cm^{-1}.
^1H NMR: Two singlets at δ 1.0 and 2.0 in the intensity ratio 3 : 1.

Explaining the spectral data, deduce the structure of the compound.

4. A compound with molecular formula $C_5H_7NO_2$ gave the following spectral data:

UV: No significant absorption above 210 nm.
IR: Significant bands at 1745, 2270 and 2950 cm^{-1}.
PMR: δ 1.3 (3H, *t*), 3.5 (2H, *s*) and 4.3 (2H, *q*).
^{13}C NMR (Proton-noise decoupled): Two singlets, two triplets and one quartet. One of the singlets appear at δ 165.
MS: Prominent peaks at *m/e* 113, 86, 68 and 40.

Rationalizing the spectral data, deduce the structural formula of the compound.

5. An organic compound with molecular mass 160 show UV absorption at 212 nm, ε_{max} 60. Its IR spectrum reveals bands at 2940-2860, 1740, 1460, 1260 and 1050 cm^{-1}. In the PMR spectrum of the compound, three signals are observed: δ 1.3 (triplet, $J = 7.2$ cps), 2.5 (singlet) and 4.2 (quartet, $J = 7.2$ cps) in the intensity ratio 3 : 1 : 2, respectively.
Deduce the structure of the compound and explain the spectral data.

6. An organic compound with molecular mass 150 gave the following spectral data:

UV: λ_{max} 205, ε_{max} 75.
IR: Significant absorption bands at 3460, 3035, 2650, 1720 and 1265 cm^{-1}.
PMR: δ 3.6 (singlet), 4.5 (singlet) and 11.0 (singlet) in the intensity ratio 1 : 1 : 1.

Explaining the spectral data, determine the structure of the compound.

7. An organic compound with molecular weight 54 is transparent to UV spectrum above 200 nm. In the IR spectrum, it shows bands at 3290, 2128 and 620 cm^{-1}. In its ^1H NMR spectrum, the following three signals are observed: δ 1.2 (triple, $J = 7.1$ cps), 1.5 (quartet, $J = 7.1$ cps) and 2.4 (singlet) in the intensity ratio 3 : 2 : 1, respectively.
Derive the structure of the compound consistent with the given data.

8. An organic compound C_3H_6O absorbs at 176 nm, ε_{max} 15000. It shows IR bands at 3520, 1650 and 1280 cm^{-1}. In its PMR spectrum four signals are observed at

δ 5.2 (double doublet), 5.7 (multiplet-complicated), 3.8 (singlet) and 2.1 (doublet) in the intensity ratio 2 : 1: 1 : 2, respectively.
Derive the structural formula of the compound and write its name.

9. An organic compound with molecular weight 58 is transparent above 200 nm in its UV spectrum. It shows absorption bands at 2940-2860 and 1460 cm^{-1} in its IR spectrum. The PMR spectrum of the compound exhibits signals at δ 4.75 (t, J = 7.1 cps) and 2.75 (quintet, J = 7.1 cps) in the intensity ratio 2 : 1. Derive the structure of the compound.

10. A hydrocarbon with molecular formula C_5H_8 exists in two isomeric forms. One of the isomers gives the following spectral data:

IR: Significant absorption bands at 3320 and 2130 cm^{-1}.
^{13}C NMR (Off-resonance): δ 83 (one singlet), 67 (one doublet), 20-30 (two triplets) and 15 (one quartet).

Explaining the spectral data, derive the structure of the compound and give the possible structure of its other isomer also.

11. An organic compound containing C, H and O gave the following spectral data:

UV: λ_{max} 220 nm, ε_{max} 1800 (EtOH).
IR: Significant absorption bands at 3075, 2975, 1745, 1605, 1500 and 1450 cm^{-1}.
PMR: δ 2.02 (singlet), 2.93 (triplet, J = 7 Hz), 4.30 (triplet, J = 7 Hz) and 7.30 (singlet) in the intensity ratio 3 : 2 : 2 : 5, respectively.
Mass: Prominent peaks at m/z 164 (M^+), 91,60, 73 and 43.

Explaining the spectral data, derive the structural formula of the compound.

12. An organic compound $C_6H_{14}O$ gave the following spectral data:

UV: No significant absorption above 210 nm.
IR: Significant bands at 2850-2960 and 1080 cm^{-1}.
PMR: δ 1.1 (doublet) and 3.6 (septet) in the ratio 6 : 1.
MS: Prominent peaks at m/z 102, 87, 59, 45, and 43.

Deduce the structure of the compound and explain the spectra data.

13. An organic compound with molecular mass 152 shows UV absorption at 223 nm, ε_{max} 100. It shows bands at 2900-3125, 2690, 2600, 1715 and 1440 cm^{-1} in its IR spectrum. In ^1H NMR spectrum the following signals were observed: δ 1.83 (d, J = 7.2 cps), 4.52 (q, J = 7.2 cps) and 11.93 (s) in the intensity ratio 3 : 1 : 1. The mass spectrum of the compound shows molecular ion peak at m/z 152 and another peak of equal intensity at m/z 154.
Explaining the spectral data, deduce the structure of the compound.

14. An organic compound with molecular weight 130 shows IR absorption bands at 2860-3080, 1825, 1755 and 1455 cm^{-1}. In PMR spectrum, it shows two signals: one triplet (J = 7.1 cps) and one quartet (J = 7.1 cps) at δ 1.3 and 2.2, respectively in the intensity ratio 3 : 2. ^{13}C NMR (proton-noise decoupled) reveals three signals.
Explain the spectral data and deduce the structure of the compound.

15. An organic compound C_4H_9NO gave the following spectral data:

UV: λ_{max} 220 nm, ε_{max} 63.
IR: Significant bands at 3500, 3400, 1682 and 1610 cm^{-1}.
^1H NMR: δ 1.0 (doublet), 2.1 (septet) and 8.08 (singlet) in the intensity ratio 6 : 1 : 2.

Rationalizing the spectral data, derive the structural formula of the compound.

16. An organic compound with molecular mass 71 is transparent in the UV spectrum. In IR spectrum, it shows bands at 2860-2940, 2250 and 1460 cm^{-1}. Its PMR spectrum shows two singlets at δ 4.22 and 3.49 in the intensity ratio 2 : 3. Deduce the structure of the compound and explain the spectral data.

17. Rationalizing the spectral data, derive the structure of an organic compound with molecular mass 72:

UV : Absorption at 274 nm, ε_{max} 17.
IR: Significant absorption bands at 2860-2940, 1715 and 1460 cm^{-1}.
^1H NMR: δ 2.48 (q, J = 7.2 cps), 2.12 (s) and 1.07 (t, J = 7.3 cps) in the intensity ratio 2 : 3 : 3.

18. An organic compound $C_6H_{12}O_2$ is transparent above 210 nm in its UV spectrum. It shows IR absorption bands at 2925, 1745 and 1455 cm^{-1}. Its PMR spectrum reveals two singlets at δ 1.97 and 1.45 in the intensity ratio 1 : 3. Derive the structure of the compound and explain the spectral data.

19. An organic compound containing C, H and O is not acidic in nature but can be easily oxidized to a crystalline compound (m.p. 122°C). It gives the following spectral data:

UV: λ_{max} 255 nm, ε_{max} 202.
IR: Significant absorption bands at 3400, 3065, 2290, 1500 and 1455 cm^{-1}.
^1H NMR: δ 3.90 (singlet), 4.60 (singlet) and 7.26 (singlet) in the intensity ratio 1 : 2 : 5, respectively.

^{13}C NMR (proton-noise decoupled): Shows five signals.
MS: Prominent peaks at 108 (M^+), 107, 105, 79, 77 and 51.

Explaining the given data, deduce the structure of the compound.

20. Explaining the spectral data, deduce the structure of the compound from its UV, IR, ^1H NMR, ^{13}C NMR and mass spectra given in Fig. P9.1 (a-e).

21. Explain the spectral data and deduce the structure of the compound from its UV,

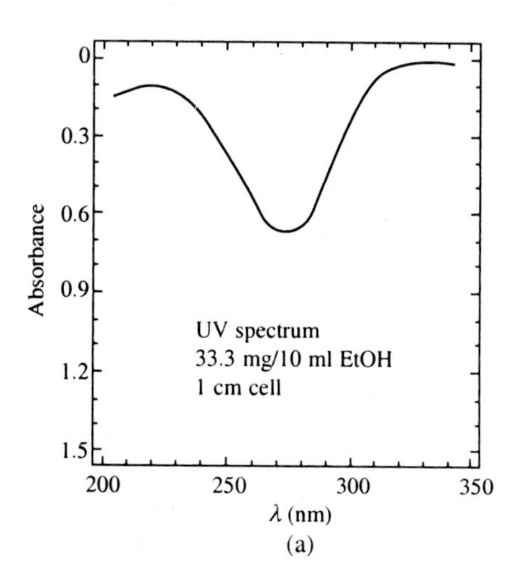

(a)

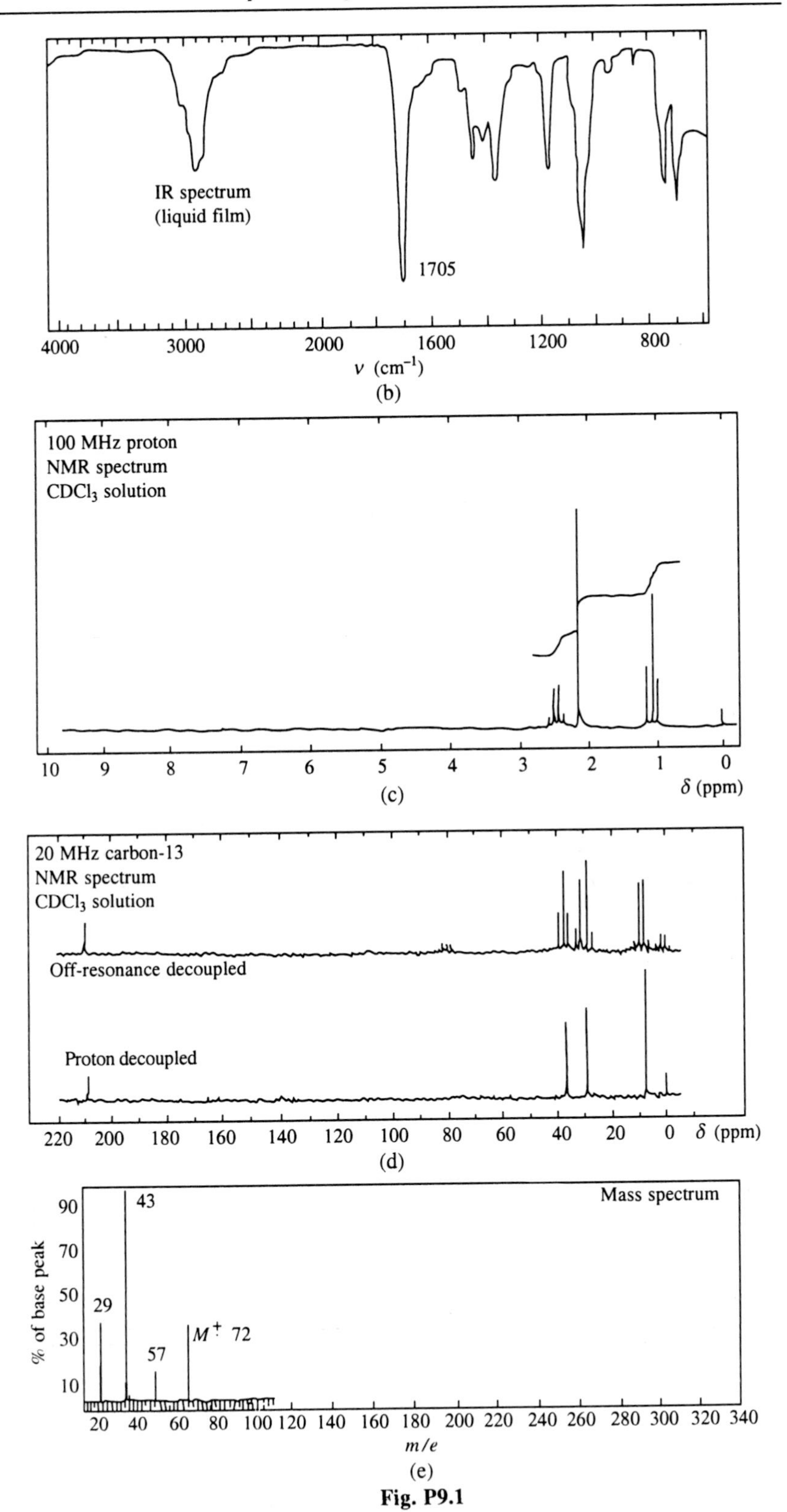
IR spectrum
(liquid film)
1705
4000
3000
2000
1600
1200
800
ν (cm⁻¹)
(b)
100 MHz proton
NMR spectrum
CDCl3 solution
10
9
8
7
6
5
4
3
2
1
0
δ (ppm)
(c)
20 MHz carbon-13
NMR spectrum
CDCl3 solution
Off-resonance decoupled
Proton decoupled
220
200
180
160
140
120
100
80
60
40
20
0
δ (ppm)
(d)
Mass spectrum
% of base peak
90
70
50
30
10
43
29
57
M+ 72
20
40
60
80
100
120
140
160
180
200
220
240
260
280
300
320
340
m/e
(e)

Fig. P9.1

IR, ^{1}H NMR, ^{13}C NMR and mass spectra given in Fig. P9.2 (a-e).

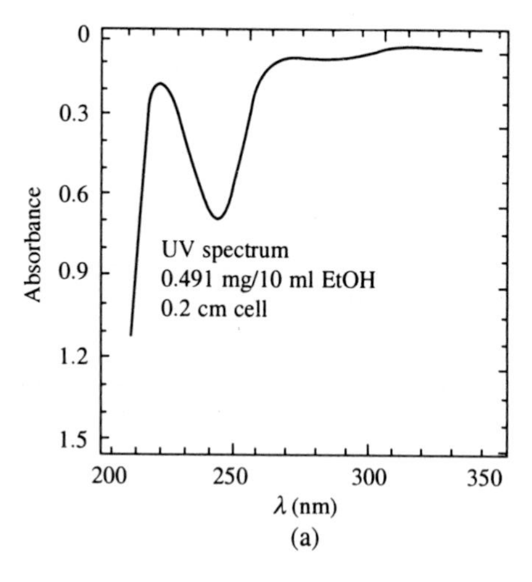

(a)

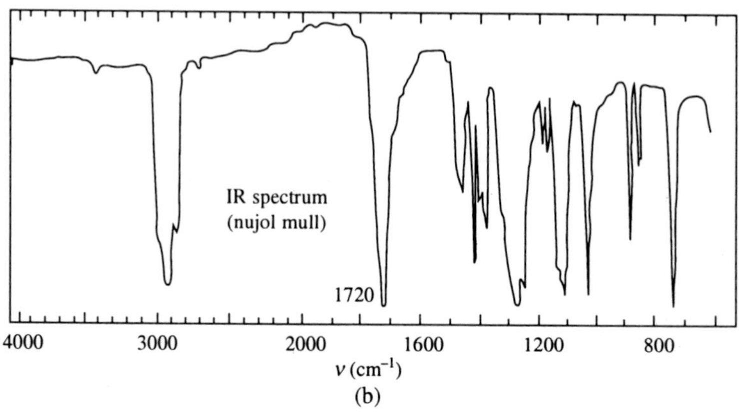

(b)

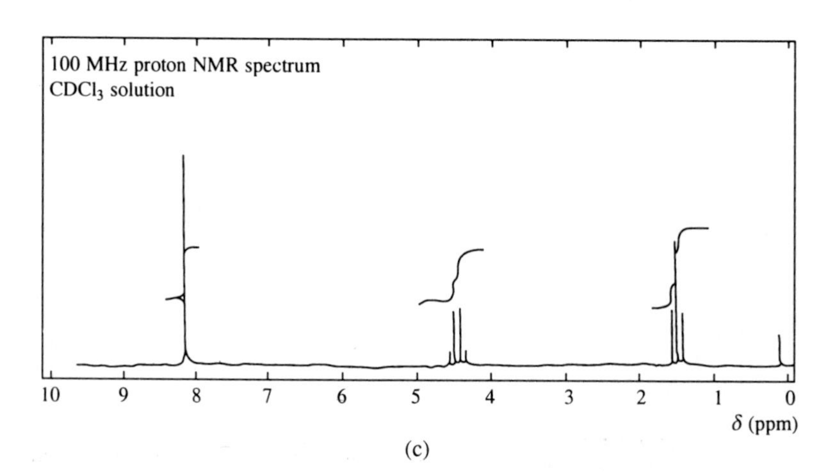

(c)

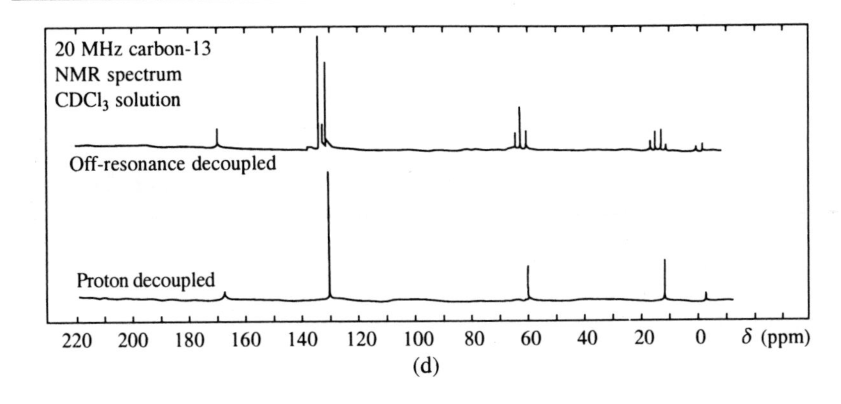

20 MHz carbon-13
NMR spectrum
$CDCl_3$ solution
Off-resonance decoupled
Proton decoupled
220 200 180 160 140 120 100 80 60 40 20 0 δ (ppm)

(d)

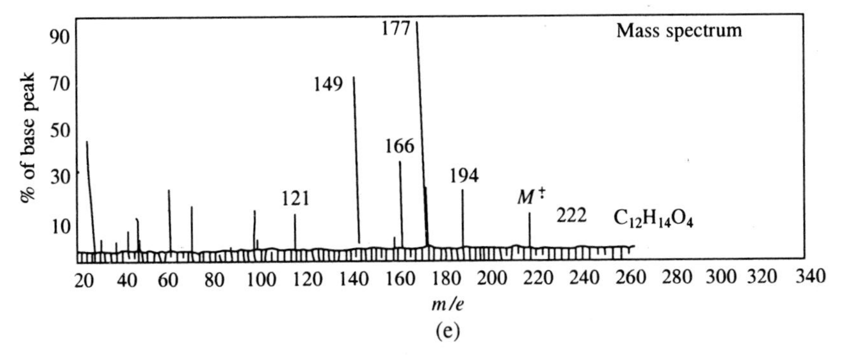

Mass spectrum
% of base peak
90
70
50
30
10
177
149
166
194
121
M^{+}
222
$C_{12}H_{14}O_4$
20 40 60 80 100 120 140 160 180 200 220 240 260 280 300 320 340
m/e

(e)

Fig. P9.2

Answers to Problems

Chapter 1

3. Gamma rays > X-rays > UV > Visible > IR > Radio waves
6. (a) (i) 4000 cm^{-1}
   (ii) 35087.7 cm^{-1}
   (iii) 3355.7 cm^{-1}
   (b) 7.495×10^8 MHz to 3.748×10^8 MHz
7. (a) 7.15 kcal/mole
   (b) 1.499×10^{15} cps to 0.7495×10^{15} cps
8. (a) 47.67 kcals $mole^{-1}$ or 199.45 kilo Joule $mole^{-1}$
   (b) 4000 cm^{-1} to 666.67 cm^{-1}
   (c) 24.176×10^{13} cps (Hz) or 24.176×10^7 MHz

Chapter 2

2. $\sigma \to \sigma^* > n \to \sigma^* > \pi \to \pi^* > n \to \pi^*$
5. (a) $n \to \pi^* < \pi \to \pi^* < n \to \sigma^*$
8. A, 1,4-Pentadiene; B, *cis*-1,3-pentadiene and C, *trans*-1,3-pentadiene
9. (i) 249 nm (ii) 342 nm (iii) 226 nm
   (iv) 418 nm (v) 280 nm (vi) 239 nm
12. (i) 254 nm (ii) 269 nm (iii) 285 nm
15. a < d < b < c
17. (a) 259 nm (b) 315 nm (c) 254 nm
19. (d)
20. *A* is 1,3-cyclohexadiene and *B* is 1,4-cyclohexadiene.
21. (b)
22. 285 nm and 257 nm
24. (a) 2-propanol, ethanol, heptene, water and dioxane
26. Structure (a) is correct.
28. *p*-benzoquinone
29. (i) π-π^* (ii) n-π^*
30. (i) 229 nm (ii) 323 nm
    (iii) 351 nm (iv) 281 nm

Chapter 3

1. (b) Rotational < Vibrational < Electronic
4. (i) 9 (ii) 21 (iii) 7 (iv) 12
   (v) 1
5. (i) Esters
   (ii) Acid halides

(iii) Amides
(iv) Anhydrides

9. (a) *p*-aminoacetophenone < acetophenone < *p*-nitroacetophenone
   (b) cyclohexanone < cyclopentanone < cyclobutanone
12. (A) CH_3COCH_3 (B) $CH_2{=}CH{-}CH_2OH$
14. (b) > (a)
15. (i) b < a < c (ii) (b) < (c) < (a)
16. $CH_3CH_2CONH_2$
19. trichloroacetic acid < chloroacetic acid < acetic acid < ethanol
20. (i) 3300 cm^{-1} (ν_{O-H}, hydrogen bonded)
    (ii) 3050 cm^{-1} (ν_{C-H}, aromatic ring)
    (iii) 2990 cm^{-1} (ν_{C-H} methyl group)
    (iv) 1700 cm^{-1} ($\nu_{C=O}$, hydrogen-bonded ester)
    (v) 1540 and 1590 cm^{-1} (ν_{C-O}, aromatic ring)
21. $C_6H_5CH_2OH$
22. CH_3COCH_3
24. $C_6H_5CHO < CH_3COCH_3 < CH_3CHO < CH_3COCl$
26. (i) Inactive (ii) Active (iii) Inactive
    (iv) Active (v) Inactive (vi) Inactive
28. Butanone
30. *m*-cresol
31. Benzaldehyde

Chapter 4

5. 5401 Å
6. 22.6
7. The energy of radiation (274.49 kJ) is lesser than the dissociation energy of H_2, hence it cannot dissociate.
9. CS_2 has a center of symmetry, whereas N_2O has no center of symmetry and thus the structures must be S—C—S and N—N—O, respectively.
10. The compound has *trans* and planar structure:

```
H       Cl
 \     /
  C = C
 /     \
Cl      H
```

15. 214, 311, 454 and 758 cm^{-1}.
17. (a), (b), (c) and (e) will exhibit both the vibrational and rotational Raman spectra; (d) will not exhibit rotational Raman spectrum.
18. $H_2 > HD > D_2$.
20. (a), (b) and (d) will exhibit rotational Raman spectra; (c) will not exhibit rotational Roman spectrum.
21. (i) Active (ii) Inactive (iii) Inactive (iv) Active (v) Active
22. The molecule has a *trans*, planar geometry:

```
    X
    |
Y — Z — Y
    |
    X
```

23. 6688, 7033, 7043 and 7839 Å
24. $H_3C—C{\equiv}C—CH_3$

Chapter 5

1. For ^{14}N, ^{2}H, ^{35}Cl and ^{31}P NMR spectroscopy is possible.
3. (a) 2(5 : 3) (b) 3(3 : 2 : 1) (c) 3(3 : 2 : 2)
6. (a) 3; one singlet, one triplet and one quartet
   (b) 3; one singlet, one triplet and one quintet
   (c) 1; singlet
   (d) 1; singlet
10. Chemical shift positions are δ 5.75 and δ 1.05; $J_{AX} = 10$ Hz.
11. (a) $ClCH_2CCl_2CH_3$ (b) $C_6H_5—(CH_3)_3$
    (c) $C_6H_5—CH_2CH\ (CH_3)_2$
12. (a) 1; 3 (3 : 2 : 1) (b) 3 (5 : 2 : 3); 2 (2 : 3)
    (c) 4 (3 : 2 : 2 : 3); 2 (3 : 2)
14. $C_6H_5C(CH_3)_2CH_2Cl$
18. $BrCH_2CH_2CH_2Br$
21. (i) 7.52 δ (ii) 4.78 δ (iii) 5.43 δ
22. $(CH_3)_2CHOH$
25. (a) $(CH_3)_3COH$ (b) $CH_3OCH_2CH_2OCH_3$ (c) $C_6H_5—C(CH_3)_2CH_2Cl$
26. 3.5 δ; 6.5τ
27. $C_6H_5CH_2CH_3$
28. $ClCH_2CH_2COOH$
29. *p*-nitrophenol
31. $C_6H_5CH_2CH_2OH$

Chapter 6

1. (a) 2, (b) 5, (c) 2, (d) 1.
4. (a) Two peaks; one singlet and one doublet
   (b) Three peaks; one singlet, one doublet and one quartet
   (c) One peak; one triplet
   (d) Three peaks; one doublet, one triplet and one quartet
5. (a) $(CH_3)_2CHC{\equiv}CH$ (b) $(CH_3CH_2)_2CHCH_3$ (c) $ClCH_2CHClCCl_3$
8. $(CH_3)_2CHCOCH_3$
9. (a) Six peaks; two singlets, three doublets, and one quartet; the singlet due to ketonic carbon will have the highest and the quartet the lowest δ value.
   (b) Four peaks, one singlet, two triplets and one quartet; the singlet will have the highest and the quartet the lowest δ value.
   (c) Four peaks; one singlet, one triplet and two quartets; the singlet will have the highest and the quartet the lowest δ value.
   (d) Three peaks; one quartet and two triplets; the triplet due to C-3 will have the highest and the quartet the lowest δ value.
10. (a) $(CH_3CH_2)_3CH$ (b) $CH_3CH_2CHOHCH_3$
    (c) $CH_3C{\equiv}CCH_2CH_3$
14. (a) 5; one singlet, three triplets and one quartet
    (b) 9; four singlets, four doublets and one quartet
    (c) 2; one doublet and one triplet
    (d) 4; one singlet and three triplets

17. $(CH_3)_2CHCH_2CH_2Br$

18. 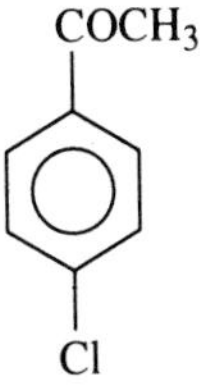 p-chlorophenyl methyl ketone: benzene ring bearing $COCH_3$ at the top and Cl at the para position

Chapter 7

1. (a), (c)
3. (a) 12, (b) 14, (c) 10, (d) 7
5. (a) 3; 1 : 1 : 1 (b) 5; 1 : 4 : 6 : 4 : 1 (c) 3; 1 : 2 : 1 (d) 6; 1 : 5 : 10 : 10 : 5 : 1
6. (a) 7
   (b) 8. A, $CH(COOH)_2$; B, CH_2COOH.
9. (a) 35 (b) 8
10. (a) 7; 1 : 6 : 15 : 20 : 15 : 6 : 1
    (b) 4; 1 : 3 : 3 : 1
    (c) 7; 1 : 6 : 15 : 20 : 15 : 6 : 1
    (d) 3; 1 : 1 : 1
12. 0.3319 T
13. (c)
14. $C_6H_6^-$
15. Yes; 2 lines in intensity ratio 1 : 1
16. 5/2

Chapter 8

2. (i) HC≡C—Br (ii) $HCOOCH_3$
9. $C_2H_5\overset{+}{O}H$, $CH_3CH{=}\overset{+}{O}H$, $CH_2{=}\overset{+}{O}H$, $C_2H_5^{+\cdot}$, $HC{\equiv}\overset{+}{O}$, $\overset{+}{O}H$ and $\overset{+}{C}H_3$
12. CH_3I
14. *n*-heptane

Chapter 9

1. 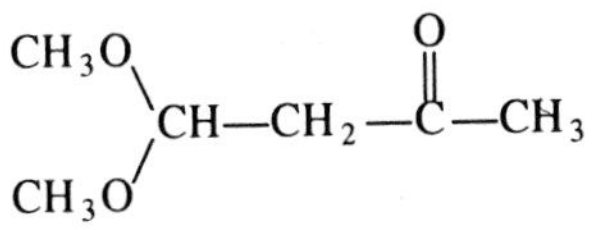 $(CH_3O)_2CH—CH_2—C(=O)—CH_3$

2. $H_3C—C(=O)—CH_2—CH_2—COOH$

3. $H_3C—C(=O)—C(CH_3)_3$

4. $N{\equiv}C—CH_2—C(=O)—O—CH_2CH_3$

5. $H_2C(COOCH_2CH_3)_2$ — H_2C bonded to two $COOCH_2CH_3$ groups

6. $CHOHCOOH$ bonded to $CHOHCOOH$ ($HOOCCH(OH)CH(OH)COOH$)

7. $CH_3CH_2C{\equiv}CH$

8. $CH_2{=}CH{-}CH_2OH$ (allyl alcohol or 2-propen-1-ol)

9. 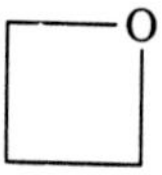 Four-membered ring: three CH_2 corners and one O (oxetane)

10. $CH_3CH_2CH_2C{\equiv}CH$, the isomer may be $CH_3CH_2C{\equiv}C{-}CH_3$ (other structures are also possible)

11. 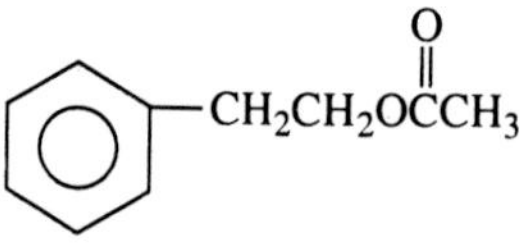 $C_6H_5{-}CH_2CH_2O\overset{O}{\overset{\|}{C}}CH_3$

12. $(CH_3)_2CH{-}O{-}CH\ (CH_3)_2$

13. $CH_3{-}CH(Br){-}COOH$ (Br on the middle C)

14. $CH_3CH_2CO{-}O{-}COCH_2CH_3$

15. $(CH_3)_2\ CHCONH_2$

16. $H_3C{-}O{-}CH_2{-}C{\equiv}N$

17. $CH_3CH_2\overset{O}{\overset{\|}{C}}CH_3$

18. $(CH_3)_3\overset{O}{\overset{\|}{C}}{-}OCH_3$

19. 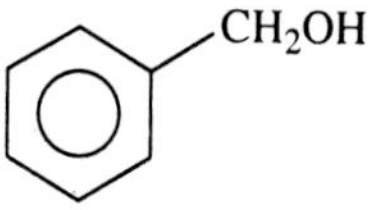 $C_6H_5{-}CH_2OH$

20. $CH_3COCH_2CH_3$

21. 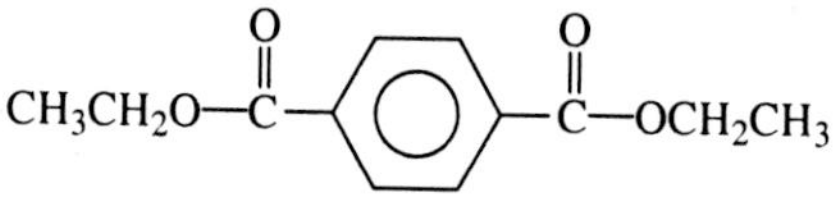 $CH_3CH_2O{-}\overset{O}{\overset{\|}{C}}{-}C_6H_4{-}\overset{O}{\overset{\|}{C}}{-}OCH_2CH_3$ (para)

Index